Toxicological Effects of Methylmercury

Committee on the Toxicological Effects of Methylmercury

Board on Environmental Studies and Toxicology

Commission on Life Sciences

National Research Council

NATIONAL ACADEMY PRESS
Washington, DC

NATIONAL ACADEMY PRESS 2101 Constitution Ave., N.W. Washington, D.C. 20418

NOTICE: The project that is the subject of this report was approved by the Governing Board of the National Research Council, whose members are drawn from the councils of the National Academy of Sciences, the National Academy of Engineering, and the Institute of Medicine. The members of the committee responsible for the report were chosen for their special competences and with regard for appropriate balance.

This project was supported by Grant Agreement No. X 827238-01 between the National Academy of Sciences and the Environmental Protection Agency. Any opinions, findings, conclusions, or recommendations expressed in this publication are those of the author(s) and do not necessarily reflect the view of the organizations or agencies that provided support for this project.

Library of Congress Card Number 00-108382

International Standard Book Number 0-309-07140-2

Additional copies of this report are available from:

National Academy Press
2101 Constitution Ave., NW
Box 285
Washington, DC 20055

800-624-6242
202-334-3313 (in the Washington metropolitan area)
http://www.nap.edu

Printed in the United States of America

THE NATIONAL ACADEMIES

National Academy of Sciences
National Academy of Engineering
Institute of Medicine
National Research Council

The **National Academy of Sciences** is a private, nonprofit, self-perpetuating society of distinguished scholars engaged in scientific and engineering research, dedicated to the furtherance of science and technology and to their use for the general welfare. Upon the authority of the charter granted to it by the Congress in 1863, the Academy has a mandate that requires it to advise the federal government on scientific and technical matters. Dr. Bruce M. Alberts is president of the National Academy of Sciences.

The **National Academy of Engineering** was established in 1964, under the charter of the National Academy of Sciences, as a parallel organization of outstanding engineers. It is autonomous in its administration and in the selection of its members, sharing with the National Academy of Sciences the responsibility for advising the federal government. The National Academy of Engineering also sponsors engineering programs aimed at meeting national needs, encourages education and research, and recognizes the superior achievements of engineers. Dr. William A. Wulf is president of the National Academy of Engineering.

The **Institute of Medicine** was established in 1970 by the National Academy of Sciences to secure the services of eminent members of appropriate professions in the examination of policy matters pertaining to the health of the public. The Institute acts under the responsibility given to the National Academy of Sciences by its congressional charter to be an adviser to the federal government and, upon its own initiative, to identify issues of medical care, research, and education. Dr. Kenneth I. Shine is president of the Institute of Medicine.

The **National Research Council** was organized by the National Academy of Sciences in 1916 to associate the broad community of science and technology with the Academy's purposes of furthering knowledge and advising the federal government. Functioning in accordance with general policies determined by the Academy, the Council has become the principal operating agency of both the National Academy of Sciences and the National Academy of Engineering in providing services to the government, the public, and the scientific and engineering communities. The Council is administered jointly by both Academies and the Institute of Medicine. Dr. Bruce M. Alberts and Dr. William A. Wulf are chairman and vice chairman, respectively, of the National Research Council.

Committee on the Toxicological Effects of Methylmercury

Robert A. Goyer (*Chair*), University of Western Ontario (Professor, Emeritus), Chapel Hill, North Carolina
H. Vasken Aposhian, University of Arizona, Tucson, Arizona
Lenore Arab, University of North Carolina, Chapel Hill, North Carolina
David C. Bellinger, Harvard Medical School, Boston, Massachusetts
Thomas M. Burbacher, University of Washington, Seattle, Washington
Thomas A. Burke, The Johns Hopkins University, Baltimore, Maryland
Joseph L. Jacobson, Wayne State University, Detroit, Michigan
Lynda M. Knobeloch, State of Wisconsin Bureau of Environmental Health, Madison, Wisconsin
Louise M. Ryan, Dana-Farber Cancer Institute, Boston, Massachusetts
Alan H. Stern, New Jersey Department of Environmental Protection, Trenton, New Jersey

Staff

Carol A. Maczka, Director, Toxicology and Risk Assessment Program
Michelle C. Catlin, Research Associate
Ruth E. Crossgrove, Editor
Mirsada Karalic-Loncarevic, Information Specialist
Judith L. Estep, Senior Program Assistant
Laura T. Holliday, Senior Program Assistant
Stephanie K. Parker, Graphics and Layout

Sponsor

U.S. Environmental Protection Agency

Commission on Life Sciences

OTHER REPORTS OF THE BOARD ON ENVIRONMENTAL STUDIES AND TOXICOLOGY

Strengthening Science at the U.S. Environmental Protection Agency: Research Management and Peer Review Practices (2000)
Scientific Frontiers in Developmental Toxicology and Risk Assessment (2000)
Modeling Mobile-Source Emissions (2000)
Copper in Drinking Water (2000)
Ecological Indicators for the Nation (2000)
Waste Incineration and Public Health (1999)
Hormonally Active Agents in the Environment (1999)
Research Priorities for Airborne Particulate Matter: I. Immediate Priorities and a Long-Range Research Portfolio (1998); II. Evaluating Research Progress and Updating the Portfolio (1999)
Ozone-Forming Potential of Reformulated Gasoline (1999)
Risk-Based Waste Classification in California (1999)
Arsenic in Drinking Water (1999)
Brucellosis in the Greater Yellowstone Area (1998)
The National Research Council's Committee on Toxicology: The First 50 Years (1997)
Toxicologic Assessment of the Army's Zinc Cadmium Sulfide Dispersion Tests (1997)
Carcinogens and Anticarcinogens in the Human Diet (1996)
Upstream: Salmon and Society in the Pacific Northwest (1996)
Science and the Endangered Species Act (1995)
Wetlands: Characteristics and Boundaries (1995)
Biologic Markers (5 reports, 1989-1995)
Review of EPA's Environmental Monitoring and Assessment Program (3 reports, 1994-1995)
Science and Judgment in Risk Assessment (1994)
Ranking Hazardous Waste Sites for Remedial Action (1994)
Pesticides in the Diets of Infants and Children (1993)
Issues in Risk Assessment (1993)
Setting Priorities for Land Conservation (1993)
Protecting Visibility in National Parks and Wilderness Areas (1993)
Dolphins and the Tuna Industry (1992)
Hazardous Materials on the Public Lands (1992)
Science and the National Parks (1992)
Animals as Sentinels of Environmental Health Hazards (1991)
Assessment of the U.S. Outer Continental Shelf Environmental Studies Program, Volumes I-IV (1991-1993)
Human Exposure Assessment for Airborne Pollutants (1991)
Monitoring Human Tissues for Toxic Substances (1991)

Rethinking the Ozone Problem in Urban and Regional Air Pollution (1991)
Decline of the Sea Turtles (1990)

Copies of these reports may be ordered from
the National Academy Press
(800) 624-6242
(202) 334-3313
www.nap.edu

PREFACE

IN 1997, the U.S. Environmental Protection Agency (EPA) issued two reports to the U.S. Congress on mercury (Hg) and its effects on public health. The first of these reports, the *Mercury Study Report to Congress*, assessed the source and amount of Hg emissions in the United States, the detrimental effects of Hg on humans and wildlife, and the feasibility of control technologies. The second report, the *Utility Hazardous Air Pollutant Report to Congress*, looked specifically at emissions from utility companies and cited Hg as a major contaminant, especially in emissions from coal-fired power plants. Once in the environment, Hg can be converted to methylmercury (MeHg), which bioaccumulates up the food chain. Such bioaccummulation can lead to high concentrations of MeHg in predatory fish. Because of concerns about MeHg exposure levels in the United States from the consumption of contaminated fish, particularly among sensitive populations, questions have arisen among federal agencies over what is an acceptable level of exposure to MeHg. Because of gaps in the scientific data regarding Hg toxicity, particularly MeHg, the potentially widespread implications for human health, and the high financial costs and feasibility problems associated with further regulating Hg emissions, Congress directed EPA in the House Appropriations Report for EPA's Fiscal 1999 funding to contract with the National Research Council (NRC) to prepare recommendations on the appropriate reference dose for Hg exposure.

In this report, the Committee on the Toxicological Effects of Methylmercury of the NRC independently reviewed the reference dose for MeHg. The committee reviewed the available toxicological, epidemiological, and exposure data (from food and water) and determined the appropriateness of the critical study, end points of toxicity, and uncertainty factors used by EPA in the derivation of the reference dose for MeHg. The committee was also asked to identify data gaps and make recommendations for future research.

This report has been reviewed in draft form by individuals chosen for their diverse perspectives and technical expertise in accordance with procedures approved by the NRC's Report Review Committee for reviewing NRC and Institute of Medicine reports. The purpose of this independent review is to provide candid and critical comments that will assist the NRC in making the published report as sound as possible and to ensure that the report meets institutional standards for objectivity, evidence, and responsiveness to the study charge. The review comments and draft manuscripts remain confidential to protect the integrity of the deliberative process. The committee wishes to thank the following individuals, who are neither officials nor employees of the NRC, for their participation in the review of this report: Melvin Andersen, Colorado State University; Michael Aschner, Wake Forest University; Kenny Crump, ICF Consulting; Kim Dietrich, University of Cincinnati; Johanna Dwyer, New England Medical Center; John Emmerson, Eli Lilly (retired); Susan Miller, University of California at San Francisco; Charles Poole, University of North Carolina; Jonathan Samet, Johns Hopkins University; Ellen Silbergeld, University of Maryland; Christopher Whipple, Environ International Corporation; James Woods, University of Washington.

The individuals listed above have provided many constructive comments and suggestions. It must be emphasized, however, that responsibility for the final content of this report rests entirely with the authoring committee and the NRC.

The committee gratefully acknowledges the following individuals for providing background information and for making presentations to the committee: Richard Duffy of the office of Senator Patrick Leahy (Vermont); Lee Alman of the office of Congressman Alan Mollohan (West Virginia); George Lucier, National Institute of Environmental Health Sciences; William Farland, EPA; Michael Bolger, Food and Drug Administration; Christopher DeRosa, Agency for Toxic Substances and Disease Registry; E. Spencer Garrett, National Oceanic and Atmospheric Administration, Fran Sharples, Office of Science and Technology; Michael Bender, Mercury Policy Project; Jane Williams, California

Communities Against Toxics; Eric Uram, Sierra Club-Great Lakes Program; Greg Schaefer, Arch Coal, Inc.; Leonard Levin, Electric Power Research Institute; and David Michaud, Wisconsin Electric Power Company. The committee also heard from a number of researchers actively investigating issues related to MeHg exposure. Those researchers are Tord Kjellstrom, University of Auckland, New Zealand; Donna Mergler, University of Quebec at Montreal; Kenny Crump, ICF Kaiser; Ellen Silbergeld, University of Maryland; Philippe Grandjean, University of Southern Denmark; Neils Keiding and Esben Budtz-Joergensen, both from the University of Copenhagen; and Thomas Clarkson, Christopher Cox, Gary Myers, Philip Davidson, and Mark Moss, all from the University of Rochester. In addition, the committee wants to give special thanks to individuals and groups who provided further analyses and information at the request of the committee. Those are Wayne Rosamond, University of North Carolina; Philippe Grandjean; Neils Keiding; Esben Budtz-Jørgensen; Thomas Clarkson; Christopher Cox; Tord Kjellstrom; Harvey Clewell III; Jeffrey Swartout; Cynthia Van Landingham; and Kenny Crump. The committee also gratefully acknowledges input from individuals representing the Environmental Working Group, the Dental Amalgam Mercury Syndrome (DAMS) organization and the Mercury Free Press.

The committee is grateful for the assistance of the NRC staff in preparing the report. Staff members who contributed to this effort are Carol A. Maczka, senior program director for the Toxicology and Risk Assessment Program; Michelle Catlin, research associate; Ruth E. Crossgrove, editor; Laura Holliday and Judy Estep, senior project assistants; and Mirsada Karalic-Loncarevic, information specialist.

Finally, I would like to thank all the members of the committee for their dedicated efforts throughout the development of this report.

Robert A. Goyer
Chair, Committee on the Toxicological
Effects of Methylmercury

CONTENTS

EXECUTIVE SUMMARY 1

1 INTRODUCTION 13
Sources of Hg, 15
Fate and Transport, 16
Health Effects, 16
Exposure Events and Studies, 18
Summary of Risk Assessments for MeHg, 21
Scientific Controversies and Sources of Uncertainty, 26
Organization of the Report, 26
References, 27

2 CHEMISTRY, EXPOSURE, TOXICOKINETICS, AND TOXICODYNAMICS 31
Physical and Chemical Properties, 31
Methods of Chemical Analysis, 37
Exposures to MeHg in the U.S. Population, 38
Toxicokinetics, 42
Mobilization of Body Hg, 51
Chemical Forms of Hg in Toxicity, 52
Toxic Effects and Target Organs, 53
Biochemical Mechanisms of Toxicity, 54
Summary and Conclusions, 58
Recommendations, 60
References, 60

3 BIOLOGICAL VARIABILITY 72
Age-Related Susceptibility, 72
Gender Differences, 73
Genetics, 74
Mechanisms of Nutritional Influence on MeHg Health Effects, 75
Toxicokinetic Variability, 83
Conclusions, 95
Recommendations, 96
References, 98

4 DOSE ESTIMATION 105
Dietary Assessment, 105
Biomarkers of Exposure, 111
Analytical Error in Biomarker Measurements, 127
Exposure and Dose Assessment in the Seychelles, Faroe Islands, and New Zealand Studies, 129
Summary and Conclusions, 136
Recommendations, 139
References, 140

5 HEALTH EFFECTS OF METHYLMERCURY 147
Carcinogenicity, 149
Genotoxicity, 154
Immunotoxicity, 156
Reproductive Effects, 161
Renal Toxicity, 164
Cardiovascular Effects, 168
Hematological Effects, 173
Developing Central-Nervous-System Toxicity, 174
Adult Central-Nervous-System Toxicity, 221
Conclusions, 228
Recommendations, 231
References, 232

6 COMPARISON OF STUDIES FOR USE IN RISK ASSESSMENT 250
Assessment of Prenatal Hg Exposure: Cord Blood Versus Maternal Hair and Timing of Exposure, 252
Differences in the Neurobehavioral End Points Assessed and the Children's Ages at Assessment, 255
Stable Versus Episodic Pattern of Exposure, 258

Study Differences in Control for Confounders, 259
Population Differences in Vulnerability, 264
Random Variation in the Detectability of Effects at Low Exposures, 266
Conclusions, 267
Recommendations, 269
References, 269

7 DOSE-RESPONSE ASSESSMENT **271**
Risk Assessment for Non-Cancer End Points, 271
Benchmark-Dose Calculations for Continuous Outcomes, 273
Some Specific Considerations for MeHg, 277
Comparing Benchmark Doses, 281
Choosing a Critical Dose for a Point of Departure, 283
An Integrative Analysis, 289
Model Choice Issues, 293
Summary and Conclusions, 298
Recommendations, 300
References, 301

8 RISK CHARACTERIZATION AND PUBLIC HEALTH IMPLICATIONS **304**
The Current EPA Reference Dose, 305
Evaluating the RfD–End Points of MeHg Toxicity, 307
Selection of the End Point for the RfD, 311
Examination of Critical Studies for the RfD, 311
BMD Considerations: Selecting a Point of Departure, 314
Selection of the Critical Study and Point of Departure for the Revised RfD, 317
Sources of Uncertainty: Consideration for Uncertainty Factors, 318
Implications for Public Health and Risk Management, 322
Committee Findings and Recommendations, 326
References, 329

APPENDIX TO CHAPTER 7 **333**

GLOSSARY **337**

Toxicological Effects of Methylmercury

EXECUTIVE SUMMARY

MERCURY (Hg) is widespread and persistent in the environment. Its use in many products and its emission from combustion processes have resulted in well-documented instances of population poisonings, high-level exposures of occupational groups, and worldwide chronic, low-level environmental exposures. In the environment, Hg is found in its elemental form and in various organic compounds and complexes. Methylmercury (MeHg), one organic form of Hg, can accumulate up the food chain in aquatic systems and lead to high concentrations of MeHg in predatory fish,[1] which, when consumed by humans, can result in an increased risk of adverse effects in highly exposed or sensitive populations. Consumption of contaminated fish is the major source of human exposure to MeHg in the United States.

In recent years, the U.S. Environmental Protection Agency (EPA) has issued two major reports on Hg to the U.S. Congress on Hg—the *Mercury Study Report to Congress* (issued in December 1997) and the *Utility Hazardous Air Pollutant Report to Congress* (issued in March 1998). In those reports, fossil-fuel power plants, especially coal-fired utility boilers, were identified as the source category that generates the greatest Hg emissions, releasing approximately 40 tons annually in the United States. EPA is currently considering rule-making for supplemental controls on Hg emissions from utilities. However, because of gaps in the

[1]In this report, the term fish includes shellfish and marine mammals, such as pilot whales, that are consumed by certain populations.

scientific data regarding Hg toxicity, Congress directed EPA, in the appropriations report for EPA's fiscal 1999 funding, to request the National Academy of Sciences to perform an independent study on the toxicological effects of MeHg and to prepare recommendations on the establishment of a scientifically appropriate MeHg exposure reference dose (RfD).[2]

THE CHARGE TO THE COMMITTEE

In response to the request, the National Research Council (NRC) of the National Academies of Sciences and Engineering convened the Committee on Toxicological Effects of Methylmercury, whose members have expertise in the fields of toxicology, pharmacology, medicine, epidemiology, neurophysiology, developmental psychology, public health, nutrition, statistics, exposure assessment, and risk assessment. Specifically, the committee was assigned the following tasks:

1. Evaluate the body of evidence that led to EPA's current RfD for MeHg. On the basis of available human epidemiological and animal toxicity data, determine whether the critical study, end point of toxicity, and uncertainty factors used by EPA in the derivation of the RfD for MeHg are scientifically appropriate. Sensitive subpopulations should be considered.
2. Evaluate any new data not considered in the 1997 *Mercury Study Report to Congress* that could affect the adequacy of EPA's MeHg RfD for protecting human health.
3. Consider exposures in the environment relevant to evaluation of likely human exposures (especially to sensitive subpopulations and especially from consumption of fish that contain MeHg). The evaluation should focus on those elements of exposure relevant to the establishment of an appropriate RfD.
4. Identify data gaps and make recommendations for future research.

---

[2]A reference dose is defined as an estimate of a daily exposure to the human population (including sensitive subpopulations) that is likely to be without a risk of adverse effects when experienced over a lifetime.

THE COMMITTEE'S APPROACH TO ITS CHARGE

To gather background information relevant to MeHg toxicity, the committee heard presentations from various government agencies, trade organizations, public interest groups, and concerned citizens. Representatives from the offices of Congressman Alan Mollohan (West Virginia) and Senator Patrick Leahy (Vermont) also addressed the committee.

The committee evaluated the body of evidence that provided the scientific basis for the risk assessments conducted by EPA and other regulatory and health agencies. The committee also evaluated new findings that have emerged since the development of EPA's current RfD and met with the investigators of major ongoing epidemiological studies to examine and compare the methods and results.

The committee was not charged to calculate an RfD for MeHg. Instead, in its report, the committee provides scientific guidance to EPA on the development of an RfD. To develop such guidance, the committee reviewed the health effects of MeHg to determine the target organ, critical study, end point of toxicity, and dose on which to base the RfD. Because various biomarkers of exposure (i.e., concentrations of Hg in hair and umbilical-cord blood) have been used to estimate the dose of MeHg ingested by individuals, the committee evaluated the appropriateness of those biomarkers for estimating dose and the extent to which individual differences can influence the estimates. Other sources of uncertainty in the MeHg data base that should be considered when deriving an RfD were also evaluated. To estimate the appropriate point of departure[3] to use in calculating an RfD, the committee statistically analyzed available dose-response data. A margin-of-exposure[4] analysis was also performed to assess the public-health implications of MeHg.

---

[3]The point of departure represents an estimate or observed level of exposure or dose which is associated with an increase in adverse effect(s) in the study population. Examples of points of departure include NOAELs, LOAELs, BMDs, and BMDLs.

[4]A margin-of-exposure analysis compares the levels of MeHg to which the U.S. population is exposed with the point of departure to characterize the risk to the U.S. population. The larger the ratio, the greater degree of assumed safety for the population.

THE COMMITTEE'S EVALUATION

Health Effects of Methylmercury

MeHg is rapidly absorbed from the gastrointestinal tract and readily enters the adult and fetal brain, where it accumulates and is slowly converted to inorganic Hg. The exact mechanism by which MeHg causes neurotoxic effects is not known, and data are not available on how exposure to other forms of Hg affects MeHg toxicity.

MeHg is highly toxic. Exposure to MeHg can result in adverse effects in several organ systems throughout the life span of humans and animals. There are extensive data on the effects of MeHg on the development of the brain (neurodevelopmental effects) in humans and animals. The most severe effects reported in humans were seen following high-dose poisoning episodes in Japan and Iraq. Effects included mental retardation, cerebral palsy, deafness, blindness, and dysarthria in individuals who were exposed in utero and sensory and motor impairment in exposed adults. Chronic, low-dose prenatal MeHg exposure from maternal consumption of fish has been associated with more subtle end points of neurotoxicity in children. Those end points include poor performance on neurobehavioral tests, particularly on tests of attention, fine-motor function, language, visual-spatial abilities (e.g., drawing), and verbal memory. Of three large epidemiological studies, two studies—one conducted in the Faroe Islands and one in New Zealand—found such associations, but those effects were not seen in a major study conducted in the Seychelles islands.

Overall, data from animal studies, including studies on nonhuman primates, indicate that the developing nervous system is a sensitive target organ for low-dose MeHg exposure. Results from animal studies have reported effects on cognitive, motor, and sensory functions.

There is also evidence in humans and animals that exposure to MeHg can have adverse effects on the developing and adult cardiovascular system (blood-pressure regulation, heart-rate variability, and heart disease). Some research demonstrated adverse cardiovascular effects at or below MeHg exposure levels associated with neurodevelopmental effects. Some studies demonstrated an association between MeHg and cancer, but, overall, the evidence for MeHg being carcinogenic is inconclusive. There is also evidence in animals that the immune and reproductive systems are sensitive targets for MeHg.

On the basis of the body of evidence from human and animal studies, the committee concludes that neurodevelopmental deficits are the most sensitive, well-documented effects and currently the most appropriate for the derivation of the RfD.

Determination of the Critical Study for the RfD

The standard approach for developing an RfD involves selecting a critical study that is well conducted and identifies the most sensitive end point of toxicity. The current EPA RfD is based on data from a poisoning episode in Iraq. However, MeHg exposures in that study population were not comparable to low-level, chronic exposures seen in the North American population, and there are a number of uncertainties associated with the Iraqi data. In light of those considerations and more recent epidemiological studies, the committee concludes that the Iraqi study should no longer be considered the critical study for the derivation of the RfD.

Results from the three large epidemiological studies—the Seychelles, Faroe Islands, and New Zealand studies—have added substantially to the body of knowledge on brain development following long-term exposure to small amounts of MeHg. Each of the studies was well designed and carefully conducted, and each examined prenatal MeHg exposures within the range of the general U.S. population exposures. In the Faroe Islands and New Zealand studies, MeHg exposure was associated with poor neurodevelopmental outcomes, but no relation with outcome was seen in the Seychelles study.

Differences in the study designs and in the characteristics of the study populations might explain the differences in findings between the Faroe and the Seychelles studies. Differences include the ways MeHg exposure was measured (i.e., in umbilical-cord blood versus maternal hair), the types of neurological and psychological tests administered, the age of testing (7 years versus 5.5 years of age), and the patterns of MeHg exposure. When taking the New Zealand study into account, however, those differences in study characteristics do not appear to explain the differences in the findings. The New Zealand study used a research design and entailed a pattern of exposure similar to the Seychelles study, but it reported associations with Hg that were similar to those found in the Faroe Islands.

The committee concludes that there do not appear to be any serious flaws in the design and conduct of the Seychelles, Faroe Islands, and New Zealand studies that would preclude their use in a risk assessment. However, because there is a large body of scientific evidence showing adverse neurodevelopmental effects, including well-designed epidemiological studies, the committee concludes that an RfD should not be derived from a study, such as the Seychelles study, that did not observe any associations with MeHg.

In comparing the studies that observed effects, the strengths of the New Zealand study include an ethnically mixed population and the use of end points that are more valid for predicting school performance. The advantages of the Faroe Islands study over the New Zealand study include a larger study population, the use of two measures of exposure (i.e., hair and umbilical-cord blood), extensive peer review in the epidemiological literature, and re-analysis in response to questions raised by panelists at a 1998 NIEHS workshop and by this committee in the course of its deliberations.

The Faroe Islands population was also exposed to relatively high levels of polychlorinated biphenyls (PCBs). However, on the basis of an analysis of the data, the committee concluded that the adverse effects found in the Faroe Islands study, including those seen in the Boston Naming Test,[5] were not attributable to PCB exposure and that PCB exposure did not invalidate the use of the Faroe Islands study as the basis of risk assessment for MeHg.

The committee concludes that, given the strengths of the Faroe Islands study, it is the most appropriate study for deriving an RfD.

Estimation of Dose and Biological Variability

In epidemiological studies, uncertainties and limitations in estimating

---

[5]The Boston Naming Test is a neuropsychological test that assesses an individual's ability to retrieve a word that appropriately expresses a particular concept.

exposures can make it difficult to quantify dose-response associations and can thereby lead to inaccuracies when deriving an RfD. An individual's exposure to MeHg can be estimated from dietary records or by measuring a biomarker of exposure (i.e., concentration of Hg in the blood or hair).

Dietary records, umbilical-cord-blood Hg concentrations, and maternal-hair Hg concentrations all provide different kinds of exposure information. Dietary records can provide information on Hg intake but depend on accurate knowledge of Hg concentrations in fish. The records also might be subject to problems with estimating portion size and capturing intermittent eating patterns. Umbilical-cord-blood Hg concentrations would be expected to correlate most closely with fetal-brain Hg concentrations during late gestation and correlate less well with Hg intake than do the other measures (e.g., dietary records and maternal-hair Hg concentration). Maternal-hair Hg concentrations can provide data on Hg exposure over time, but they might not provide as close a correlation with fetal-brain Hg concentrations as umbilical-cord-blood Hg concentrations, at least during the latter period of gestation. Use of data from two or more of these measurement methods increases the likelihood of uncovering true dose-response relationships. The use of either umbilical-cord-blood or maternal-hair Hg concentrations as biomarkers of exposure is adequate for estimating a dose received by an individual.

Individual responses to MeHg exposure are variable and a key source of uncertainty. Factors that might influence the responses include genetics, age, sex, health status, nutritional supplements, nutritional influences, including dietary interactions, and linking the time and intensity of MeHg exposure to the critical periods of brain development. In addition, people exposed to the same amount of MeHg can have different concentrations of Hg at the target organ because of individual variability in the way the body handles MeHg. Individual differences that affect the estimation of dose can be addressed in the derivation of the RfD by applying an uncertainty factor to the estimated dose. If an RfD is based on a Hg concentration in maternal-hair or umbilical-cord blood, adjusting by an uncertainty factor of 2-3 would account for individual differences in the estimation of dose in 95% to 99% of the general population.

Modeling the Dose-Response Relationships

An important step in deriving an RfD is choosing an appropriate dose to be used as the "point of departure" (i.e., the dose to which uncertainty factors will be applied to estimate the RfD). The best available data for assessing the risk of adverse effects for MeHg are from the Faroe Islands study. Because those data are epidemiological, and exposure is measured on a continuous scale, there is no generally accepted procedure for determining a dose at which no adverse effects occur. The committee concludes, therefore, that a statistical approach (i.e., calculation of a benchmark dose level, BMDL[6]) should be used to determine the point of departure for MeHg instead of identifying the dose at which no adverse effects occur or the lowest dose at which adverse effects occur. The committee cautions, however, that the type of statistical analysis conducted (i.e., the model choice – K power, logarithmic, or square root) can have a substantial effect on the estimated BMDL. The committee recommends the use of the K-power model with the constraint of K ≥ 1, because it is the most plausible model from a biological perspective and also because it tends to yield the most consistent results for the Faroe Islands data. It should be noted that, for the data from the Faroe Islands study, the results of the K-power model with the constraint of K ≥ 1 are equivalent to the results of the linear model.

The adverse effects observed in the Faroe Islands study were most sensitively detected when using cord blood as the biomarker. Based on cord-blood analyses from the Faroe Islands study, the lowest BMD for a neurobehavioral end point the committee considered to be sufficiently reliable is for the Boston Naming Test. Thus, on the basis of that study and that test, the committee's preferred estimate of the BMDL is 58 parts per billion (ppb)[7] of Hg in cord blood. To estimate this BMDL, the

---

[6]A benchmark dose level is the lowest dose, estimated from the modeled data, that is expected to be associated with a small increase in the incidence of adverse outcome (typically in the range of 1% to 10%).

[7] The BMDL of 58 ppb is calculated statistically and represents the lower 95% confidence limit on the dose (or biomarker concentration) that is estimated to result in a 5% increase in the incidence of abnormal scores on the Boston Naming Test.

committee's calculations involved a series of steps, each involving one or more assumptions and related uncertainties. Alternative assumptions could have an impact on the estimated BMDL value. In selecting a single point of departure, the committee followed established public-health practice of using the lowest value for the most sensitive, relevant end point.

In addition to deriving a BMDL based on the Faroe Islands study, the committee performed an integrative analysis of the data from all three studies to evaluate the full range of effects of MeHg exposure. The values obtained by the committee using that approach are consistent with the results of the benchmark analysis of the Boston Naming Test from the Faroe Islands study. Because an integrative analysis is not a standard approach at present, the committee does not recommend that it be used as the basis for an RfD.

Public-Health Implications

The committee's margin-of-exposure analysis based on estimates of MeHg exposures in U.S. populations indicates that the risk of adverse effects from current MeHg exposures in the majority of the population is low. However, individuals with high MeHg exposures from frequent fish consumption might have little or no margin of safety (i.e., exposures of high-end consumers are close to those with observable adverse effects). The population at highest risk is the children of women who consumed large amounts of fish and seafood during pregnancy. The committee concludes that the risk to that population is likely to be sufficient to result in an increase in the number of children who have to struggle to keep up in school and who might require remedial classes or special education. Because of the beneficial effects of fish consumption, the long-term goal needs to be a reduction in the concentrations of MeHg in fish rather than a replacement of fish in the diet by other foods. In the interim, the best method of maintaining fish consumption and minimizing Hg exposure is the consumption of fish known to have lower MeHg concentrations.

In the derivation of an RfD, the benchmark dose is divided by uncertainty factors. The committee identified two major categories of uncertainty, based on the body of scientific literature, that should be considered when revising the RfD: (1) biological variability when estimating dose and (2) data-base insufficiencies. On the basis of the available scientific data, the committee concludes that a safety factor of 2-3 will account for biological variability in dose estimation. The choice of an uncertainty factor for data-base insufficiencies is, in part, a policy decision. However, given the data indicating possible long-term neurological effects not evident at childhood, immunotoxicity, and cardiovascular effects, the committee supports an overall composite uncertainty factor of no less than 10.

RESEARCH NEEDS

To better characterize the health effects of MeHg, the committee recommends further investigation of the following:

- The impacts of MeHg on the prevalence of hypertension and cardiovascular disease in the United States. Such data should be considered in a re-evaluation of the RfD as they become available.
- The relationships between low-dose exposure to MeHg throughout the life span of humans and animals and carcinogenic, reproductive, neurological, and immunological effects.
- The potential for delayed neurological effects resulting from Hg remaining in the brain years after exposure.
- The emergence of neurological effects later in life following low-dose prenatal MeHg exposure.
- The mechanisms underlying MeHg toxicity.

To improve estimates of dose and to clarify the impact of biological variability and other factors on MeHg dose-response relationships, the committee recommends the following:

- The analysis of hair samples to evaluate the variability in short-term exposures, including peak exposures. Hair that has been stored from the Seychelles and the Faroe Islands studies should be analyzed to determine variability in exposures over time.
- The collection of information on what species of fish are eaten at

specific meals to improve estimates of dietary intakes and temporal variability in MeHg intake.
- The assessment of factors that can influence individual responses to MeHg exposures in humans and animals. Such factors include age, sex, genetics, health status, nutritional supplement use, and diet. Food components considered to be protective against MeHg toxicity in humans also deserve closer study (e.g., wheat bran and vitamin E).

To determine the most appropriate methods for handling model uncertainty in benchmark analysis, the committee recommends that further statistical research be conducted.

To better characterize the risk to the U.S. population from current MeHg exposures, the committee recommends obtaining data on the following:

- Regional differences in MeHg exposure, populations with high consumptions of fish, and trends in MeHg exposure. Characterization should include improved nutritional and dietary exposure assessments and improved biomonitoring of subpopulations.
- Exposure to all chemical forms of Hg, including exposure to elemental Hg from dental amalgams.

RECOMMENDATIONS

On the basis of its evaluation, the committee's consensus is that the value of EPA's current RfD for MeHg, 0.1 μg/kg per day, is a scientifically justifiable level for the protection of public health. However, the committee recommends that the Iraqi study no longer be used as the scientific basis of the RfD. The RfD should still be based on the developmental neurotoxic effects of MeHg, but the Faroe Islands study should be used as the critical study for the derivation of the RfD. Based on cord-blood analyses from the Faroe Islands study, the lowest BMD for a neurobehavioral end point the committee considered to be sufficiently reliable is for the Boston Naming Test. For that end point, dose-response data based on Hg concentrations in cord blood should be modeled using the K-power model ($K \geq 1$). That approach estimates a BMDL of 58 ppb of Hg in cord blood (corresponding to a BMDL of 12 ppm of Hg in hair) as a reasonable point of departure for deriving the RfD. To calculate the RfD, the BMDL should be divided by uncertainty factors that take into consideration biological variability when estimating dose and MeHg data-base insufficiencies. As stated earlier, given those considerations, an uncertainty factor of at least 10 is supported by the committee.

The committee further concludes that the case of MeHg presents a strong illustration of the need for harmonization of efforts to establish a common scientific basis for exposure guidance and to reduce current differences among agencies, recognizing that risk-management efforts reflect the differing mandates and responsibilities of the agencies.

1

INTRODUCTION

MERCURY (Hg) is a persistent substance that comes from natural and anthropogenic sources. Hg that enters our oceans, lakes, and rivers is converted to methylmercury (MeHg) by aquatic biota and bioaccumulates in aquatic food webs including fish and shellfish. Humans and wildlife are exposed to MeHg primarily through the consumption of contaminated fish,[1] particularly large predatory fish species such as tuna, swordfish, shark, and whale. In humans, MeHg is known to be neurotoxic. The fetus is more sensitive to those effects than the adult (EPA 1997a).

In 1997, the U.S. Environmental Protection Agency (EPA) issued two reports on Hg and its effects on public health to the U.S. Congress. The first of these reports, the *Mercury Study Report to Congress* (EPA 1997a,b,c), assessed the source and amount of Hg emissions in the United States, the detrimental effects of Hg on humans and wildlife, and the feasibility of control technologies. The second report, the *Study of Hazardous Air Pollutant Emissions from Electric Utility Steam Generating Units. Final Report to Congress* (EPA 1998), looked specifically at emissions from utility companies and cited Hg as a major contaminant,

---

[1]In this report, the term fish includes shellfish and marine mammals, such as the pilot whale, that are consumed by certain populations.

especially in emissions from coal-fired power plants. Because concerns have been raised about Hg exposure levels in the United States, particularly among sensitive populations, questions have arisen among federal agencies over what is an acceptable level of exposure to MeHg.

Due to disagreement over the appropriate level of concern for MeHg exposure, the potentially widespread implications for human health, and the challenges associated with further regulating Hg emissions, Congress directed EPA in the House Appropriations Report for EPA's Fiscal 1999 funding to contract with the National Research Council (NRC) to prepare recommendations on the appropriate value for a Hg exposure reference dose (RfD). In response, the NRC convened the Committee on the Toxicological Effects of Mercury, whose membership includes experts in toxicology, pharmacology, medicine, epidemiology, developmental psychology, neurophysiology, neuropsychology, public health, nutrition, statistics, exposure assessment, and risk assessment. The committee was charged with the following specific tasks:

1. Evaluate the body of evidence that led to the EPA-derived MeHg RfD. Human epidemiological and animal toxicity data should be the basis of the evaluation. The evaluation should determine the appropriateness of the critical study, end point of toxicity, and uncertainty factors used by EPA in deriving the RfD for MeHg. Sensitive populations should be considered.
2. Evaluate any new data (e.g., mechanistic data) that were not considered in EPA's 1997 Hg report that are relevant to EPA's MeHg RfD for protecting human health.
3. Consider exposure pathways (especially from the consumption of MeHg in fish) in evaluating likely human exposures, especially exposures of sensitive subpopulations. The evaluation should focus on those elements of exposure relevant to the establishment of an appropriate RfD.
4. Identify data gaps and make recommendations for future research.

Although the committee name, the Committee on the Toxicological Effects of Mercury, does not limit the scope of this report to MeHg, the committee focused on the health effects of this organic form of Hg because the toxicity due to this form is of greatest concern. In addition, the committee did not attempt to establish an RfD for MeHg. Instead, the committee provides guidance to EPA on the data sets, exposure-assessment approaches, modeling techniques, and statistical analysis that should be considered in deriving an appropriate Hg RfD.

SOURCES OF Hg

In the environment, Hg comes from natural and anthropogenic sources. Mercuric sulfide, or Hg in cinnabar, is the natural form of Hg. The concentration of cinnabar varies greatly with the location of deposits. Hg can be released into the air through weathering of rock containing Hg ore or through human activities, principally incineration and burning of fossil fuels. Hg is a global pollutant, that once released to the air can travel long distances and impact distant sites. Water contamination can occur from run-off water, contaminated by either natural or anthropogenic sources, or from air deposition. Potential sources of general population exposure to Hg include inhalation of Hg vapors in ambient air, ingestion of drinking water and foodstuffs contaminated with Hg, and exposure to Hg from dental amalgams and medical treatments. Dietary intake is one of the most important sources of non-occupational exposure to Hg, fish and other seafood products being the dominant source of Hg in the diet. Most of the Hg consumed in fish or other seafood is the highly absorbable MeHg form. The substantial variation in human MeHg exposure is based on the differences in frequency and amount of fish consumed and Hg concentration in the fish. MeHg exposure is a major problem in some populations, especially subsistence fish eaters who consume large amounts of fish (EPA 1997a). Intake of elemental Hg from dental analgams is another major contributing source to the total Hg body burden in humans in the general population (IPCS 1990, 1991).

The World Health Organization (WHO) has estimated that anthropogenic sources, mainly the combustion of fossil fuels, contribute 25% of the overall (natural and anthropogenic) Hg emissions to the atmosphere (ATSDR 1999). EPA has estimated that those sources account for 50% to 75% of the total yearly input of Hg into the atmosphere (EPA 1997a). In the United States, the majority of Hg emissions are from combustion sources. Medical and municipal waste incinerators and coal-fired utility boilers account for greater than 80% of the Hg emitted from point sources (EPA 1997b; ATSDR 1999).

FATE AND TRANSPORT

Hg has three valence states (Hg^0, Hg^{1+}, Hg^{2+}) and is found in the environment in the metallic form and in various inorganic and organic complexes. The natural global bio-geochemical cycling of Hg is characterized by degassing of the element from soils and surface waters, atmospheric transport, deposition of Hg back to land and surface water, sorption of the compound onto soil or sediment particles, and revolatilization from land and surface water (see Figure 1-1). This emission, deposition, and revolatilization creates difficulties in tracing the movement of Hg to its sources (ATSDR 1999). Once in the environment, interconversion between the different forms of Hg can occur. Particulate-bound Hg can be converted to insoluble Hg sulfide and precipitated or bioconverted into more volatile or soluble forms that re-enter the atmosphere or are bioaccumulated in aquatic and terrestrial food chains. Conversion of inorganic Hg to MeHg occurs primarily in microorganisms especially in aquatic systems. Once in its methylated form, Hg bioaccumulates up the food chain; the microorganisms are consumed by fish, and the smaller fish are consumed by larger fish. Such bioaccumulation can result in very high concentrations of MeHg in some fish, which are one of the main sources of human and piscivorus wildlife exposure to MeHg.

HEALTH EFFECTS

Human exposure to MeHg from contaminated fish and seafood can pose a variety of health risks. A spectrum of adverse health effects has been observed following MeHg exposure, with the severity depending largely on the magnitude of the dose. Fatalities and devastating neurological damage were observed in association with the extremely high exposures that occurred during the Minamata and Iraqi poisoning

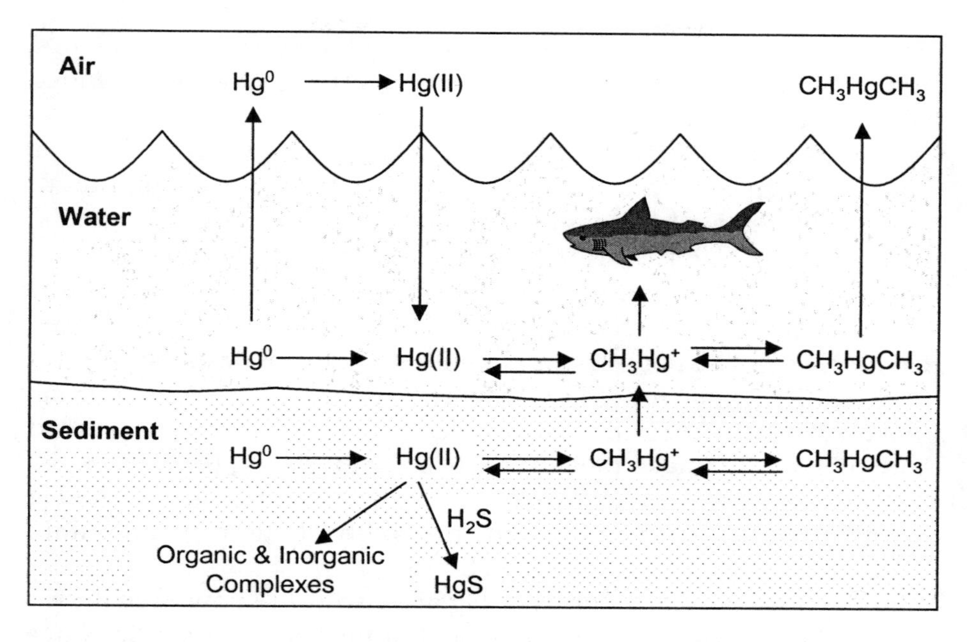

FIGURE 1-1 Cycling of Hg in aquatic system. CH_3Hg^+, methylmercury ion; CH3HgCH$_3$, dimethylmercury; Hg(ll), mercuric mercury; Hg^0, elemental mercury; H_2S, hydrogen sulfide; HgS, cinnabar. Source: Adapted from EPA 1997b.

episodes. The fetus is considered much more sensitive than the adult. Prenatal exposures interfere with the growth and migration of neurons and have the potential to cause irreversible damage to the developing central nervous system (EPA 1997a). Infants exposed in utero to MeHg during the Minamata and Iraqi episodes were born with severe disabilities, such as mental retardation, seizure disorders, cerebral palsy, blindness, and deafness. At much lower doses that result from chronic maternal fish consumption, infants might appear normal during the first few months of life but might later display deficits in subtle neurological end points (e.g., IQ deficits, abnormal muscle tone, decrements in motor function, attention, and visuospatial performance).

Exposures that occur during childhood and adulthood can also cause damage to the central nervous system, as evidenced by human poisoning incidents in Japan, Iraq, and the United States, in which the first signs of toxicity often appear several months after exposure has ended (EPA 1997b, Davis et al. 1994).

There is evidence that MeHg also effects other systems. In 1995, researchers in Finland found a correlation between consumption of MeHg-contaminated fish and the risk of acute myocardial infarction (Salonen et al., 1995). This prospective study of 1,833 fishermen was intended to confirm previous studies in which fish consumption was associated with a reduced risk of heart disease. Instead, they discovered that hair Hg levels above 2 parts per million (ppm), or daily ingestion of more than 30 grams (g) of fish, increased the risk of acute myocardial infarction (AMI) or cardiovascular death 2- to 3-fold. The estimated daily dietary Hg intake ranged from 1.1 μg to 95.3 μg (mean, 7.6 μg). The investigators theorized that the cardiovascular effects of MeHg might be caused, at least in part, by the ability of Hg to enhance lipid peroxidation via a Fenton-type reaction.

Inorganic and organic forms of Hg are also well-known renal toxicants. Human case investigations and animal feeding studies have repeatedly confirmed that effect. Human exposures to organic Hg have resulted in symptoms of polyuria and albuminuria (Jalili and Abbasi 1961; Cinca et al. 1979). Autopsies of patients who died following ingestion of alkyl Hg revealed nephritis and tubular degeneration (Al-Saleem 1976; Cinca et al. 1979). Animal studies have shown that MeHg damages the proximal tubules in the kidney (Mitsumori et al. 1990).

During the past decade, researchers have studied the effects of MeHg on immune function and blood-pressure regulation. After administering MeHg to mice for 12 weeks, Ilbäck (1991) noted changes in the thymus and natural killer-cell activity. Sørensen et al. (1999) found an association between prenatal exposure to MeHg and childhood blood pressure. Diastolic and systolic blood pressures, measured at age 7, increased 13.9 millimeters (mm) and 14.6 mm, respectively, as cord-blood Hg concentrations rose from 1 to 10 micrograms per liter (μg/L).

EXPOSURE EVENTS AND STUDIES

Between 1950 and 1975, several MeHg poisoning incidents occurred in Japan and Iraq. Scientists who investigated those events identified developmental neurotoxicity as the health effect of greatest concern following high-level episodic exposures. Individuals poisoned by MeHg through consumption of contaminated fish in Japan exhibited paresthesia, ataxia, sensory disturbances, tremors, impairment of hearing, and difficulty walking (Harada 1995). In Iraq, exposure was due to the consumption of home-made bread that was made with grain treated with MeHg as a fungicide. In that outbreak, the most common symptom in adults was paresthesia; the most severely affected individuals exhibited ataxia, blurred vision, slurred speech, hearing difficulties, blindness, deafness, and death (Marsh et al. 1987). In both Iraq and Japan, the effects in offspring who were exposed to MeHg in utero were more serious, and in some cases seen at lower doses, than in adults. Both exposure episodes have been studied to determine the doses and the effects resulting from exposure to MeHg. Although the doses that produced those effects in the Japanese and Iraqi populations were undoubtedly quite high, precise dose-response relationships have not been established, and the exposure scenarios are not comparable to the low-dose chronic exposure that the general population in North America might experience.

In an attempt to establish dose-response relationships, three large prospective epidemiological studies have evaluated subtle end points of neurotoxicity. One study was conducted in the Republic of the Seychelles, a nation of islands located in the Indian Ocean off the coast of East Africa (Davidson et al. 1995, 1998). Another major study was conducted in the Faroe Islands (part of Denmark), which are located in the North Sea between Scotland and Iceland (Grandjean et al. 1997, 1998, 1999). The other major study was conducted in New Zealand (Kjellström et al. 1986, 1989). The populations of the Seychelles, Faroe Islands, and New Zealand were chosen for study, because their dietary dependence on fish and marine mammals provides an ongoing source of exposure to MeHg. Prenatal MeHg exposures in those populations were within the range of at least some U.S. population exposures. All three studies evaluated large numbers of subjects.

The 66-month study of 711 children in the Seychelles islands assessed the effects of prenatal MeHg in tests of global intelligence and developmental milestones. No adverse effects were seen that could be attributed to MeHg. Maternal hair samples collected at birth contained Hg concentrations that ranged from 0.5 to 27 ppm (mean, 6.8 ppm). Meanwhile, scientists working in the Faroe Islands found that children whose prenatal exposures were similar to those observed in the Seychelles population had subtle developmental dose-related deficits that were apparent at 7 years of age. Abnormalities were seen in tests of memory, attention, and language and, to a lesser extent, in neurophysiological end points. Measurements of blood pressure, heart rate, and heart-rate variability were also taken when the children reached 7 years of age. Researchers found that diastolic and systolic blood pressures increased, and heart-rate variability decreased as cord-blood Hg concentrations rose from 1 to 10 μg/L.

A prospective study carried out in New Zealand (Kjellström et al. 1986, 1989) examined the effects in offspring exposed in utero to MeHg via maternal consumption of fish. Scores on the Denver Developmental Screening Test (DDST), a standardized test for childhood mental and motor development, were compared in groups of children 4 years of age categorized by maternal Hg exposure (as measured in parts per million in maternal hair) (Kjellström et al. 1986). At 6 years of age, a battery of specific cognitive tests was administered (Kjellström et al. 1989). At both ages, the researchers found significant decrements in test performance in the children exposed to moderate-to-high doses of MeHg prenatally (more than 6 ppm).

A correlation was demonstrated between hair Hg concentrations and neurophysiological effects in a study of an adult population in the Amazon, where gold-mining activities have resulted in fish highly contaminated with Hg (Lebel et al. 1996). In that study population, it is likely that the adult population was also exposed to MeHg in utero.

The studies of the Iraqi, Amazon, Seychelles, and Faroe Islands populations were reviewed by an expert panel that met in Raleigh, North Carolina, at the Workshop on the Scientific Issues Relevant to Assessment of Health Effects from Exposure to MeHg. A report of that workshop has been published (NIEHS 1998). In suggesting possible explanations for the discrepant findings of the Seychelles and Faroe studies, the panel pointed to differences in sources of exposures or exposure measures, differences in the neurobehavioral tests used or their administration or interpretation, influences of confounders and covariates, and biostatistical issues involved in the analysis of the data. The differences between those studies are discussed further in Chapter 6.

SUMMARY OF RISK ASSESSMENTS FOR MeHg

State and national governments as well as international organizations have recommended acceptable levels of Hg exposure that are thought to be protective against adverse effects (see Table 1-1). General risk assessment approaches used by the various agencies are described in NRC (1983) and NRC (1984). In this report, information on how EPA derives an RfD can be found in the section on Risk Assessment for Noncancer End Points in Chapter 7. Specific details on the derivation of EPA's MeHg RfD can be found in the section on The Current EPA Reference Dose in Chapter 8. In the United States, responsibility for regulating Hg is shared by two federal agencies: the Food and Drug Administration (FDA) and EPA. FDA is responsible for ensuring that Hg concentrations in commercially sold fish and seafood do not exceed what the agency defines as an action level for this contaminant (FDA 1979). EPA monitors Hg concentrations in the environment and regulates industrial releases to air and surface water. Although not a regulatory agency, the Agency for Toxic Substances and Disease Registry (ATSDR) evaluates the potential for humans to be exposed to MeHg and investigates reported health effects. Currently, each of these agencies uses a different guideline to assess exposure to toxicants.

The differences in guidelines among the agencies are due to the use of different risk-assessment methods, data sets, and uncertainty factors and the different mandates of each agency (EPA 1984, 2000; FDA 1979; ATSDR 2000). For example, EPA used data from the 1971 Iraqi poisoning incident to derive an RfD of 0.1 microgram per kilogram (µg/kg) of body weight per day for MeHg (EPA 1997a). The reference dose was calculated using a benchmark dose of 1.1 µg/kg per day. That benchmark dose was divided by uncertainty factors (UF) to account for the variability in the human population (UF of 3) and for the lack data on reproductive effects, sequelae, and adult paresthesia (UF of 3). Although MeHg is classified by the agency as a possible human carcinogen, no uncertainty factor was used to protect against that effect. The RfD calculated by EPA was in the range of other values obtained by EPA using similar analysis of other data sets.

In 1998, ATSDR used the Seychelles study (Davidson et al. 1998) as the starting point for estimating a minimal risk level for exposure to MeHg (ATSDR 1999). In this study, the investigators examined

TABLE 1-1 Summary of Risk Assessments for Methylmercury

Agency	Key Studies	End Points	Biomarker and Exposure Level	Critical Dose	Uncertainty Factors	Acceptable Level
EPA[a]	Iraqi study (Marsh et al. 1987)	Combined instance of neurological effects following in utero exposure[b]	Maternal hair, 11 ppm; equivalent to intake of 1.1 μg/kg/d	Benchmark dose, 1.1 μg/kg/d[c]	UF, 10[d]	RfD, 0.1 μg/kg/d (based on fetal effects)
ATSDR	Seychelles study (Davidson et al. 1998)	Developmental neurotoxicity measured by neurological evaluation, behavioral, psychological tests	Maternal hair, 15.3 ppm; equivalent to intake of 1.3 μg/kg/d	NOAEL, 1.3 μg/kg/d	UF, 4.5[e]	MRL, 0.3 μg/kg/d
FDA	Japanese data (Friberg et al. 1971)	Overt neurological symptoms in adults	Adult blood, 0.2 ppm; equivalent to intake of 300 μg/d	LOAEL, 4.3 μg/kg/d	SF, 10[f]	Action level in fish, 1 ppm in edible portion[g] (equivalent to 0.5 μg/kg/d)
JECFA[h]	Japanese data (Friberg et al. 1971)	Overt neurological symptoms in adults	Adult blood, 0.2 ppm; equivalent to intake of 300 μg/d	LOAEL, 4.3 μg/kg/d	SF of 10[i]	pTWI, 3.3 μg/kg/wk (equivalent to 0.5 μg/kg/d)
Health Canada	Seychelles study (Davidson et al. 1998); Faroe Islands (Grandjean et al. 1997); New Zealand (Kjellstrom 1986, 1989)	Developmental neurotoxicity	Maternal hair, 10 ppm; equivalent to intake of 1 μg/kg/d	Benchmark dose, 1 μg/kg/d[j]	UF, 5[k]	pTDI, 0.2 μg/kg/d (for women of childbearing ages, infants, and young children)[l]

North Carolina	Seychelles study (Davidson et al. 1998)	Developmental neurotoxicity	Maternal hair, 6.8 ppm; equivalent to intake of 0.5 μg/kg/d	Arithmetic mean,[m] 0.5 μg/kg/d	Modifying factor, 3	RfD, 0.2 μg/kg/d (for women of childbearing age and developing fetuses)[n]
Washington State	Faroe Islands (Grandjean et al. 1997)	Impaired neurological development and long-term or delayed sequelae in children	Maternal hair, 4.3-10 ppm; equivalent to intake of 0.35-0.8 μg/kg/d	Daily intake range,[o] 0.35-0.8 μg/kg/d	UF, 10[p]	TDI, 0.035-0.08 μg/kg/d

[a]EPA report to Congress states that "a number of additional studies of human populations generally support the dose range of the benchmark dose level for perinatal effects." The agency is awaiting the results of this NRC report before updating its RfD based on more recent data.

[b]The data for delayed onset of walking and talking, neurological scores of less than 3, mental symptoms, and seizures were grouped together for this analysis.

[c]EPA carried out the analysis using the polynomial model and the Weibull model. The results of the two models were within 3% of each other. EPA based its analysis on the Weibull model due to goodness of fit and history of use. The Benchmark dose is an estimate of an experimental dose associated with a specified low incidence of adverse effects.

[d]According to the Integrated Risk Information System (IRIS), the following uncertainty factors were applied: 3 for the variability in human population (variability in the half-life of methylmercury and in hair-to-blood ratio) and 3 for the lack of a two-generation reproductive study and data on the effect of exposure duration on sequelae of the developmental neurotoxicity effects and on adult paresthesia.

[e]The following uncertainty factors were applied: 1.5 for human pharmacokinetic variability, 1.5 for human pharmacodynamic variability and 1.5 to account for domain-specific findings in the Faroe study.

[f]Arbitrary value; the *Federal Register* states that, in cases in which human data are available, the safety factor used is 10.

[g]This conversion was calculated using fish consumption data from the National Marine Fisheries Service.

[h]The Joint Food and Agriculture Organization/World Health Organization Expert Committee on Food Additives (JECFA) concluded in 1999 that "the information available was insufficient for evaluating the neurodevelopmental effects on offspring of mothers with low intakes of methylmercury."

TABLE 1-1 *(Continued)*

[i]In the absence of relevant toxicodynamic and toxicokinetic data, the committee uses a safety factor of 10 when the pTWI is based on human data.

[j]An approximate benchmark dose determined by qualitatively looking at the data.

[k]Arbitrary value.

[l]Health Canada also maintains the provisional TDI for adults of 0.47 microgram per kilogram (μg/kg) of body weight per day that was established by JECFA.

[m]Arithmetic mean exposure value for entire Seychelles cohort.

[n]North Carolina also maintains an RfD for nonsensitive populations of 0.5 μg/kg of body weight per day.

[o]Obtained using the algorithm relating Hg levels in blood to a daily intake level as described by ATSDR (1997). The geometric average maternal hair level of 4.3 ppm was used because the regression relationship between MeHg exposure and adverse effects was described from the entire cohort and the average value reflects that cohort (not withstanding that the regression may be driven by values close or below the average value), while 10 ppm represents the cutoff value used in the bivariate categorical analyses which showed a significant difference for MeHg above and below that value.

[p]Accounts for interindividual pharmacokinetic variability associated with determining a tolerable intake level based on hair mercury concentrations, for uncertainty associated with toxicodynamic variations within the populations, and for the lack of ability to address long-term or delayed sequelae.

Abbreviations: EPA, Environmental Protection Agency; ATSDR, Agency for Toxic Substances and Disease Registry; FDA; Food and Drug Administration; NOAEL; no-observed-adverse-effect level; LOAEL, lowest-observed-adverse-effect level; UF, uncertainty factor; SF, safety factor; RfD, reference dose (an amount of a substance that is anticipated to be without adverse health effects in humans, including sensitive populations, when ingested daily over a lifetime; MRL, minimal risk level (an estimate of daily human exposure to a hazardous substance that is likely to be without an appreciable risk of adverse noncancer health effects over a specified route and duration of exposure); pTDI, provisional tolerable daily intake (maximum daily exposure level to a contaminant; provisional meaning that it is considered temporary until more data are available, especially the completed Seychelles study); JECFA, Joint FAO/WHO Expert Committee on Food Additives; pTWI, provisional tolerable weekly intake.

the correlation between subtle neurological effects and low-dose chronic exposure to MeHg. No correlation between Hg concentrations and neurological effects was seen. ATSDR determined a minimal risk level of 0.3 μg/kg per day, based on a dose of 1.3 μg/kg per day, which reflects the average concentration of the upper quintile of the exposed population but does not necessarily correspond to a no-observed-adverse-effect level (NOAEL). The agency used two uncertainty factors of 1.5 each to account for pharmacokinetic and pharmacodynamic variability within the human population. A modifying factor of 1.5 was applied to account for the possibility that domain-specific tests used in the Faroe Islands study might have allowed detection of subtle neurological effects that were not evaluated in the Seychelles cohort. Although the conventional risk-assessment approach is to multiply uncertainty factors, the agency summed these factors to develop an overall safety factor of 4.5.

According to Tollefson and Cordle (1986), FDA used data from the Minamata Bay poisoning episode to determine the action level of 1 ppm (in the edible portion of fish), which corresponds to a daily intake of 0.5 μg/kg (Friberg et al. 1971). FDA followed the approach taken by the Joint Food and Agriculture Organization/World Health Organization Expert Committee on Food Additives (JECFA), who had determined a provisional tolerable weekly intake (pTWI) of 0.5 μg/kg in adults and stated that the fetus and children might be more sensitive but that the data are insufficient to determine a safe intake in these populations (JECFA 1972). That pTWI was recently confirmed at the JECFA meeting in June 1999 (JECFA 1999). Canadian recommendations are based on the JECFA pTWI in adults; however, Canada also has a provisional tolerable daily intake of 0.2 μg/kg per day for children and women of child-bearing years, an intake based on a qualitative assessment of available data (M.-T. Lo, Food Directorate, Health Canada, personal commun., June 1999). The effect on public health of using one dose rather than another to set acceptable exposure levels might be substantial, leaving open the question of which value best ensures public safety. Differences in acceptable levels can affect many government programs, including state fish advisories, and regulation of such industries as commercial fishing and electric power plants (Renner 1999).

SCIENTIFIC CONTROVERSIES AND SOURCES OF UNCERTAINTY

Many controversies surround the determination of what is an acceptable level of exposure to MeHg. Some of these controversies stem from the science underlying the toxicity data base for MeHg. For example, there is disagreement over which studies and which end points of concern should be used to derive an acceptable level. There is emerging evidence of potential effects on both the immune and cardiovascular systems at low doses. The contradictory findings from the Seychelles and Faroe Islands studies have made it difficult to determine an appropriate point of departure for risk assessment. Scientists also do not agree on whether Hg in hair or blood is the more appropriate biomarker or measurement of exposure. There is debate over the assumptions on the disposition and metabolism of MeHg that are used to extrapolate from a measured biomarker value to a corresponding Hg exposure level. In addition, there is debate over the assumptions on fish intake and the concentration of Hg in the fish that are used to determine a safe amount of fish for consumption. The choice of dose-response model and uncertainty factors, if any, is also controversial.

ORGANIZATION OF THE REPORT

The remainder of this report is organized into six chapters and an appendix. In Chapter 2, information on the chemistry, toxicokinetics, toxicodynamics, and exposure of MeHg is presented. Chapter 3 presents a discussion on toxicokinetic variability and other factors that influence variation in human sensitivity to MeHg. Those factors include age, genetics, and nutrition. In Chapter 4, issues involved in assessing MeHg exposure and dose are presented. The focus is on the selection and interpretation of dose metrics and the implications of the possible dose metrics for dose-response assessment and nutritional assessment. The health effects associated with the ingestion of MeHg are discussed in Chapter 5. Emphasis is placed on the more-recent studies with respect to the choice of end points, possible confounders, and sensitive subpopulations. Evidence from experimental animal studies is also discussed. In Chapter 6, a comparison of studies that are appropriate for risk assessment for MeHg is presented. Chapter 7 provides an evaluation of the various data sets and statistical approaches for deriving an acceptable Hg exposure level. Further details of one approach are provided in the appendix. In Chapter 8, the risks from ingestion of MeHg and the sources of uncertainty are characterized and the adequacy of the EPA MeHg RfD for protecting human health is evaluated. The public-health implications of exposure to MeHg, including the implications of choosing one Hg exposure level over another, and how these relate to state and federal concerns, such as fish advisories and consumption, are also addressed.

REFERENCES

Al-Saleem, T. 1976. Levels of mercury and pathologic changes in patients with organomercury poisoning. Bull. WHO 53(Suppl.):99-104.

ATSDR (Agency for Toxic Substances and Disease Registry). 1997. Toxicological Profile for Mercury. (Update). Draft. U.S. Department of Health and Human Services, Agency for Toxic Substances and Disease Registry, Atlanta, GA.

ATSDR (Agency for Toxic Substances and Disease Registry). 1999. Toxicological Profile for Mercury. (Update). U.S. Department of Health and Human Services, Agency for Toxic Substances and Disease Registry, Atlanta, GA.

Cinca, I., I. Dumetrescu, P. Onaca, A. Serbanescu, and B. Nestorescu. 1979. Accidental ethyl mercury poisoning with nervous system, skeletal muscle, and myocardium injury. J. Neurol. Neurosurg. Psychiatry 43(2):143-149.

Davidson, P.W., G.J. Myers, C. Cox, C.F. Shamlaye, D.O. Marsh, M.A. Tanner, M. Berlin, J. Sloane-Reeves, E. Cernichiari, O. Choisy, A. Choi, and T.W. Clarkson. 1995. Longitudinal neurodevelopmental study of Seychellois children following in utero exposure to methylmercury from maternal fish ingestion: outcomes at 19 and 29 months. Neurotoxicology 16(4):677-688.

Davidson, P.W., G.J. Myers, C. Cox, C. Axtell, C. Shamlaye, J. Sloane-Reeves, E. Cernichiari, L. Needham, A. Choi, Y. Wang, M. Berlin, and T.W. Clarkson. 1998. Effects of prenatal and postnatal methylmercury exposure from fish consumption on neurodevelopment: Outcomes at 66 months of age in the Seychelles Child Development Study. JAMA 280(8):701-707.

Davis, L.E., M. Kornfeld, H.S. Mooney, K.J. Fiedler, K.Y. Haaland, W.W. Orrison, E. Cernichiari, and T.W. Clarkson. 1994. Methylmercury poisoning: Long-term clinical, radiological, toxicological, and pathological studies of an affected family. Ann. Neurol. 35(6):680-688.

EPA (U.S. Environmental Protection Agency). 1997a. Mercury Study for

Congress. Volume I: Executive Summary. EPA-452/R-97-003. U.S. Environmental Protection Agency, Office of Air Quality Planning and Standards and Office of Research and Development.

EPA (U.S. Environmental Protection Agency). 1997b. Mercury Study for Congress. Volume V: Health Effects of Mercury and Mercury Compounds. EPA-452/R-97-007. U.S. Environmental Protection Agency, Office of Air Quality Planning and Standards and Office of Research and Development.

EPA (U.S. Environmental Protection Agency). 1997c. Mercury Study Report to Congress. Volume VII: Characterization of Human Health and Wildlife Risks from Mercury Exposure in the United States. EPA-452/R-97-009. U.S. Environmental Protection Agency, Office of Air Quality Planning and Standards and Office of Research and Development.

EPA (U.S. Environmental Protection Agency). 1998. Study of Hazardous Air Pollutant Emissions from Electric Utility Steam Generating Units. Final Report to Congress. EPA-453/R-98-004a,-b. U.S. Environmental Protection Agency, Office of Air Quality Planning and Standards and Office of Research and Development.

FDA (U.S. Food and Drug Administration). 1979. Action level for mercury in fish, shellfish, crustaceans and other aquatic animals. Withdrawal of proposed rulemaking. Dept of Health, Education and Welfare. Fed. Regist. 44(14):3990-3993. Jan. 19.

Friberg, L. (Swedish Expert Group). 1971. Methylmercury in fish: A toxicological-epidemiologic evaluation of risks report from an expert group. Nord. Hyg. Tidskr. 4(Suppl.):19-364.

Grandjean, P., E. Budtz-Jørgensen, R.F. White, P. Weihe, F. Debes, and N. Keiding. 1999. Methylmercury exposure biomarkers as indicators of neurotoxicity in children aged 7 years. Am. J. Epidemiol. 150(3):301-305.

Grandjean, P., P. Weihe, R.F. White, F. Debes, S. Araki, K. Yokoyama, K. Murata, N. Sørensen, R. Dahl, and P.J. Jørgensen. 1997. Cognitive deficit in 7-year-old children with prenatal exposure to methylmercury. Neurotoxicol. Teratol. 19(6):417-428.

Grandjean, P., P. Weihe, R.F. White, N. Keiding, E., Budtz-Jørgensen, K. Murato, and L. Needham. 1998. Prenatal exposure to methylmercury in the Faroe Islands and neurobehavioral performance at age seven years. Response to workgroup questions for presentation on 18-20 Nov. 1998. In Scientific Issues Relevant to Assessment of Health Effects from Exposure to Methylmercury. Appendix II-B. Faroe Islands Studies. National Institute for Environmental Health Sciences. Available: http://ntp-server.niehs.nih.gov/Main_Pages/PUBS/MethMercWkshpRpt.html

Harada, M. 1995. Minamata disease: Methylmercury poisoning in Japan

caused by environmental pollution. Crit. Rev. Toxicol. 25(1):1-24.
Ilbäck, N.G. 1991. Effects of methylmercury exposure on spleen and blood natural killer cell activity in the mouse. Toxicology 67(1):117-124.
IPCS (International Programme on Chemical Safety). 1990. Environmental Health Criteria Document 101: Methylmercury. Geneva: World Health Organization.
IPCS (International Programme on Chemical Safety). 1991. Environmental Health Criteria Document 118: Inorganic Mercury. Geneva: World Health Organization.
Jalili, H.A., and A.H. Abbasi. 1961. Poisoning by ethyl mercury toluene sulphonanilide. Br. J. Indust. Med. 18(Oct.):303-308.
JECFA (Joint FAO/WHO Expert Committee on Food Additives). 1972. Evaluation of Certain Food Additives and the Contaminants Mercury, Lead, and Cadmium. World Health Organization Technical Series No. 505. Geneva: World Health Organization.
JECFA (Joint FAO/WHO Expert Committee on Food Additives). 1999. Joint FAO/WHO Expert Committee on Food Additives. 53rd meeting. Rome, 1-10 June, 1999. Online. Available: http://www.who.int/pes/jecta/jecta.htm
Kjellström, T., P. Kennedy, S. Wallis, and C. Mantell. 1986. Physical and Mental Development of Children with Prenatal Exposure to Mercury from Fish. Stage I: Preliminary tests at age 4. National Swedish Environmental Protection Board Report 3080. Solna, Sweden.
Kjellström, T., P. Kennedy, S. Wallis, A. Stewart, L., Friberg, B. Lind, T. Wutherspoon, and C. Mantell. 1989. Physical and Mental Development of Children with Prenatal Exposure to Mercury from Fish. Stage II: Interviews and psychological tests at age 6. National Swedish Environmental Protection Board Report 3642. Solna, Sweden.
Lebel, J., D. Mergler, M. Lucotte, M. Amorim, J. Dolbec, D. Miranda, G. Arantes, I. Rheault, and P. Pichet. 1996. Evidence of early nervous system dysfunction in Amazonian populations exposed to low-levels of methylmercury. Neurotoxicology 17(1):157-168.
Marsh, D.O., T.W. Clarkson, C. Cox, G.J. Myers, L. Amin-Zaki, and S. Al-Tikriti. 1987. Fetal methylmercury poisoning: Relationship between concentration in single strands of maternal hair and child effects. Arch. Neurol. 44(10): 1017-1022.
Mitsumori, K., M. Hirano, H. Ueda, K. Maita, and Y. Shirasu. 1990. Chronic toxicity and carcinogenicity of methylmercury chloride in B6C3F1 mice. Fundam. Appl. Toxicol. 14(1):179-190.
NIEHS (National Institute of Environmental Health Sciences). 1998. Scientific Issues Relevant to Assessment of Health Effects from Exposure to Methyl-

mercury. Report of the Workshop on Scientific Issues Relevant to Assessment of Health Effects from Exposure to Methylmercury, Nov. 18-10, 1998, Raleigh, NC.

Renner, R. 1999. Consensus on health risks from mercury exposure eludes federal agencies. Environ. Sci. Technol. 33(13):269A-270A.

Salonen, J.T., K. Seppänen, K. Nyyssönen, H. Korpela, J. Kauhanen, M. Kantola, J. Tuomilehto, H. Esterbauer, F. Tatzber, and R. Salonen. 1995. Intake of mercury from fish, lipid peroxidation, and the risk of myocardial infarction and coronary, cardiovascular, and any death in Eastern Finnish men. Circulation 91(3):645-655.

Sørensen, N., K. Murata, E. Budtz-Jørgensen, P. Weihe, and P. Grandjean. 1999. Prenatal methylmercury exposure as a cardiovascular risk factor at seven years of age. Epidemiology 10(4):370-375.

Tollefson, L., and F. Cordle. 1986. Methylmercury in fish: A review of residue levels, fish consumption and regulatory action in the United States. Environ. Health Perspect. 68:203-208.

2

CHEMISTRY, EXPOSURE, TOXICOKINETICS, AND TOXICODYNAMICS

THIS chapter presents background information that serves as a foundation for understanding the toxicology of MeHg. The chemical, toxicokinetic, and toxicodynamic properties of MeHg are presented. There is extensive literature on MeHg, and this review is not meant to be exhaustive. Although the primary emphasis of this report is on MeHg, this chapter includes discussions of other Hg species to provide a general review of the sources of exposure and toxicological properties of different Hg species. The emphasis is on human Hg data. Animal data are also discussed.

PHYSICAL AND CHEMICAL PROPERTIES

Chemical species of Hg that are of toxicological importance include the inorganic forms, elemental or metallic Hg (Hg^0), mercurous Hg (Hg^{1+}), and mercuric Hg (Hg^{2+}), and the organic forms, MeHg and ethylmercury. Although there are many organic Hg compounds, the emphasis in this chapter is on MeHg. The structure, chemical formula, and physical and chemical properties of some Hg-containing compounds are shown in Table 2-1. A more complete table of physical and chemical properties of some Hg compounds can be found in the Agency of Toxic Substances and Disease Registry (ATSDR) *Toxicological Profile for Mercury (Update)* (ATSDR 1999). Table 2-2 summarizes the informa-

TABLE 2-1 Physical and Chemical Properties of Some Toxicologically Relevant Mercury Compounds

Chemical Name	Elemental Mercury[a]	Mercuric Chloride	Mercurous Chloride[b]	Methylmercuric Chloride[c]	Dimethylmercury
Molecular formula	Hg^0	$HgCl_2$	Hg_2Cl_2	CH_3HgCl	C_2H_6Hg
Molecular structure		Cl–Hg–Cl	Cl–Hg–Hg–Cl	CH_3–Hg–Cl	CH_3–Hg–CH_3
Molecular weight	200.59	271.52	472.09	251.1	230.66
Solubility	5.6×10^{-5} g/L at 25°C	69 g/L at 20°C	2.0×10^{-3} g/L at 25°C	0.100 g/L at 21°C	1 g/L at 21°C
Density	13.534 g/cm^3 at 25°C	5.4 g/cm^3 at 25°C	7.15 g/cm^3 at 19°C	4.06 g/cm^3 at 20°C	3.1874 g/cm^3 at 20°C
Oxidation state	+1, +2	+2	+1	+2	+2

[a]Also known as metallic mercury.
[b]Also known as calomel.
[c]Methylmercuric chloride is used experimentally to investigate the effects of methylmercury.

tion on some toxicologically relevant Hg compounds discussed later in this chapter.

At 25° C, elemental Hg has a water solubility of 5.6×10^{-5} g/L. Mercuric chloride is considerably more soluble, having a solubility of 69 g/L at 20° C. In comparison, an organic Hg compound, such as methylmercury chloride, is much less water soluble, having a solubility of 0.100 g/L at 21° C. Dimethylmercury, a very toxic by-product of the chemical synthesis of MeHg (Nierenberg et al. 1998), also has a relatively low water solubility (1.0 g/L at 21° C). Due to its low water solubility, MeHg chloride is considered to be relatively lipid soluble. As discussed later in this chapter, the solubility of the different forms of Hg might play a role in their differential toxicity.

TABLE 2-2 Summary Table Comparing Toxicologically Relevant Mercury Species

Methylmercury (CH_3Hg^+)	Elemental Mercury (Hg^0)	Mercuric Mercury (Hg^{2+})
Sources of Exposure		
Fish, marine mammals, crustaceans, animals and poultry fed fish meal	Dental amalgams, occupational exposure, Caribbean religious ceremonies, fossil fuels, incinerators	Oxidation of elemental mercury or demethylation of MeHg; deliberate or accidental poisoning with $HgCl_2$
Biological Monitoring		
Hair, blood, cord blood	Urine, blood	Urine, blood
Toxicokinetics		
Absorption		
Inhalation: Vapors of MeHg absorbed	Inhalation: Approximately 80% of inhaled dose of Hg^0 readily absorbed	Inhalation: Aerosols of $HgCl_2$ absorbed
Oral: Approximately 95% of MeHg in fish readily absorbed from GI tract	Oral: GI absorption of metallic Hg is poor; any released vapor in GI tract converted to mercuric sulfide and excreted	Oral: 7-15% of ingested dose of $HgCl_2$ absorbed from the GI tract; absorption proportional to water solubility of mercuric salt; uptake by neonates greater than adults
Dermal: In guinea pigs, 3-5% of applied dose absorbed in 5 hr	Dermal: Average rate of absorption of Hg^0 through human skin, 0.024 ng/cm^2 for every 1 mg/m^3 in air	Dermal: In guinea pigs, 2-3% of applied dose of $HgCl_2$ absorbed
Distribution		
Distributed throughout body since lipophilic; approximately 1-10% of absorbed oral dose of MeHg distributed to blood; 90% of blood MeHg in RBCs	Rapidly distributed throughout the body since it is lipophilic	Highest accumulation in kidney; fraction of dose retained in kidney dose dependent

TABLE 2-2 *(Continued)*

Methylmercury (CH_3Hg^+)	Elemental Mercury (Hg^0)	Mercuric Mercury (Hg^{2+})
MeHg-cysteine complex[a] involved in transport of MeHg into cells		
Half-life in blood, 50 d; 50% of dose found in liver; 10% in head.	Half-life in blood, 45 d (slow phase); half-life appears to increase with increasing dose	Half-life in blood, 19.7-65.6 d; 1st phase, 24 d, 2nd phase, 15-30 d
Readily crosses blood-brain and placental barriers	Readily crosses blood-brain and placental barriers	Does not readily penetrate blood-brain or placental barriers
		In neonate, mercuric Hg not concentrated in kidneys; therefore, more widely distributed to other tissues
		In fetus and neonate, blood-brain barrier incompletely formed, so mercuric Hg brain concentrations higher than those in adults
Biotransformation		
MeHg slowly demethylated to mercuric Hg (Hg^{2+})	Hg^0 in tissue and blood oxidized to Hg^{2+} by catalase and hydrogen peroxide (H_2O_2); H_2O_2 production the rate-limiting step	Hg^0 vapor exhaled by rodents following oral administration of mercuric Hg
Tissue macrophages, intestinal flora, and fetal liver are sites of tissue demethylation		Mercuric Hg not methylated in body tissues but GI microorganisms can form MeHg

Mechanisms of demethylation unknown; free radicals demethylate MeHg in vitro; bacterial demethylation enzymes studied extensively, none has been characterized or identified in mammalian cells		
Does not bind or induce metallothionein		Binds and induces metallothionein
Excretion		
Daily excretion, 1% of body burden; major excretory route is bile and feces; 90% excreted in feces as Hg^{2+}; 10% excreted in urine as Hg^{2+}	Excreted as Hg^0 in exhaled air, sweat, and saliva, and as mercuric Hg in feces and urine	Excreted in urine and feces; also excreted in saliva, bile, sweat, exhaled air, and breast milk
Lactation increases clearance from blood; 16% of Hg in breast milk is MeHg		
Half-Life elimination		
(Whole body) 70-80 d; dependent on species, dose, sex, and animal strain	58 d	1-2 mo
Toxicodynamics		
Critical target organ		
Brain, adult and fetal	Brain and kidney	Kidney
Causes of Toxicity		
Demethylation of MeHg to Hg^{2+} and the intrinsic toxicity of MeHg	Oxidation of Hg^0 to Hg^{2+}	Hg^{2+} binding to thiols in critical enzyme (e.g., cysteine) and structural proteins

TABLE 2-2 *(Continued)*

Methylmercury (CH_3Hg^+)	Elemental Mercury (Hg^0)	Mercuric Mercury (Hg^{2+})
Latency period		
In Iraq, from weeks to month; in Japan, more than a year; differences suggested to be caused by Se in fish; no toxic signs during latency period		
Mobilization		
DMPS, DMSA	After oxidation to Hg^{2+}: DMPS, DMSA	DMPS, DMSA
Possible Antagonists		
Selenium, garlic, zinc		

[a]MeHg-cysteine complex is structurally analogous to methionine.

Abbreviations: $HgCl_2$, mercuric chloride; DMPS, 2,3-dimercapto-1-propane sulfonate; DMSA, meso 2,3-dimercaptosuccinic acid; GI, gastrointestinal tract; RBC, red blood cells.

METHODS OF CHEMICAL ANALYSIS

The methods used for analyzing Hg in biological samples include atomic absorption spectrometry (AAS), atomic fluorescence spectrometry (AFS) (Vermeir et al. 1991a, b), X-ray fluorescence (XRF) (Marsh et al. 1987), gas chromatography (GC)-electron capture (Cappon and Smith 1978), and neutron activation analysis (NAA) (Fung et al. 1995). Anodic stripping voltammetry (ASV) has also been used (Liu et al. 1990). Of those procedures, GC-electron capture is able to distinguish MeHg from other species, but only cold vapor (CV)-AAS will detect Hg at parts per billion. CV-AAS, AFS, XRF, and NAA have all been used to analyze Hg content in hair (Zhuang et al. 1989).

To measure total Hg in biological samples, the Hg must first be reduced to the elemental form. CV-AAS is most frequently used to measure Hg in urine (Magos and Cernik 1969) and blood (Magos and Clarkson 1972). For example, CV-AAS, the most commonly used method for analyzing Hg in biological samples, involves reduction of the Hg in the sample with stannous chloride to elemental Hg. To measure inorganic Hg, the analysis is carried out without chemical reduction of the sample. The difference between the total Hg concentration and the inorganic Hg concentration represents the concentration of organic Hg that was present in the sample.

Biological samples containing MeHg can also be analyzed using *Pseudomonas putida* strain FB1. That bacteria converts MeHg to methane gas and elemental Hg (Baldi and Filippelli 1991). This method is one of the most reliable and specific methods for MeHg quantification, because chemical interference is negligible. It can detect 15 ng of MeHg in 1 g of biological tissue with a coefficient of variation of 1.9%.

New methods for analyzing Hg in biological samples have been developed such as inductively coupled plasma-mass spectrometry (ICP-MS) (Kalamegham and Ash 1992). Most of the new methods are expensive and beyond the reach of most laboratories. The cost is approximately $150,000-250,000 for the instrument and more than $35,000 a year for gases and maintenance costs.

Regardless of the analytical method used, care must be taken to eliminate or prevent contamination of the sample by Hg during preparation and analysis. All glassware and plasticware used for collection and analysis of the specimen must be acid washed. In addition, care must be taken to avoid losses due to volatilization of elemental Hg and MeHg, especially when preserving or concentrating the samples.

Many procedures require the digestion of the sample before reduction. When attempting to quantify Hg content, especially in biological samples, data are needed to validate the procedures and their use in a given laboratory. All the methods of analysis are prone to large variations.

Biological monitoring of inorganic Hg, including elemental Hg, requires measurement of Hg concentrations in blood, urine, or both (Clarkson et al. 1988). Biological monitoring for MeHg usually involves measuring Hg content in scalp hair, blood, or both. The MeHg incorporated into hair is stable and can be used for longitudinal timing (historical record) of exposure to MeHg by analyzing segments of hair (Phelps et al. 1980; IPCS 1990; Grandjean et al. 1992; Suzuki et al. 1992). One source of error in hair Hg analysis is the presence of Hg on the hair surface due to external deposition. Adequate washing of the hair sample before analysis minimizes that error (Francis et al. 1982).

An excellent summary of the analytical methods for determining various species of Hg in biological specimens, including blood, urine, hair, breath, and tissues, as well as in environmental samples can be found in Table 6-1 in *Toxicological Profile for Mercury (Update)* (ATSDR 1999) and in the World Health Organization (WHO) report *Methylmercury* (IPCS 1990).

EXPOSURES TO MeHg IN THE U.S. POPULATION

The major source of MeHg exposure in humans is consumption of fish, marine mammals, and crustaceans. Because exposure to MeHg occurs almost entirely through fish consumption and varies according to the types of fish consumed, variations in exposure to MeHg in the U.S. population are based on individual characteristics of fish consumption. Exposure also varies according to the characteristic amounts and types of fish consumed in different regions of the United States. Hg concentrations in commercial fish and seafood in the United States span about two orders of magnitude. For example, herring contains Hg at approximately 0.01 ppm and shark contains Hg at greater than 1 ppm (EPA 1997a). Limited data suggest that coastal regions generally have higher rates of fish consumption (Rupp et al. 1980). In addition, specific ethnic and cultural subgroups, as well as recreational fishermen, can have increased exposures (EPA 1997a). Population-based estimates of MeHg exposure in the United States have been made on the basis of dietary assessment studies, which provide information on fish consumption by species and by portion size. The combination of intake frequency by species and portion size by species for each individual consumer provides an estimate of the average mass of fish consumed (in grams per day). Summaries of such studies giving national data are provided in EPA's report to Congress (EPA 1997a). Another such dietary assessment study was conducted in New Jersey (Stern et al. 1996). To estimate population-based MeHg exposure from such studies, the gram-per-day amount of each species consumed by each individual is multiplied by the characteristic MeHg concentration of each species (microgram per gram) and then is summed across species to give the average intake of MeHg by each individual (microgram/day). The distribution of individual intakes for the study sample can then provide an estimate of MeHg intake in the underlying population. Uncertainties in such assessments include those in recall and recording of intake frequency and portion size, misidentification of the species consumed, extrapolation of short-term dietary studies to long-term average exposure, and the outdated and incomplete national database on average MeHg concentrations of different fish species. Estimates also typically vary depending on the length of time over which the fish-intake data was obtained (e.g., 1-day recall versus 1-week recall). These uncertainties are discussed by EPA (1997a) and Stern et al. (1996). Table 2-3 presents the EPA (1997c) analysis of MeHg intake for the general population and for the population of women of childbearing age based on fish-consumption data for month-long consumption. Estimates based on intake from such data are generally lower than those based on 1-day dietary data. Table 2-3 also presents data from New Jersey based on a 7-day recall survey. These data, along with the study by Rupp et al. 1980, suggest that the population in that region of the United States has higher intakes than the U.S. population in general. Estimates of population exposure and risk based on the average exposure of the U.S. population might, therefore, underestimate exposure to large subpopulations. Upon completion, data from Continuing Survey of Food Intakes by Individuals (CFSII) and National Health and Nutrition Examination

TABLE 2-3 Estimated Average MeHg Intake for the U.S. Population and for New Jersey Fish Consumers

	Average Daly Intake of MeHg (μg/day)[a]			
Percentiles of	General Population		Women of Childbearing Age	
the Population	U.S.[b,c]	New Jersey[d]	U.S.[b,e]	New Jersey[c,f]
50th	1.4	3.1	0.6	3.2
75th	3.5	5.8	1.8	5.4
90th	9.1	13.1	4.8	10.8
95th	15.6	21.1	7.8	15.7
99th		49.9	22.2	26.5

[a]Assuming body weight of 70 kg for the general population and 60 kg for women of childbearing age.

[b]Data from EPA 1997a.

[c]Unweighted average across ethnic groups.

[d]Data from Stern et al. 1996.

[e]Women 15-45 years old.

[f]Women 18-40 years old.

Survey (NHANES IV) might provide information on regional fish consumption. NHANES IV is also designed to provide information on MeHg exposure in U.S. populations.

Consumption of animals or poultry fed fish meal might increase the exposure to MeHg, but data are not available. The use of organic Hg compounds as preservatives in vaccines and medical preparations is also a source of exposure and is of particular importance in young children who might be more sensitive to those mercurials than adults. As many as 219 such products are in use (FDA 1999). Thimerosal (TM) (sodium ethylmercurithiosalicylate) and phenylmercuric acetate (PMA) are the most frequently used compounds, at concentrations of 0.01% and 0.0002%, respectively. The FDA estimates that 75-80 kg of Hg compounds are used annually by the manufacturers of those vaccines and medical preparations. The risks associated with thimerosal use in vaccines have been discussed in an interim report to clinicians (American Academy of Pediatrics 1999).

Small amounts of MeHg can be formed in the gut by intestinal bacteria. A.O. Summers (University of Georgia, personal commun., Dec. 1999) estimated that 9 μg of MeHg can be formed per day in the gut of humans. That estimate is based on the bacterial species reported to occur in the human gut and assumes that there are 454 g of feces in the lower bowel of an adult human. However, not all the MeHg that is synthesized would be absorbed. Some of the methylation would occur in the colon, where absorption is less. In addition, intestinal flora can demethylate MeHg to inorganic Hg, which is poorly absorbed by the GI tract (Nakamura et al. 1977; Rowland et al. 1980).

The major source of exposure to elemental Hg in the general U.S. population is due to Hg vapor released from dental amalgams (Goering et al. 1992; Halbach 1994; Lorscheider et al. 1995). Approximately 300 metric tons of Hg are used annually by dentists for amalgams (Arenholt-Bindslev and Larsen 1996). Most amalgams used in the United States contain approximately 50% Hg (IPCS 1991; Aposhian et al. 1992a; Lorscheider et al. 1995). In a study of college students who have dental amalgams, two-thirds of the Hg excreted in the urine appeared to be derived from the Hg vapor released from their amalgams (Aposhian et al. 1992a). Evidence shows that Hg vapor from dental amalgams enters tissues, including the brain, where it is oxidized to inorganic Hg. Pregnant sheep given amalgam fillings labeled with radioactive Hg accumulated radioactivity in maternal and fetal tissues within a few days (Vimy et al. 1990). Significant positive correlations between the number of amalgams in the mouth and the mercury content of human tissues, including the brain, are also seen (Drasch et al. 1994). The mean concentration of total Hg in whole blood (in the absence of consumption of fish with high concentrations of MeHg) is probably of the order of 5-10 μg/L (IPCS 1991; Mahaffey and Mergler 1998). This concentration is most likely due to exposure to Hg vapors from amalgams, because retention of inorganic Hg is very low compared with retention of organic and elemental Hg. Furthermore, exposure to MeHg from non-fish sources is also very low (IPCS 1991).

Occupational exposure to elemental Hg has occurred because of accidents in chloralkali plants (Bluhm et al. 1992). However, there are other potential occupational exposures to elemental Hg. In addition, some Caribbean religions use elemental Hg in religious ceremonies (Wendroff 1995). Children have been known to play with elemental Hg because of its fascinating physical properties (i.e., liquid silver), possibly severely contaminating living and play areas (ATSDR 1999). In wastewaters, the main sources of elemental Hg are dental offices, hospitals, and laboratories (Arenholt-Bindsley and Larsen 1996). Exposure of humans to mercuric Hg has occurred because of intentional or accidental (e.g., occupational exposures) poisonings with mercuric chloride (Clarkson et al. 1988).

TOXICOKINETICS

Absorption and Distribution

Methylmercury

Most fish contain MeHg. Many freshwater fish in the United States contain more than 2-3 ppm of Hg (Northeast States for Coordination of Air Use Management (NESCAUM 1998). Populations worldwide that eat fish regularly can have concentrations of more than 10 ppm in their hair (Cernichiari et al. 1995). About 95% of the MeHg in fish ingested by humans (Aberg et al. 1969; Miettinen 1973) or about 95% of methylmercuric nitrate given orally to volunteers (Aberg et al. 1969) was found to be absorbed from the gastrointestinal (GI) tract. Although MeHg toxicity following ingestion is the primary focus of this report, it should be noted that MeHg also is readily absorbed through the skin and lungs. The extent of absorption following inhalation exposure is believed to be high.

Once absorbed into the bloodstream, MeHg enters the red blood cells. More than 90% of the MeHg that is found in blood is bound to hemoglobin in red blood cells (Kershaw et al. 1980). Aberg et al. (1969) studied the distribution of Hg compounds in three healthy male volunteers administered [^{203}Hg]-methylmercuric nitrate orally. ^{203}Hg was found in the blood 15 min after administration and peaked within 3-6 hr. The concentration in red blood cells was 10 times greater than that in plasma. MeHg binds to cysteine residue number 104, of the α chain and numbers 93 and 112 of the β chain of hemoglobin. Numbers 104 and 112 are cysteine residues in the contact junction of the hemoglobin molecule. Number 93 is out of the junction and binds to MeHg easily because it is on the external surface of the hemoglobin molecule. The number and position of the junctional and external cysteine residues on hemoglobin differ in animal species. An extensive table, including the data for the hemoglobin of eight animal species, can be found in Doi (1991). Some MeHg is also bound to plasma proteins. In humans exposed orally to large amounts of MeHg daily, the percentage of the total Hg found as inorganic Hg in whole blood, plasma, breast milk, liver, and urine was 7%, 22%, 39%, 16-40%, and 73% respectively (IPCS 1990). Matsuo et al. (1989) reported autopsy data on Japanese subjects. Kidney and liver contained total Hg concentrations on the order of hundreds of ng/g. Cerebrum, cerebellum, heart, and spleen contained total Hg concentrations on the order of tens of ng/g. Approximately 80% of the Hg in those organs was in the form of MeHg. In the liver, kidney medulla and kidney cortex 33%, 15% and 11% of the mercury was methylmercury, respectively. Consumption of high concentrations of MeHg in fish results in only about 5% inorganic Hg in whole blood and about 20% inorganic Hg in scalp hair (Phelps et al. 1980). It should be emphasized that the exact form(s) in which MeHg exists in the body is still unknown. MeHg ion is hydrated in aqueous solutions. There are pH-dependent reactions giving rise to Hg-substituted oxonium ions (Figure 2-1). Cotton and Wilkinson (1988), state that "the types of complexes formed by the two ions differ markedly; Hg^{2+} compounds of amino acids containing SH groups are polymeric and polar, whereas the CH_3HgR species are nonpolar and monomeric. For example the cysteinate is with a linear C—Hg—S." The chemistry and formation of Hg-substituted oxonium ion complexes may affect MeHg transport, but investigators who study such transport largely ignore them.

About 10% of the body burden of MeHg is found in the brain where it is slowly demethylated to inorganic mercuric Hg (see Figure 2-2). MeHg is also readily transferred to the fetus and the fetal brain. Evidence from rat experiments suggests that MeHg transport across the blood-brain barrier occurs via a MeHg-L-cysteine complex, which is transported by the L-system (leucine preferring) amino acid carrier (Kerper et al. 1992). MeHg-cysteine is released in vitro from a MeHg-glutathione complex by the action of γ-glutamyltransferase and dipeptidases (Naganuma et al. 1988). That action suggests that glutathione might play an indirect role in the transport of MeHg into endothelial cells. The MeHg-cysteine or MeHg-glutathione complex would be expected to be water soluble. That would not support the hypothesis

$$CH_3Hg(OH_2)^+ + OH^- \rightleftharpoons CH_3HgOH + H_2O$$

$$CH_3Hg(OH_2)^+ + CH_3HgOH \rightleftharpoons (CH_3Hg)_2OH^+ + H_2O$$

$$CH_3HgOH + (CH_3Hg)_2OH^+ \rightleftharpoons (CH_3Hg)_3O^+ + H_2O$$

FIGURE 2-1 Hg-substituted oxonium ions formed in aqueous solution. Source: Cotton and Wilkinson 1988.

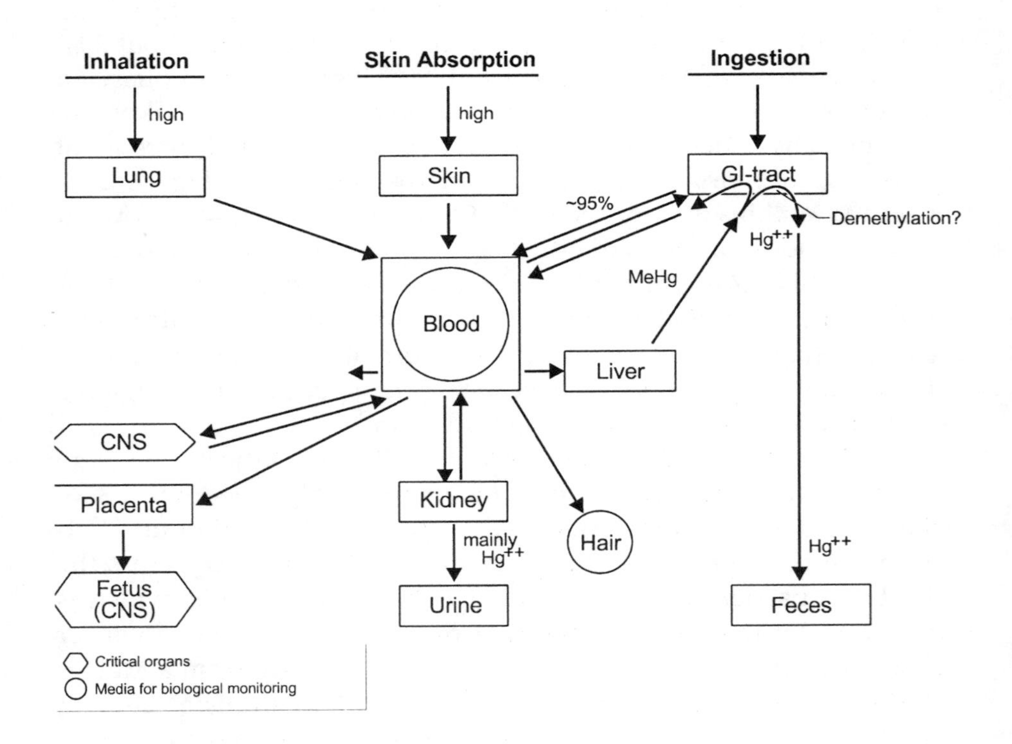

FIGURE 2-2 Methylmercury kinetics. Source: Elinder et al 1988. Reprinted with permission from *Biological Monitoring of Toxic Metals*; copyright 1988, Plenum Publishing Corporation.

that the rapid uptake of MeHg by the brain is due to lipid solubility in body tissues and fluids. Recently, Fujiyama et al. (1994) proposed that the MeHg-glutathione complex is the mechanism by which MeHg can efflux rat astroglia. Aschner et al. (1991), however, proposed that the MeHg-cysteine complex is the mechanism by which MeHg is exported from astroglia.

A case study of family members that developed classic signs of MeHg poisoning due to the consumption of contaminated pork indicates that the cerebrum and the cerebellum are particularly sensitive to MeHg (Davis et al. 1994). Analyses of various regions of the brain of one female member upon autopsy, several years later, revealed that the extent of brain damage correlated with regional-brain Hg concentrations. Inorganic Hg comprised 82-100% of the total Hg, suggesting that most of the MeHg had been converted to inorganic Hg during the period. The highest levels of Hg were found in the cerebrum and cerebellum. Magnetic Resonance Imaging (MRI) studies showed brain damage in the calcarine cortices, parietal cortices, and cerebellum of other family members. The damage in those areas is believed to underlie many of their persistent clinical signs, because those areas of the brain are responsible for coordination, balance, and sensations (see Chapter 5).

Dimethylmercury

Dimethylmercury is a supertoxic form of Hg (Gosselin et al. 1984) that has been fatal after accidental exposure. At Dartmouth College, a chemistry professor died 298 days after several drops of dimethylmercury fell on her latex gloves. The gloves did not appear to act as a barrier, and the compound was rapidly absorbed through her skin. Six to 7 months after her exposure, her blood Hg concentration was 1,000 μg/L (Nierenberg et al. 1998). Typical blood concentrations of Hg are in the range of 1 to 8 μg/L. Mouse studies suggest that the extremely toxic dimethlymercury must be metabolized to MeHg before it can enter the brain (Ostlund 1969).

Elemental Mercury

Absorption of elemental Hg vapor via the lungs is rapid. In humans,

75-85% of an inhaled dose is absorbed (Kudsk 1965; Okawa et al. 1982; Hursh 1985; Hursh et al. 1985). Elemental Hg in liquid or vapor form is not well absorbed from the GI tract (less than 0.01%) (Bornmann et al. 1970). In humans exposed to elemental Hg vapor, 97% of the absorption occurred via the lungs, and less than 3% of the total amount absorbed was via the skin (Hursh et al. 1989).

Because elemental Hg is very lipid soluble, its diffusion across the lungs and dissolution in blood lipids is rapid (Berlin 1986). The fact that it is uncharged with intermediate molecular weight and size might be another reason why it passes readily from air to blood. It is distributed throughout the body, and readily crosses the placenta and the blood-brain barrier (Vimy et al. 1997; Fredricksson et al. 1992, 1996; Drasch et al. 1994) (see Figure 2-3). Elemental Hg is oxidized to mercuric Hg. Eventually, the Hg ratio of red blood cells to plasma is 1:1.

Inorganic Mercury

Approximately 7-15% of an ingested dose of mercuric chloride is absorbed from the GI tract (WHO 1976; Miettinen 1973). Absorption is proportional to the water solubility of the mercuric salt. Mercuric Hg has a high affinity for sulfhydryl groups in the red blood cells and plasma. The half-life in the blood is reported to be 19.7-65.6 days (Hall et al. 1995).

The highest accumulation of mercuric Hg is in the kidneys. The major fraction of inorganic Hg in rat kidney is bound to metallothionein (Jakubowski et al. 1970; Wisniewska et al. 1970; Komsta-Szumska et al. 1976). In contrast MeHg has a low affinity for metallothionein (Chen et al. 1973). Because of its ionic charge, mercuric Hg does not readily penetrate the blood-brain barrier or the placenta.

Biotransformation

MeHg is converted in tissues to mercuric Hg (Magos and Butler 1972; Dunn and Clarkson 1980). The rate of demethylation in rats and most other species is very slow. The mechanisms involved in conversion of MeHg to mercuric Hg are controversial. The enzymes in mammalian

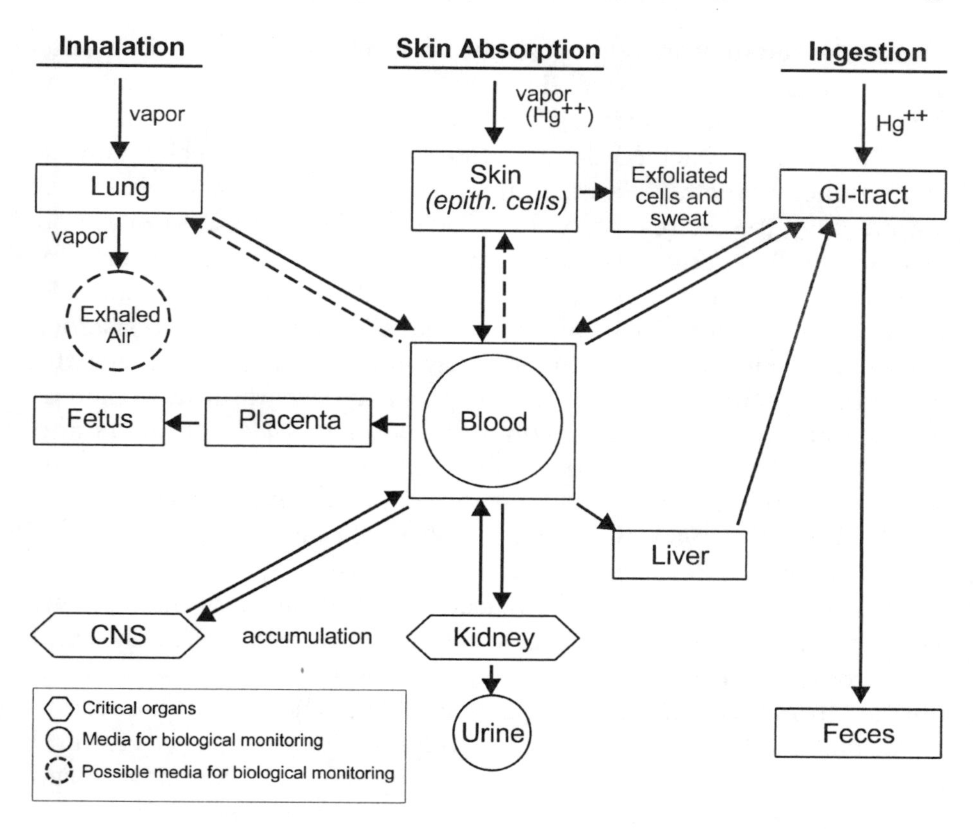

FIGURE 2-3 Inorganic mercury kinetics. This diagram is complicated by the fact that inhaled vapor is oxidized to Hg^{++} so that both species are present. The inhaled vapor is highly mobile, readily crosses cell membranes, the blood-brain barrier, and the placenta. The Hg^{++} species is much less mobile, crossing the blood-brain barrier and placenta much more slowly than dissolved vapor. Source: Elinder et al. 1988. Reprinted with permission from *Biological Monitoring of Toxic Metals;* copyright 1988, Plenum Publishing Corporation.

tissues believed to be responsible for the biotransformation have never been identified. Greater emphasis has been placed on investigating the possible role of a free radical mechanism (Suda and Hirayama 1992). In addition, γ-globulin and serum albumin have been shown to have similar degradation activity that can be stimulated further by glutathione (Gage 1975). Intestinal flora, tissue macrophages, and fetal liver are all sites of tissue demethylation.

Experiments in bacteria demonstrate many different mechanisms to detoxify heavy metals. For example, some metals are actively transported out of the cell (e.g., arsenite) (Perry and Silver 1982; Mobley and Rosen 1982; Silver and Keach 1982), and other metals are sequestered by protein binding in the cell (Kägi and Nordberg 1979). Organic Hg compounds are detoxified by a microbial organomercurial resistance system (see Figure 2-4). An organomercurial lyase catalyzes the protonolysis of the carbon-Hg bond to give a hydrocarbon and a mercuric ion (Summers 1985; Robinson and Tuovinen 1984; Summers and Silver 1978). Mercuric reductase then catalyzes the reduction of mercuric Hg to elemental Hg. NADPH is the coenzyme in that reaction (Fox and Walsh 1982, 1983; Brown et al. 1983). Because elemental Hg is volatile, it evaporates from the bacterial culture.

GI absorption of MeHg is decreased by intestinal flora that convert MeHg to inorganic Hg (mercuric ion) (Nakamura et al. 1977; Rowland et al. 1980), which is poorly absorbed. Organomercurial lyase has been purified from *Escherichia coli* (Begley et al. 1986). The enzyme is encoded on the plasmid R831. No cofactors are required for enzyme activity, and the enzyme structure does not contain any metals. The enzyme can catalyze protonolysis of the C-Hg bond in primary, secondary, and tertiary alkyl, vinyl, allyl, and aryl organomercurial salts to the hydrocarbon and mercuric ion. A thiol must be present for activity, cysteine being the most active thiol compound, for demethylation of organic mercurials.

Equation 1: $CH_3HgCl + H^4 \xrightarrow{\text{organomercurial lyase}} CH_4 + Hg^{2+} + Cl^-$

Equation 2: $Hg^{2+} \xrightarrow[\text{NADPH}]{\text{mercuric reductase}} Hg^0$

FIGURE 2-4 Organomercurical detoxification pathway in bacteria.

Enzymes similar to those found in bacteria have not been found in mammals. Demethylation of MeHg is thought to occur via a free-radical mechanism in the mammalian brain, possibly eliminating the need for those enzymes. It is also possible that the enzymes have not yet been identified.

Lefevre and Daniel (1973) examined rat, mouse, and guinea-pig liver homogenates for activity that would degrade organic Hg compounds. Although a minimum level of activity was found, phenylmercuric acetate and methoxyethylmercury chloride were degraded, but not MeHg. Fang and Fallin (1974) were able to show cleavage of phenylmercuric acetate (PMA) and ethylmercury chloride in the kidney and liver of rats, but no activity was seen against MeHg.

Elemental Hg vapor is oxidized to mercuric mercury by catalase and hydrogen peroxide (H_2O_2) in blood and tissues (Berlin 1986). H_2O_2 production is the rate-limiting step.

When mercuric Hg is administered orally to rodents, elemental Hg vapor has been detected in the expired air, indicating that some metabolism to elemental Hg must have occurred. Mammals do not methylate mercuric Hg; however, intestinal flora can methylate Hg^{2+} to a small extent (Rowland et al. 1977).

Excretion

Approximately 1% of the human body burden of MeHg is excreted daily (Clarkson et al. 1988). In humans, the major routes of excretion are via the bile and feces. About 90% of a given dose of MeHg is eventually excreted in the feces as mercuric Hg in humans and other species. Approximately 10% is excreted as mercuric Hg via the urine. Much of the biliary MeHg is reabsorbed; MeHg complexed with glutathione is eliminated via the bile.

Following oral administration of [^{203}Hg] methylmercuric nitrate, only about 33% of the administered dose was excreted in 49 days, fecal excretion being the main route of excretion (Miettinen 1973). There was a 0.18% to 0.27% excretion of the dose in the urine in 10 days and 3.3% excretion in 49 days. The extent of urinary excretion continued to increase up to 71 days after ingestion. A maximum of 0.12% of the administered dose of Hg was found per gram of hair. That amount was found 40-50 days after ingestion. Using whole-body measurements, the half-life of MeHg was 70-74 days. No methylmercuric chloride was found in the sperm, but about 50% of the body content was found in the liver and about 10% was found in the head.

In humans, the whole-body half-life of MeHg was estimated to be 70-80 days (Aberg et al. 1969; Miettinen 1973; Bernard and Purdue 1984; EPA 1997b).

The half-life in blood for MeHg as measured in blood and hair of humans ranged from 48 to 53 days (Miettinen et al. 1971; Kershaw et al. 1980; Sherlock et al. 1984; Cox et al. 1989). Elimination rates for MeHg are dependent upon species, dose, sex, and animal strain (Nielsen 1992).

It is pertinent to note that neonatal rats and monkeys are limited in their ability to excrete MeHg into the bile (Ballatori and Clarkson 1982). Therefore, it takes them longer than mature animals to excrete MeHg (Thomas et al. 1982). In addition, their intestinal flora might also be less able to demethylate MeHg during this suckling period (Rowland et al. 1977; Sundberg et al. 1998; Grandjean et al 1994). If those two phenomena are true for humans, then neonates might be particularly sensitive to exposure to MeHg. GSH may be the major cellular defense against MeHg toxicity. GSH complexation with MeHg is a major mechanism for MeHg excretion from the cell, thus protecting against MeHg toxicity (Kromidas et al. 1990).

MeHg has been measured in the breast milk of rats, humans, and guinea pigs (Sundberg and Oskarsson 1992; Yoshida et al. 1992). Therefore, breast milk is considered a route of excretion, but it is also an important route of exposure to suckling neonates. In human breast milk, 16% of the Hg was found to be MeHg (Skerfving 1988). That percent is much lower than the percent of Hg found as MeHg in whole blood. In animals, the total Hg content of breast milk was found to be proportional to the total Hg content of the plasma (Skerfving 1988; Sundberg and Oskarsson 1992).

A small amount of elemental Hg vapor is excreted unchanged in exhaled air, sweat, saliva, feces, and urine (Cherian et al. 1978). Only small amounts of elemental Hg can be detected in the urine (Stopford et al. 1978) and exhaled air (Hursh et al. 1976). Excretion via sweat and saliva is usually minimal. The half-life for whole-body Hg excretion was 58 days in humans (Hursh et al. 1976). Elemental Hg is also oxidized in the body to mercuric Hg, which is then excreted in the feces and urine. That is demonstrated by the observation that, after exposure to Hg vapor, the mercuric Hg content in the feces increases and is four times greater than that in the urine.

Following oral administration of mercuric Hg to humans, about 85% was excreted in the feces within a few days (Miettinen 1973). Fecal excretion of mercuric Hg occurs as the result of secretion through the small intestine epithelium and colon, and bile secretion (Berlin 1986). Mercuric Hg is also excreted in the urine, sweat, lung (Clarkson et al. 1988), and breast milk (Yoshida et al. 1992). Urinary excretion is useful for biological monitoring of inorganic Hg. Absorbed inorganic Hg has been estimated to have a half-life of 40 days (Rahola et al. 1973) or 67 days (Hall et al. 1995). WHO (IPCS 1990) reported a half-life of 35 days. Humans occupationally exposed to inorganic Hg excrete in their urine three forms of this element: a metallic form, a Hg-cysteine complex, and a large unidentified complex (Henderson et al. 1974).

MOBILIZATION OF BODY Hg

Synthetic chelating or complexing agents that compete with endogenous ligands for mercuric or organic Hg increase the urinary excretion of inorganic Hg and organic Hg and reduce the body burden (Aposhian 1983; Aposhian and Aposhian 1990). Compounds that have been used therapeutically are 2,3-dimercapto-1-propane sulfonate (DMPS, Dimaval, and Unithiol) and meso 2,3-dimercaptosuccinic acid (DMSA, succimer). DMPS and DMSA are water-soluble, less-toxic chemical analogs of 2,3-dimercapto-1-propanol (British Anti-Lewisite BAL dimercaprol). BAL is lipid soluble and must be given by deep intramuscular injection. DMPS and DMSA can be taken orally. There is an injectable preparation of DMPS. About 55% of patients administered BAL have one or more adverse reactions to it (Klaassen 1996), although most of the reactions are not serious. However, because BAL redistributes Hg, increasing brain Hg concentrations when given to Hg-intoxicated animals, its continued therapeutic use is questionable (Berlin 1986).

DMPS was introduced into the official Soviet drug armamentarium in 1958 (Klimova 1958) and to the western world in 1978. A number of reviews of DMPS and other chelating agents have appeared during the last 18 years (Aposhian 1983; Aposhian et al. 1992b, 1995; Aaseth et al. 1995). It is approved for use by the German and Chinese equivalents of the U.S. Food and Drug Administration (FDA). Clarkson et al. (1981) used DMPS to treat the MeHg-poisoned humans in Iraq and showed that it is more potent than D-penicillamine *N*-acetyl-DL-penicillamine or a thiolated resin for decreasing inorganic Hg in the blood. Elsewhere, including the United States, DMPS has been used by alternative-medicine physicians concerned with dental amalgam toxicity. It was used recently to increase the urinary excretion of Hg in eight humans exposed to mercurous Hg (Gonzalez-Ramirez et al. 1998). In contrast to BAL, DMPS does not increase brain Hg concentration in rats (Aposhian et al. 1996).

Although it has not been approved for this use, DMSA has been used to treat Hg intoxication (Aposhian 1983; Aposhian and Aposhian 1990). It was approved by the FDA in 1990 for the treatment of children with blood Pb concentrations greater than or equal to 45 μg per deciliter.

CHEMICAL FORMS OF Hg IN TOXICITY

There is ample evidence from studies of humans (Takizawa 1979; Matsuo et al. 1987; Takeuchi et al. 1989) and experimental studies using animal models (Vahter et al. 1994) that MeHg is slowly biotransformed to inorganic Hg in the brain. Although the rate of demethylation of MeHg in the brain appears to be dose related, many questions remain concerning the mechanisms by which the brain biotransforms MeHg to inorganic Hg and the slow rate at which it occurs. Davis et al. (1994) reported that a New Mexico patient who died approximately 21 years after eating MeHg-tainted pork had greatly increased total Hg concentrations in various regions of the brain (71 to 300 times greater than controls). A minimum of 82% of the Hg in the patients brain was in the inorganic form. In most regions of her brain, 100% of the Hg was in the inorganic form. Similar results were found in a Minamata patient who died 18 years after the exposure (Takeuchi and Komyo 1977).

Experimental studies have also reported a slow increase in the concentration of inorganic Hg in the brain in a number of species after administering MeHg and analyzing the brain for total and inorganic Hg from days to years after exposure (Friberg and Mottet 1989). When monkeys were exposed daily to high doses of MeHg for long periods of time, there were significant concentrations of inorganic Hg found in the brain (Lind et al. 1988). By 69-166 weeks, 10-33% of the brain Hg was inorganic. Female monkeys (*Macaca fascicularis*) received daily doses of MeHg for up to 18 months (Vahter et al. 1994, 1995). When the brains were examined for Hg species, inorganic Hg made up about 9% of the total Hg after 6-12 months of exposure and 18% after 18 months of exposure. Six months after a 12-month exposure ended, it was 74%. The authors stated that they believed that inorganic Hg "was formed by demethylation of MeHg in the brain."

The extent to which demethylation of MeHg produces toxicity in the brain is not known. Studies by Norseth and Clarkson (1970) and Syversen (1974) indicated that MeHg itself mediates the toxicity following MeHg exposure. In addition, Magos et al. (1985) provided direct evidence that the extent of brain damage correlates better with the brain concentration of intact organic Hg than inorganic Hg when MeHg or ethyl Hg is administered to rats. A possible hypothesis is that the long half-life of inorganic Hg in the brain once demethylation occurs might be responsible for the latent or long-term MeHg effects that have been reported. No direct evidence to support that hypothesis is available at this time.

In addition to the questions regarding whether inorganic or organic Hg mediates MeHg toxicity at the cellular level, questions have also been raised regarding the species responsible for the Iraqi poisonings. In the Iraqi poisoning episode, some of the grain seeds appeared to contain phenyl Hg instead of MeHg. There is no doubt, however, that gas chromatography identified MeHg in the blood of most of the exposed population, and phenyl Hg would have been quickly converted to inorganic Hg in the blood (T.W. Clarkson, University of Rochester, personal commun., Nov. 1999). In addition, the phenyl Hg was in the barley seeds and no barley seeds were used to make bread (T.W. Clarkson, University of Rochester, personal commun., Nov. 1999).

TOXIC EFFECTS AND TARGET ORGANS

Currently, there is a general consensus that the critical organ for MeHg toxicity is the brain. Both the adult and fetal brains are susceptible to MeHg toxicity (see Figure 2-2), although the developing nervous system appears to be more sensitive. Studies of the Minamata disaster in Japan indicate that prenatal exposure causes damage throughout the fetal brain and, at high doses, results in effects in the offspring that are largely indistinguishable from cerebral palsy caused by other factors (Harada 1995). Exposure of adults to MeHg resulted in focal lesions (Clarkson 1997). The neurotoxicity of chronic MeHg exposure at lower levels is not immediately evident. A latent period of 1 month or more usually occurs (Bakir et al. 1973; IPCS 1990). Other adverse effects (e.g., cardiovascular and immunological effects) have been reported to occur at MeHg doses lower than those producing adverse effects in the nervous system. Those effects, however, are not as well studied as the neurotoxic effects. The health effects of MeHg are discussed in more detail in Chapter 5.

The target organs of elemental Hg are the brain and kidney. The toxicity of elemental Hg is believed to be due to mercuric Hg. Inhaled elemental Hg vapor readily crosses the blood-brain barrier and is then oxidized to mercuric Hg. The latter becomes firmly bound to macromolecules in the brain. There does not seem to be any endogenous mechanism for the rapid removal of mercuric Hg from such sites. In humans occupationally exposed to elemental Hg vapor, signs of severe exposure include tremor, psychiatric disturbances, gingivitis, and altered behavior.

The target organ of mercuric Hg toxicity is the kidney due to Hg accumulation there. The earliest signs of renal injury due to Hg compounds are increased urinary excretion of *N*-acetyl-β-glucoseaminidase, β_2-microglobulin and retinol-binding protein. Although the exact mechanism of renal toxicity is not known, it is known that mercuric Hg has a strong affinity for sulfhydryl moieties. The formation constants of Hg sulfhydryl complexes are very high (approximately $K_f = 10^{30}$) (Divine et al. 1999). The formation constant for mercuric Hg and the anionic form of a sulfhydryl group, RS^-, is greater than or equal to 10^{10}-fold higher than that for the carboxyl or amino groups (Ballatori 1991; Divine et al. 1999). Since there is a wide distribution of sulfhydryl groups in the body, especially in proteins, mercuric Hg is believed to cause toxicity by combining with the active centers of critical enzymes and structural proteins.

BIOCHEMICAL MECHANISMS OF TOXICITY

Experimental studies of the possible biochemical mechanisms of MeHg neurotoxicity have been reviewed in detail (Atchison and Hare

1994; Chang and Verity 1995; ATSDR 1999). Mitochondrial changes, induction of lipid peroxidation, microtuble disruption, and disrupted protein synthesis have all been proposed as possible mechanisms. In developmental toxicity, disruption of cell-surface recognition has also been proposed as a possible mechanism (Baron et al. 1998; Dey et al. 1999). To date, no definitive data are available that point to any one mechanism as the proximate cause for the neurotoxic symptoms associated with MeHg exposure in adults.

Exposure of rats to MeHg has long been known to cause biochemical and ultrastructural changes in the mitochondria, but the evidence is not convincing that those changes are the primary mechanism for MeHg toxicity (Denny and Atchison 1994; Yoshino et al. 1966). Sarafian and Verity (1991) showed that MeHg causes membrane peroxidation in nerve cells. Because antioxidants, such as vitamin E and selenium, offer some protection in vivo against MeHg neurotoxicity (Chang et al. 1978; Magos and Webb 1980), free-radical-induced lipid peroxidation might be involved in the cellular damage caused by MeHg. However, lipid peroxidation does not appear to be the critical mechanism that causes cell lethality for many reasons, as summarized by Atchison and Hare (1994).

MeHg disrupts protein synthesis, and disruption has been proposed as the primary mechanism of MeHg neurotoxicity. In the rat, inorganic Hg, however, was 10 times more potent an inhibitor of cell-free protein synthesis than MeHg (Sugano et al. 1975). Stimulation of protein synthesis by MeHg was also reported (Burbaker et al. 1973). Mitotic arrest is one of the most sensitive indicators of MeHg exposure in mice. A single 4-mg/kg dose MeHg on postnatal day 2 resulted in a brain Hg concentration of only 1.8 μg/g of tissue. The ratio of late mitotic figures to total mitotic figures was significantly reduced in the cerebellum of exposed mice, indicating mitotic arrest (Sager et al. 1984).

Oxidative stress might also be involved in MeHg toxicity. Glutathione is the major antioxidant of the cell. After exposure to MeHg, glutathione concentrations decline and then increase. Cells that are made resistant to MeHg toxicity had an increase in the rate of efflux of MeHg and had 4-fold higher glutathione concentrations than normal cells (Miura and Clarkson 1993).

Another proposed mechanism underlying MeHg toxicity is disruption of microtubules in the neuronal cytoskeleton (Miura and Imura 1987). Hg binds to thiols in the tubulin, the protein monomers that form microtubules, and blocks the depolymerization and repolymerization of microtubules. Because the breakdown and assembly of microtubules are required for many cell functions, including cell division and migration, disruption of microtubule assembly could disrupt cellular processes. In vitro, MeHg has been shown to disrupt cell-cycle progression in primary rat brain cells (Ponce et al. 1994). The developing nervous system would be particularly sensitive to those effects due to the extensive cell division and migration that occurs during its development.

The ability to exchange between thiols forms the basis of therapeutic techniques following both MeHg exposure and exposure to Hg vapor. The neurotoxic effects of combined exposure to MeHg and Hg vapor have been reported to be similar in nature but more severe than those observed following exposure to each alone (Fredriksson et al 1996). There are many similarities in the biochemistry of the $MeHg^{+}$ and the inorganic Hg cation (Hg^{2+}), which is responsible for the toxicity following Hg vapor exposure (Clarkson 1997). Both cations exhibit a high affinity for SH groups, and association and dissociation reactions are rapid (Carty and Malone 1979). Both are found in tissues bound to large and small molecular-weight thiol-containing molecules (proteins, cysteine, and glutathione). The formation of Hg thiol bonds is believed to underlie the mobility and toxicity of Hg in the body (Clarkson 1997).

Although the exposure patterns and toxicokinetics and toxicodynamics of the different Hg species are usually studied separately, organic Hg and elemental Hg are eventually converted in vivo to inorganic Hg. The estimated average daily intake and retention of various forms of Hg are shown in Table 2-4. Estimates of the retention in the body of Hg from dental amalgams range from 3.1 to 17 μg per day. Estimates of MeHg retention range from 1 to 6 μg per day. The ratio of MeHg to total Hg will be different among those with high fish consumption. The data in Table 2-4 suggest that average exposure to Hg from dental amalgams might be considerably higher than exposure to Hg from MeHg. However, the available data are not adequate to permit a definitive comparison.

MeHg is very slowly but ultimately metabolized in situ in the brain to inorganic Hg. Elemental mercury is also oxidized to inorganic Hg in the brain. It is unclear whether MeHg toxicity at the cellular level is caused by the parent compound itself, due to the inorganic Hg that is its metabolite, or is caused indirectly by the free radicals generated by the

TABLE 2-4 Estimated Daily Intake and Retention[a] (micrograms per day) of Total Hg and Hg Compounds in the General Population Not Occupationally Exposed to Hg

Exposure	Elemental Hg Vapor	Inorganic Hg Compounds[b]	MeHg
Air	0.030 (0.024)[b,c]	0.002 (0.001)[b,c]	0.008 (0.0064)[b,c]
Food Sources			
Fish	0[b,c]	0.600 (0.042)[b,c]	1[d,e,f], 3[e,g,h], 6[c,g,h]
Non-Fish	0[b,c]	3.6 (0.25)[b,c]	0[b,c]
Drinking water	0[b,c]	0.050 (0.0035)[b,c]	0[b,c]
Dental amalgams	3.8-21 (3-17)[b,c]	0[b,c]	0[b,c]
Total	3.9-21 (3.1-17)	4.3 (0.3)	1-6 (1-6)

[a]Retention is assumed to be 95% of intake for MeHg, 80% for elemental Hg vapor, and 7% for inorganic Hg compounds.

[b]Data from IPCS 1991.

[c]Mean value.

[d]Data are for United States nationwide (per capita), calculated assuming a body weight of 70 kg (EPA 1997).

[e]Median value.

[f]Equivalent data for women of childbearing age: median = 1, assuming a body weight of 60 kg.

[g]Data are for the general population of reproductive age in New Jersey fish consumers, calculated assuming a body weight of 70 kg (Stern et al. 1996).

[h]Equivalent data for women of childbearing age: mean = 5, median = 3, assuming a body weight of 60 kg.

Note: Values given are the estimated average daily intake; the figures in parentheses represent the estimated amount retained in the body of an adult. Values are quoted to two significant figures.

metabolism of MeHg to inorganic Hg. If the ultimate toxic form of MeHg is indeed its inorganic Hg metabolite, that suggests that the dose of inorganic Hg to the brain from elemental Hg exposure (particularly from dental amalgams) and MeHg might be cumulative. That is the case even if oxidation of elemental Hg in the blood before absorption to the brain is considered. Risk-assessment models for MeHg, therefore, should consider additional chronic sources of exposure to Hg such as dental amalgams.

Such considerations are complicated by uncertainty about the mechanisms by which MeHg specifically exerts its neurodevelopmental toxicity. Such mechanisms might not be the same as those responsible for adult neurotoxicity. Nonetheless, the potential implications of additive toxicity from fish consumption and dental amalgams make elucidation of the mechanisms of MeHg toxicity in the brain a critical research priority.

SUMMARY AND CONCLUSIONS

- The major source of MeHg exposure in humans is consumption of fish, marine mammals, and crustaceans.
- The water solubility of mercuric chloride is greater than elemental Hg. That of elemental Hg is greater than MeHg. The solubility of the different forms of Hg might play a role in their differential toxicity.
- Elemental Hg and a portion of MeHg are converted to mercuric Hg in the body. The conversion of MeHg occurs at a very slow rate.
- Analytical methods for analyzing Hg in biological samples include AAS, AFS, NAA, ASV, ICP-MS, and XRF. Care must be taken to prevent contamination by Hg during sample preparation and analysis.
- MeHg is readily absorbed from the GI tract. After ingestion, 90% of the MeHg in blood can be found in red blood cells. It is bound primarily to red-blood-cell hemoglobin, but some is bound to plasma proteins.
- Hg in blood reflects recent exposure to MeHg and inorganic Hg. The half-life in blood for humans averages 50 days but can vary substantially. Because neonates have an immature transport system, they do not excrete MeHg as rapidly as adults.
- Hg in hair is approximately 90% MeHg. Hair measurements have the advantage of providing a historical record of MeHg exposure but do not accurately reflect exposure to inorganic Hg.
- The daily excretion of MeHg is about 1% of the human body burden. It is excreted mainly via the bile and feces as MeHg and mercuric Hg. Complexing with GSH is involved. Urine MeHg concentrations do not accurately reflect MeHg exposure.

- For elemental and inorganic Hg, the half-life in blood is 1-2 months. The whole-body half-life is slightly longer, but that does not take into account Hg in the brain, which is cleared very slowly. Excretion occurs primarily via urine and feces and, to a small extent, saliva, bile, sweat, and lungs.
- DMPS and DMSA can be used to increase Hg excretion. Dimercaprol (BAL), used in the past for chelation, is contraindicated because it redistributes Hg to the brain.
- MeHg readily crosses the blood-brain barrier. The rapid uptake of MeHg in the brain has been proposed to be due to lipid solubility, but evidence in rats suggests that the transport is due to the formation of MeHg-cysteine complexes.
- MeHg accumulates in the brain where it is slowly converted to inorganic Hg. Whether CNS damage is due to MeHg per se, to its biotransformation to inorganic Hg, or to both is still controversial. The mechanisms and cellular site for the biotransformation in humans are not well understood. Both free-radical and enzymatic biotransformation has been proposed.
- The critical organ for MeHg toxicity is the brain. Both adult and fetal brains are vulnerable. For elemental Hg, the critical organs are the brain and kidney. Both MeHg and elemental Hg are converted to mercuric Hg in the brain, where it is trapped. The biological mechanisms for removing mercuric Hg from the brain are limited. The critical organ for mercuric Hg toxicity is the kidney, where it accumulates.
- There is emerging evidence that the cardiovascular and immune systems might be major sites of MeHg toxicity (see Chapter 5).
- The high affinity of MeHg and mercuric Hg for sulfhydryl groups is believed to be a major mechanism that underlies their toxicity. If those sulfhydryl groups are in the active center of critical enzymes, severe inhibition of essential biochemical pathways occurs.
- The toxicology of the three species of Hg—elemental Hg, mercuric Hg and MeHg—are intertwined, because MeHg and elemental Hg are transformed to inorganic Hg in the brain. Risk-assessment models for MeHg in humans are complicated because of inadequate data regarding the cumulative neurotoxic effects of MeHg per se and its biotransformation product mercuric Hg, which has a very long half-live in the brain.

RECOMMENDATIONS

- As data become available, exposure to elemental Hg from dental amalgams should be considered in risk assessment of MeHg. Exposure to other chemical forms of Hg should also be considered.
- Retention of inorganic Hg in the brain for years following early MeHg intake is possibly related to the latent or long-term neurotoxic effects reported. The long half-life of inorganic Hg in the brain following MeHg intake should be considered in risk assessment of MeHg.
- The mechanisms, including any enzymes, involved in the biotransformation of MeHg to mercuric Hg in human tissues need to be investigated, especially at the subcellular level. The effects of Hg on signaling pathways and the conformation of enzymes and structural proteins should be further elucidated, because the development and function of the brain would be particularly sensitive to such effects.
- Exposure assessment of the U.S. population—including those with high fish consumption—is needed to provide a full picture of the distribution of MeHg and total Hg exposure nationally and regionally.

REFERENCES

Aaseth, J., D. Jacobsen, O. Andersen, and E. Wickstrom. 1995. Treatment of mercury and lead poisoning with dimercaptosuccinic acid and sodium dimercaptopropane-sulfonate: A review. Analyst 120(3):853-854.

Aberg, B., L. Ekman, R. Falk, U. Greitz, G. Persson, and J.O. Snihs. 1969. Metabolism of methyl mercury (^{203}Hg) compounds in man. Arch. Environ. Health 19(4):478-484.

American Academy of Pediatrics. 1999. Thimerosal in vaccines—An interim report to clinicians. Committee on Infectious Diseases and Committee on Environmental Health. Pediatrics 104(3):570-574.

Aposhian, H.V. 1983. DMSA and DMPS: Water-soluble antidotes for heavy metal poisoning. Annu. Rev. Pharmacol. Toxicol. 23:193-215.

Aposhian, H.V., and M.M. Aposhian. 1990. Meso-2,3-dimercaptosuccinic acid: Chemical, pharmacological and toxicological properties of an orally effective metal chelating agent. Annu. Rev. Toxicol. 30:279-306.

Aposhian, H.V., D.C. Bruce, W. Alter, R.C. Dart, K.M. Hurlbut, and M.M.

Aposhian. 1992a. Urinary mercury after administration of 2,3-dimercaptopropane-1-sulfonic acid: Correlation with dental amalgam score. FASEB J. 6(7):2472-2476.

Aposhian, H.V., R.M. Maiorino, M. Rivera, D.C. Bruce, R.C. Dart, K.M. Hurlbut, D.J. Levine, W. Zheng, Q. Fernando, D. Carter, and M.M. Aposhian. 1992b. Human studies with the chelating agents DMPS and DMSA. Clin. Toxicol. 30(4):505-528.

Aposhian, H.V., R.M. Maiorino, D. Gonzalez-Ramirez, M. Zuniga-Charles, Z. Xu, K.M. Hurlbut, P. Junco-Munoz, R.C. Dart, and M.M. Aposhian. 1995. Mobilization of heavy metals by newer, therapeutically useful chelating agents. Toxicology 97(1-3):23-28.

Aposhian, M.M., R.M. Maiorino, Z. Xu, and H.V. Aposhian. 1996. Sodium 2,3-dimercapto-1-propanesulfaonte (DMPS) treatment does not redistribute lead or mercury to the brain of rats. Toxicology 109(1):49-55.

Arenholt-Bindslev, D., and A.H. Larsen. 1996. Mercury levels and discharge in waste water from dental clinics. Water Air Soil Pollut. 86(1-4):93-99.

Aschner, M., N.B. Eberle, and H.K. Kimelberg. 1991. Interactions of methylmercury with rat primary astrocyte cultures: Methylmercury efflux. Brain Res. 554(1-2):10-14.

Atchison, W.D., and M.F. Hare. 1994. Mechanisms of methylmercury-induced neurotoxicity. FASEB J. 8(9):622-629.

ATSDR (Agency for Toxic Substances and Disease Registry). 1999. Toxicological Profile for Mercury. (Update). U.S. Department of Health & Human Services, Agency for Toxic Substances and Disease Registry, Atlanta, GA.

Bakir, F., S.F. Damluji, L. Amin-Zaki, M. Murthadha, A. Khalidi, N.Y. Al-Rawi, S. Tikriti, H.I. Dhahir, T.W. Clarkson, J.C. Smith, and R.A. Doherty. 1973. Methylmercury poisoning in Iraq. Science 181:230-241.

Baldi, F., and M. Filippelli. 1991. New method for detecting methylmercury by its enzymic conversion to methane. Environ. Sci. Technol. 25(2):302-305.

Ballatori, N. 1991. Mechanisms of metal transport across liver cell plasma membranes. Drug Metab. Rev. 23(1-2):83-132.

Ballatori, N., and T.W. Clarkson. 1982. Developmental changes in the biliary excretion of methylmercury and glutathione. Science 216(4541):61-63.

Baron Jr, S., N. Haykal-Coates, and H.A. Tilson. 1998. Gestational exposure to methylmercury alters the developmental pattern of trk-like immunoreactivity in the rat brain and results in cortical dysmorphology. Dev. Brain Res. 109(1):13-31.

Begley, T.P., A.E. Walts, and C.T. Walsh. 1986. Bacterial organomercurial lyase: Overproduction, isolation, and characterization. Biochemistry 25(22): 7186-7192.

Berlin, M. 1986. Mercury. Pp. 387-445 in Handbook on the Toxicology of

Metals, 2nd Ed., L. Friberg, G.F. Nordberg, and V.B. Vouk, eds. New York: Elsevier.

Bernard, S., and P. Purdue. 1984. Metabolic models for methyl and inorganic mercury. Health Phys. 46(3):695-699.

Bluhm, R.E., R.G. Bobbitt, L.W. Welch, A.J.J. Wood, J.F. Bonfiglio, C. Sarzen, A.J. Heath, and R.A. Branch. 1992. Elemental mercury vapour toxicity, treatment, and prognosis after acute, intensive exposure in chloralkali plant workers: Part I: History, neuropsychological findings and chelator effects. Hum. Exp. Toxicol. 11(3):201-210.

Bornmann, G., G. Henke, H. Alfes, and H. Mollmann. 1970. Intestinal absorption of metallic mercury. [in German]. Arch. Toxicol. 26(3):203-209.

Brown, N.L., S.J. Ford, R.D. Pridmore, and D.C. Fritzinger. 1983. Nucleotide sequence of a gene from the Pseudomonas transposon Tn501 encoding mercuric reductase. Biochemistry 22(17):4089-4095.

Burbaker, P.E., R. Klein, S.P. Herman, G.W. Lucier, L.T. Alexander, and M.D. Long. 1973. DNA, RNA and protein synthesis in brain, liver and kidneys of symptomatic methylmercury treated rats. Exp. Mol. Pathol. 18(3):263-280.

Cappon, C.J., and J.C. Smith. 1978. A simple and rapid procedure for the gas-chromatographic determination of methylmercury in biological samples. Bull Environ. Contam. Toxicol. 19(5):600-607.

Carty, A.J., and S.F. Malone. 1979. The chemistry of mercury in biological systems. Pp. 433-470 in The Biogeochemistry of Mercury in the Environment, J.O. Nriagu, ed. Amsterdam: Elsevier.

Cernichiari, E., R. Brewer, G.J. Myers, D.O. Marsh, L.W. Lapham, C. Cox, C.F. Shamlaye, M. Berlin, P.W. Davidson, and T.W. Clarkson. 1995. Monitoring methylmercury during pregnancy: Maternal hair predicts fetal brain exposure. Neurotoxicology 16(4):705-710.

Chang, L.W., and M.A. Verity. 1995. Mercury neurotoxicity: Effects and mechanisms. Pp. 31-59 in Handbook of Neurotoxicology, L.W. Chang, and R.S. Dyer, eds. New York: Marcel Dekker.

Chang, L.W., M. Gilbert, and J. Sprecher. 1978. Modification of methylmercury neurotoxicity by vitamin E. Environ. Res. 17(3):356-366.

Chen, R.W., H.E. Ganther, and K.G. Hoekstra. 1973. Studies on the binding of methylmercury by thionein. Biochem. Biophys. Res. Commun. 51(2):383-390.

Cherian, M.G., J.B. Hursh, T.W. Clarkson, and J. Allen. 1978. Radioactive mercury distribution in biological fluids, and excretion in human subjects after inhalation of mercury vapor. Arch. Environ. Health 33(3):109-114.

Clarkson, T.W. 1997. The toxicology of mercury. Crit. Rev. Clin. Lab. Sci. 34(4):369-403.

Clarkson, T.W., L. Friberg, G. Nordberg, and P.R. Sager, eds. 1988. Biological Monitoring of Toxic Metals. New York: Plenum Press.

Clarkson, T.W., L. Magos, C. Cox, M.R. Greenwood, L. Amin-Zaki, M.A. Majeed, and S.F. al-Damluji. 1981. Tests of efficacy of antidotes for removal of methylmercury in human poisoning during the Iraq outbreak. J. Pharmacol. Exp. Ther. 218(1):74-83.

Cotton, F.A., and G. Wilkinson. 1988. Advanced Inorganic Chemistry, 5th Ed. New York: John Wiley & Sons.

Cox, C., T.W. Clarkson, D.O. Marsh, L. Amin-Zaki, S. Tikriti, and G.G. Myers. 1989. Dose-response analysis of infants prenatally exposed to methyl mercury: An application of a single compartment model to single-strand hair analysis. Environ. Res. 49(2):318-332.

Davis, L.E., M. Kornfeld, H.S. Mooney, K.J. Fiedler, K.Y. Haaland, W.W. Orrison, E. Cernichiari, and T.W. Clarkson. 1994. Methylmercury poisoning: Long-term clinical, radiological, toxicological, and pathological studies of an affected family. Ann. Neurol. 35(6):680-688.

Denny, M.F., and W.D. Atchison. 1994. Elevations in the free intrasynaptosomal concentration of endogenous zinc by methylmercury. [Abstract]. Toxicologist 14:290.

Dey, P.M., M. Gochfield, and K.R. Reuhl. 1999. Developmental methylmercury administration alters cerebellar PSA--NCAM expression and Golgi sialytransferase activity. Brain Res. 845(2):139-151.

Divine, K.K., F. Ayala-Fierro, D.S. Barber, and D.E. Carter. 1999. Glutathione, albumin, cysteine, and cys-gly effects on toxicity and accumulation of mercuric chloride in LLC-PK_1 cells. J. Toxicol. Environ. Health 57(7):489-505.

Doi, R. 1991. Individual difference of methylmercury metabolism in animals and its significance in methylmercury toxicity. Pp. 77-98 in Advances in Mercury Toxicology, T. Suzuki, N. Imura, and T.W. Clarkson, eds. New York: Plenum Press.

Drasch, G., I. Schupp, H. Hofl, R. Reinke, and G. Roider. 1994. Mercury burden of human fetal and infant tissues. Eur. J. Pediatr. 153(8):607-610.

Dunn, J.D., and T.W. Clarkson. 1980. Does mercury exhalation signal demethylation of methylmercury? Health Phys. 38(3):411-414.

Elinder, C.G., L. Gerhardsson, and G. Oberdörster. 1988. Biological monitoring of toxic metals—Overview. Pp. 1-72 in Biological Monitoring of Toxic Metals, T.W. Clarkson, L. Friberg, G.F. Nordberg, and P. R. Sager, eds. New York: Plenum Press.

EPA (U.S. Environmental Protection Agency). 1997a. Mercury Study Report to Congress. Vol. IV: An Assessment of Exposure to Mercury in the United States. EPA-452/R-97-006. U.S. Environmental Protection Agency, Office of Air Quality Planning and Standards and Office of Research and Development.

EPA (U.S. Environmental Protection Agency). 1997b. Mercury Study Report for Congress. Volume V: Health Effects of Mercury and Mercury Compounds. EPA-452/R-97-007. U.S. Environmental Protection Agency, Office of Air Quality Planning and Standards and Office of Research and Development.

EPA (U.S. Environmental Protection Agency). 1997c. Mercury Study Report to Congress. Volume VII: Characterization of Human Health and Wildlife Risks from Mercury Exposure in the United States. EPA-452/R-97-009. U.S. Environmental Protection Agency, Office of Air Quality Planning and Standards and Office of Research and Development.

Fang, S.C., and E. Fallin. 1974. Uptake and subcellular cleavage of organomercury compounds by rat liver and kidney. Chem. Biol. Interact. 9(1):57-64.

FDA (Center for Drug Evaluation and Research). 1999. List of Drug and Food that Contain Intentionally Introduced Mercury Compounds. Updated November 17, 1999. Online. Available: http://www.fda.gov/cder/fdama/mercury300.htm

Fox, B., and C.T. Walsh. 1982. Mercuric reductase. Purification and characterization of a transposon-encoded flavoprotein containing an oxidation-reduction-active disulfide. J. Biol. Chem. 257(5):2498-2503.

Fox, B.S., and C.T. Walsh. 1983. Mercuric reductase: Homology to glutathione reductase and lipoamide dehydrogenase. Iodoacetamide alkylation and sequence of the active site peptide. Biochemistry 22(17):4082-4088.

Francis, P.C., W.J. Birge, B.L. Roberts, and J.A. Black. 1982. Mercury content of human hair: A survey of dental personnel. J. Toxicol. Environ. Health 10(4-5):667-672.

Fredriksson, A., L. Dahlgren, B. Danielsson, P. Eriksson, L. Dencker, and T. Archer. 1992. Behavioural effects of neonatal metallic mercury exposure in rats. Toxicology 74(2-3):151-160.

Fredriksson, A., L. Dencker, T. Archer, and Danielsson. 1996. Prenatal coexposure to metallic mercury vapour and methylmercury produce interactive behavioural changes in adult rats. Neurotoxicol. Teratol. 18(2):129-134.

Friberg, L., and N.K. Mottet. 1989. Accumulation of methylmercury and inorganic mercury in the brain. Biol. Trace Elem. Res. 21:201-206.

Fujiyama, J., K. Hirayama, and A. Yasutake. 1994. Mechanism of methylmercury efflux from cultured astrocytes. Biochem. Pharmacol. 47(9):1525-1530.

Fung, Y.K., A.G. Meade, E.P. Rack, A.J. Blotcky, J.P. Claassen, M.W. Beatty, and T. Durham. 1995. Determination of blood mercury concentrations in Alzheimer's patients. Clin. Toxicol. 33(3):243-247.

Gage, J.C. 1975. Mechanisms for the biodegradation of organic mercury compounds: The actions of ascorbate and of soluble proteins. Toxicol. Appl. Pharmacol. 32(2):225-238.

Goering, P.L., W.D. Galloway, T.W. Clarkson, F.L. Lorscheider, M. Berlin, and

A.S. Rowland. 1992. Toxicity assessment of mercury vapor from dental amalgams. Fundam. Appl. Toxicol. 19(3):319-329.

Gonzalez-Ramirez, D.M., M. Zuniga-Charles, A. Narro-Juarez, Y. Molina-Recio, K.M. Hurlbut, R.C. Dart, and H.V. Aposhian. 1998. DMPS (2,3-dimercapto-propane-1-sulfonate, dimaval) decreases the body burden of mercury in humans exposed to mercurous chloride. J. Pharmacol. Exp. Ther. 287(1):8-12.

Gosselin, R.E., R.P. Smith, H.C. Hodge. 1984. Clinical Toxicology of Commercial Products, 5th Ed. Baltimore: Williams & Wilkins.

Grandjean, P., P.J. Jørgensen, and P. Weihe. 1994. Human milk as a source of methylmercury exposure in infants. Environ. Health Perspect. 102(1):74-77.

Grandjean, P., P. Weihe, P.J. Jørgensen, T.W. Clarksen, E. Cernichiari, and T. Viderø. 1992. Impact of maternal seafood diet on fetal exposure to mercury, selenium, and lead. Arch. Environ. Health 47(3):185-195.

Halbach, S. 1994. Amalgam tooth fillings and man's mercury burden. Hum. Exp. Toxicol. 13:496-501.

Hall, L.L., P.V. Allen, H.L. Fisher, and B. Most. 1995. The kinetics of intravenously-administered inorganic mercury in humans. Pp. 265-280 in Kinetic Models of Trace Elements and Mineral Metabolism During Development, K.N.S. Subramanian, and M.E. Wastney, eds. Boca Raton, FL: CRC Press.

Harada, M. 1995. Minamata disease: Methylmercury poisoning in Japan caused by environmental pollution. Crit. Rev. Toxicol. 25(1):1-24.

Henderson, R., H.P. Shotwell, and L.A. Krause. 1974. Analyses for total, ionic and elemental mercury in urine as a basis for biological standard. Am. Ind. Hyg. Assoc. J. 35:576-580.

Hursh, J.B. 1985. Partition coefficients of mercury (^{203}Hg) vapor between air and biological fluids. J. Appl. Toxicol. 5(5):327-332.

Hursh, J.B., M.G. Cherian, T.W. Clarkson, J.J. Vostal, and P.V. Mallie. 1976. Clearance of mercury (Hg-197, Hg-203) vapor inhaled by human subjects. Arch. Environ. Health 31(6):302-309.

Hursh, J.B., T.W. Clarkson, T.V. Nowak, R.C. Pabico, B.A. McKenna, E. Miles, and F.R. Gibb. 1985. Prediction of kidney mercury content by isotope techniques. Kidney Int. 27(6):898-907.

Hursh, J.B., T.W. Clarkson, E.F. Miles, and L.A. Goldsmith. 1989. Precutaneous absorption of mercury vapor by man. Arch. Environ. Health 44(2):120-127.

IPCS (International Programme on Chemical Safety). 1990. Environmental Health Criteria Document 101: Methylmercury. Geneva: World Health Organization.

IPCS (International Programme on Chemical Safety). 1991. Environmental Health Criteria Document 118: Inorganic Mercury. Geneva: World Health Organization.

Jakubowski, M., J. Piotrowski, and B. Trojanowska. 1970. Binding of mercury in the rat: Studies using 203HgCl2 and gel filtration. Toxicol. Appl. Pharmacol. 16(3):743-753.

Kägi, J.H., and M. Nordberg. 1979. Metallothionein. International Meeting on Metallothionein and Other Low Molecular Weight Metal Binding Proteins. Basel, Switzerland: Birkhäuser.

Kalamegham, R., and K.O. Ash. 1992. A simple ICP-Ms procedure for the determination of total mercury in whole blood and urine. J. Clin. Lab. Anal. 6(4):190-193.

Kerper, L.E., N. Ballatori, and T.W. Clarkson. 1992. Methylmercury transport across the blood-brain barrier by an amino acid carrier. Am. J. Physiol. 262(5):R761-R765.

Kershaw, T.G., T.W. Clarkson, and P.H. Dhahir. 1980. The relationship between blood-brain levels and dose of methylmercury in man. Arch. Environ. Health 35(1):28-36.

Klaassen, C.D. 1996. Heavy metals and heavy-metal antagonists. Pp. 1649-1671 in The Pharmacological Basis of Therapeutics, J.G. Hardman, L.E. Limbird, P.B. Molinoff, R.W. Ruddon, and A.G. Gilman, eds. New York: McGraw-Hill.

Klimova, L.K. 1958. Pharmacology of a new Unithiol antidote [in Russian]. Farmakol. Toksikol. (Moscow) 21:53-59.

Komsta-Szumska, E., J. Chmielnicka, and J.K. Piotrowski. 1976. Binding of inorganic mercury by subcellllar fractions and proteins of rat kidneys. Arch. Toxicol. 37(1):57-66.

Kromidas, L., L.D. Trombetta, and I.S. Jamall. 1990. The protective effects of glutathione against methylmercury cytotoxicity. Toxicol. Lett. 51(1):67-80.

Kudsk, F.N. 1965. The influence of ethyl alcohol on the absorption of methyl mercury vapor from the lungs of man. Acta Pharmacol. Toxicol. 23:263-274.

Lefevre, P.A., and J.W. Daniel. 1973. Some properties of the organomercury-degrading system in mammalian liver. FEBS Lett. 35(1):121-123.

Lind, B., L. Friberg, and M. Nylander. 1988. Preliminary studies on methylmercury biotransformation and clearance in the brain of primates: II. Demethylation of mercury in brain. J. Trace Elem. Exp. Med. 1(1):49-56.

Liu, K.Z., Q.G. Wu, and H.I. Liu. 1990. Application of a Nafion-Schiff-base modified electrode in anodic-stripping voltammetry for the determination of trace amounts of mercury. Analyst 115(6):835-837.

Lorscheider, F.L., M.J. Vimy, and A.O. Summers. 1995. Mercury exposure from "silver" tooth fillings: Emerging evidence questions a traditional dental paradigm. FASEB J. 9(7):504-508.

Magos, L., and W.H. Butler. 1972. Cumulative effects of methylmercury dicyandiamide given orally to rats. Food Cosmet. Toxicol. 10(4):513-517.

Magos, L., and A.A. Cernik. 1969. A rapid method for estimating mercury in undigested biological samples. Br. J. Ind. Med. 26(2):144-149.

Magos, L., and T.W. Clarkson. 1972. Atomic absorption determination of total, inorganic, and organic mercury in blood. J. Assoc. Off. Anal. Chem. 55(5):966-971.

Magos, L., and M. Webb. 1980. The interaction of selenium with cadmium and mercury. Crit. Rev. Toxicol. 8(1):1-42.

Magos, L., A.W. Brown, S. Sparrow, E. Bailey, R.T. Snowden, and W.R. Skipp. 1985. The comparative toxicology of ethyl- and methylmercury. Arch. Toxicol. 57(4):260-267.

Mahaffey, K.R., and D. Mergler. 1998. Blood levels of total and organic mercury in residents of the upper St. Lawrence River basin, Quebec: Association with age, gender, and fish consumption. Environ. Res. 77(2):104-114.

Marsh, D.O., T.W. Clarkson, C. Cox, G.J. Myers, L. Amin-Zaki, and S. Al-Tikriti. 1987. Fetal methyl mercury poisoning: Relationship between concentration in single strands of maternal hair and child effects. Arch. Neurol. 44(10):1017-1022.

Matsuo, N, T. Suzuki, and H. Akagi. 1989. Mercury concentration in organs of contemporary Japanese. Arch. Environ. Health 44(5):298-303.

Matsuo, N., M. Takasugi, A. Kuroiwa, and H. Ueda. 1987. Thymic and splenic alterations in mercuric chloride-induced glomerulopathy. Pp. 333-334 in Toxicology of Metals: Clinical and Experimental Research, S.S. Brown, and Y. Kodama, eds. Chichester, UK: Ellis Horwood Limited.

Miettinen, J.K. 1973. Absorption and elimination of dietary (Hg^{++}) and methylmercury in man. Pp. 233-246 in Mercury, Mercurial, and Mercaptans, M.W. Miller, and T.W. Clarkson, eds. Springfield, IL: C.C. Thomas.

Miettinen, J.K., T. Rahola, T. Hattula, K. Rissanen, and M. Tillander. 1971. Elimination of ^{203}Hg-methylmercury in man. Ann. Clin. Res. 3(2):116-122.

Miura, K., and T.W. Clarkson. 1993. Reduced methylmercury accumulation in a methylmercury resistant rat pheochromocytoma PC12 cell line. Toxicol. Appl. Pharmacol. 118(1):39-45.

Miura, K., and N. Imura. 1987. Mechanism of methylmercury cytotoxicity. Crit. Rev. Toxicol. 18(3):161-188.

Mobley, H.L., and B.P. Rosen. 1982. Energetics of plasmid-mediated arsenate resistance in Escherichia coli. Proc. Natl. Acad. Sci. U.S.A. 79(20):6119-6122.

Naganuma, A., N. Oda-Urano, T. Tanaka, and N. Imura. 1988. Possible role of hepatic glutathione in transport of methylmercury into mouse kidney. Biochem. Pharmacol. 37(2):291-296.

Nakamura, I., K. Hosokawa, H. Tamura, and T. Miura. 1977. Reduced mercury excretion with feces in germfree mice after oral administration of methylmercury chloride. Bull. Environ. Contam. Toxicol. 17(5):528-533.

NESCAUM (Northeast States for Coordinated Air Use Management), NEWMOA (Northeast Waste Management Officials' Association), NEIWPCC (New England Interstate Water Pollution Control Commission), and EMAN (Canadian Ecological Monitoring and Assessment Network). 1998. Mercury in Northeastern freshwater fish: current level and ecological impacts. Pp. IV.1-IV. 21 in Northeast States/Eastern Canadian Provinces Mercury Study—A Frame Work for Action. February, 1998.

Nielsen, J.B. 1992. Toxicokinetics of mercuric-chloride and methylmercuric chloride in mice. J. Toxicol. Environ. Health 37(1):85-122.

Nierenberg, D.W., R.E. Nordgren, M.B. Chang, R.W. Siegler, M.B. Blayney, F. Hochberg, T.Y. Toribara, E. Cernichiari, and T. Clarkson. 1998. Delayed cerebellar disease and death after accidental exposure to dimethylmercury. N. Engl. J. Med. 338(23):1672-1676.

Norseth, T., and T.W. Clarkson. 1970. Studies on the biotransformation of ^{203}Hg-labeled methylmercury chloride in rats. Arch. Environ. Health 21(6):717-727.

Okawa, K., H. Saito, I. Kifune, T. Ohshina, M. Fujii, and Y. Takizawa. 1982. Respiratory tract retention of inhaled air pollutants. 1. Mercury absorption by inhaling though the nose and expiring through the mouth at various concentrations. Chemosphere 11(9):943-951.

Ostlund, K. 1969. Studies on the metabolism of methyl mercury in mice. Acta Pharmacol. Toxicol. 27(Suppl.1):1-132.

Perry, R.D., and S. Silver. 1982. Cadmium and manganese transport in Staphylococcus aureus membrane vesicles. J. Bacteriol. 150(2):973-976.

Phelps, R.W., T.W. Clarkson, T.G. Kershaw, and B. Wheatley. 1980. Interrelationships of blood and hair mercury concentrations in a North American population exposed to methylmercury. Arch. Environ. Health 35(3):161-168.

Ponce, R.A., T.J. Kavanagh, N.K. Mottet, S.G. Whittaker, and E.M. Faustman. 1994. Effects of methyl mercury on the cell cycle of primary rat CNS cells in vitro. Toxicol. Appl. Pharmacol. 127(1):83-90.

Rahola, T., T. Hattula, A. Korolainen, and J.K. Miettinen. 1973. Elimination of free and protein-bound ionic mercury (20Hg2+) in man. Ann. Clin. Res. 5(4):214-219.

Robinson, J.B., and O.H. Tuovinen. 1984. Mechanisms of microbial resistance and detoxification of mercury and organomercury compounds: Physiological, biochemical, and genetic analyses. Microbiol. Rev. 48(2):95-124.

Rowland, I., M. Davies, and J.G. Evans. 1980. Tissue content of mercury in rats given methylmercuric chloride orally: Influence of intestinal flora. Arch. Environ. Health 35(3):155-160.

Rowland, I., M. Davies, and P. Grasso. 1977. Biosynthesis of methylmercury compounds by the intestinal flora of the rat. Arch. Environ. Health 32(1):24-28.

Rupp, E.M., F.I. Miller, and C.F. Baes III. 1980. Some results of recent surveys of fish and shellfish consumption by age and region of U.S. residents. Health Phys. 39(2):165-175.

Sager, P.R., M. Aschner, and P.M. Rodier. 1984. Persistent, differential alterations in developing cerebellar cortex of male and female mice after methylmercury exposure. Brain Res. 314(1):1-11.

Sarafian, T., and M.A. Verity. 1991. Oxidative mechanisms underlying methylmercury neurotoxicity. Int. J. Dev. Neurosci. 9(2):147-153.

Sherlock, J., J. Hislop, D. Newton, G. Topping, and K. Whittle. 1984. Elevation of mercury in human blood from controlled chronic ingestion of methylmercury in fish. Hum. Toxicol. 3(2):117-131.

Silver, S., and D. Keach. 1982. Energy-dependent arsenate efflux: The mechanism of plasmid-mediated resistance. Proc. Natl. Acad. Sci. U.S.A. 79(20):6114-6118.

Skerfving, S. 1988. Mercury in women exposed to methylmercury through fish consumption, and in their newborn babies and breast milk. Bull. Environ. Contam. Toxicol. 41(4):475-482.

Stern, A.H., L.R. Korn, and B.E. Ruppel. 1996. Estimation of fish consumption and methylmercury intake in the New Jersey population. J. Expo. Anal. Environ. Epidemiol. 6(4):503-525.

Stopford, W., S.D. Bundy, L.J. Goldwater, and J.A. Bittikofer. 1978. Microenvironmental exposure to mercury vapor. Am. Ind. Hyg. Assoc. J. 39(5):378-384.

Suda, I., and K. Hirayama. 1992. Degradation of methy- and ethylmercury into inorganic mercury by hydroxyl radical produced from rat liver microsomes. Arch. Toxicol. 66(6):398-402.

Sugano, H., S. Omata, and H. Tsubaki. 1975. Methylmercury inhibition of protein synthesis in brain tissue. I. Effects of methylmercury and heavy metals on cell-free protein synthesis in rat brain and liver. Pp. 129-136 in Studies on the Health Effects of Alkylmercury in Japan, Environmental Agency, Japan.

Summers, A.O. 1985. Bacterial resistance to toxic elements. Trends Biotechnol. 3(5):122-125.

Summers, A.O., and S. Silver. 1978. Microbial transformations of metals. Annu. Rev. Microbiol. 32:637-672.

Sundberg, J., and A. Oskarsson. 1992. Placental and lactational transfer of mercury from rats exposed to methylmercury in their diet: Speciation of mercury in the offspring. J. Trace Elem. Exp. Med. 5(1):47-56.

Sundberg, J., S. Jonsson, M.O. Karlsson, I.P. Hallen, and A. Oskarsson. 1998. Kinetics of methylmercury and inorganic mercury in lactating and nonlactating mice. Toxicol. Appl. Pharmacol. 151(2):319-329.

Suzuki, T., T. Hongo, N. Matsuo, H. Imai, M. Nakazawa, T. Abbe, Y. Yamamura, M. Yoshida, and H. Aoyama. 1992. An acute mercuric mercury poisoning: Chemical speciation of hair mercury shows a peak of inorganic mercury value. Hum. Exp. Toxicol. 11(1):53-57.

Syversen, T.L. 1974. Distribution of mercury in enzymatically characterized subcellular fractions from the developing rat brain after injections of methylmercuric chloride and diethylmercury. Biochem. Pharmacol. 23(21):2999-3007.

Takeuchi, T., and E. Komyo. 1977. Pathology and pathogenesis of Minamata disease. Pp. 103-142 in Minamata Disease, T. Tsubake and K. Irukayama, eds. New York: Elsevier.

Takeuchi, T., K. Eto, and H. Tokunaga. 1989. Mercury level and histochemical distribution in a human brain with Minamata Disease following a long-term clinical course of 26 years. Neurotoxicology 10(4):651-657.

Takizawa, Y. 1979. Epidemiology of mercury poisoning. Pp. 325-366 in The Biogeochemistry of Mercury in the Environment, J.O. Nriagu, ed. Amsterdam: Elsevier/North-Holland Biomedical Press.

Thomas, D.J., H.L. Fisher, L.L. Hall, and P. Mushak. 1982. Effects of age and sex on retention of mercury by methylmercury-treated rats. Toxicol. Appl. Pharmacol. 62(3):445-454.

Vahter, M., N.K. Mottet, L. Friberg, B. Lind, D.D. Shen, and T. Burbacher. 1994. Speciation of mercury in the primate blood and brain following long-term exposure to methyl mercury. Toxicol. Appl. Pharmacol. 124(2):221-229.

Vahter, M.E., N.K. Mottet, L.T. Friberg, S.B. Lind, J.S. Charleston, and T.M. Burbacher. 1995. Demethylation of methyl mercury in different brain sites of Macaca fascicularis monkeys during long-term subclinical methyl mercury exposure. Toxicol. Appl. Pharmacol. 134(2):273-284.

Vermeir, G., C. Vandecasteele, and R. Dams. 1991a. Atomic fluorescence spectrometry combined with reduction aeration for the determination of mercury in biological samples. Anal. Chim. Acta 242(2):203-208.

Vermeir, G., C. Vandecasteele, and R. Dams. 1991b. Atomic fluorescence spectrometry for the determination of mercury in biological samples. Pp. 29-36 in Trace Elements in Health and Disease, A. Aitio, A. Aro, J. Jarvisalo, and H. Vainio, eds. Cambridge, UK: The Royal Society of Chemistry.

Vimy, M.J., D.E. Hooper, W.W. King, and F.L. Lorscheider. 1997. Mercury from maternal "silver" tooth fillings in sheep and human breast milk. A source of neonatal exposure. Biol. Trace Elem. Res. 56(2):143-152.

Vimy, M.J., Y. Takahashi, and F.L. Lorscheider. 1990. Maternal-fetal distribution of mercury (203Hg) released from dental amalgam fillings. Am. J. Physiol. 258(4):R939-R945.

Wendroff, A.P. 1995. Magico-religious mercury use and cultural sensitivity. Am. J. Public Health 85(3):409-410.

WHO (World Health Organization). 1976. Mercury. Environmental Health Criteria 1. Geneva: World Health Organization.

Wisniewska, J.M., B. Trojanowska, J. Piotrowski, and M. Jakubowski. 1970. Binding of mercury in the rat kidney by metallothionein. Toxicol. Appl. Pharmacol. 16(3):754-763.

Yoshida, M., H. Satoh, T. Kishimoto, and Y. Yamamura. 1992. Exposure to mercury via breast milk in suckling offspring of maternal guinea pigs exposed to mercury vapor after parturition. J. Toxicol. Environ. Health 35(2):135-139.

Yoshino, Y., T. Mozai, and K. Nakao. 1966. Distribution of mercury in the brain and its subcellular units in experimental organic mercury poisoning. J. Neurochem. 13:397-406.

Zhuang, G., Y. Wang, M. Zhi, W. Zhou, J. Yin, M. Tan, and Y. Cheng. 1989. Determination of arsenic, cadmium, mercury, copper and zinc in biological samples by radiochemical neutron-activation analysis. J. Radioanal. Nucl. Chem. 129(2):459-464.

3

BIOLOGICAL VARIABILITY

INDIVIDUAL responses to MeHg exposure are variable. For example, individuals receiving the same dose of MeHg in the Iraqi accident did not all have the same effects. Even in controlled animal experiments, considerable variability in response is noted (Burbacher et al. 1988; Rice and Gilbert 1990). Differences in susceptibility to MeHg might be due to differences in the uptake, storage, transport, or metabolism of MeHg. Susceptibility to the effects of MeHg can also be predetermined by genetic polymorphisms that affect the delivery of MeHg to the target organs or affect the response of the target organs to MeHg. In addition, other external factors can influence vulnerability to the effects of MeHg. Factors that deserve consideration are age, gender, health status, nutritional status, and the intake of other foods or nutrients that might influence the absorption, uptake, distribution, and metabolism of MeHg. The ability of the individual to compensate for damage caused by MeHg exposure would also affect susceptibility. This chapter discusses those factors that could underlie the variability in response to Hg exposure. The implications of that variability on the study of the toxicokinetics of Hg are discussed.

AGE-RELATED SUSCEPTIBILITY

Exposure to MeHg during the neonatal period, infancy and childhood has different effects due to the different stages of brain development and organ growth and the ratio of the MeHg concentration to body size. Age also affects the detection of toxic responses to MeHg, because some of the most sensitive end points examined— neurological development and cognitive ability—are dependent upon the age of the subject and the stage of cognitive maturation. There are also natural differences among individuals in performance on tests used. Therefore, the sensitivity of the test or assessment is dependent upon the developmental stage and age of the subject. In addition, many of the tests are carried out during periods of rapid development, which results in greater natural variation between individuals.

Data from Japanese poisoning episodes provide strong evidence that susceptibility to MeHg changes with age. Takeuchi (1968) described three distinct patterns of MeHg neuropathology termed adult, infantile, and fetal Minamata disease. In autopsy cases following fetal exposures, clear evidence of interference with brain development was observed. Disorganized cell layers and misoriented cells were observed, providing evidence of disrupted cell migration. For fetal and infantile exposures, lesions were observed throughout the cortex. A more selective pattern of lesions, localized in the calcarine and precentral cortices, particularly in the depths of the sulci, was observed in adult cases. Lesions in the granular layer of the cerebellum were observed in all cases. Reports of age-related neurological effects due to MeHg exposure in Japan and Iraq have also been described (Bakir et al. 1973; Harada 1968; Marsh et al. 1980). In both cases, mothers with few or no symptoms gave birth to infants severely affected. Studies with animal models also have reported significant age-dependent effects from MeHg exposure. As in human cases, offspring are sometimes severely affected with little or no signs of toxicity in the mother (Spyker et al. 1972; Mottet et al. 1987). Thus, age needs to be considered in the design of studies of MeHg, including in the choice of end points and the determination of how to analyze the results.

GENDER DIFFERENCES

Several reports have described gender differences in the toxicokinetics and the toxicodynamics of MeHg. Evidence of gender-dependent MeHg metabolism has been reported in humans (Miettinen 1973) and animal models (Thomas et al. 1986; Nielsen and Andersen 1991). However, this gender sensitivity does not apply in the same way for all outcomes. The Iraqi MeHg epidemic appeared to affect three times as many females as males (Magos et al. 1976). Epidemiological studies of human infants and children have reported gender specific effects on development with males exhibiting greater effects than females. (McKeown-Eyssen et al. 1983)

In general, results in animal studies indicate that females exhibit a higher body burden of Hg per given dose than males. That result might be due to higher metabolism and urinary-excretion rates for MeHg in sexually mature male mice compared with female mice (Hirayama and Yasutake 1986). Animal data also indicate gender differences in the sensitivity to MeHg toxicity. Fowler (1972) and Yasutake et al. (1990) found that females are more likely to show renal toxicity following MeHg exposure. Yasutake and Hirayama (1988) found the gender differences to be strain sensitive following oral administration of MeHg chloride at 5 mg/kg per day since male Balb/cA mice died earlier than female Balb/cA mice, but female C57B/6N mice died earlier than male C57B/6N mice. Reports regarding the neurological effects have been mixed; females were found to be more susceptible than males in some studies, and males were observed to be more susceptible than females in others (Magos et al. 1976; Tagashira et al. 1980; McKeown-Eyssen et al. 1983; Vorhees 1985; Tamashiro et al. 1986a).

GENETICS

Aside from gender differences within a population, there is evidence of differences in sensitivity within populations that result in greater damage from a given exposure in one individual than in another. The extent to which that difference is due to familial characteristics compared with nutritional and environmentally mediated susceptibility remains to be determined. Differences in enzymatic expression might result in individual differences in sensitivity to MeHg. Currently, no evidence of polymorphisms affecting the metabolism or detoxification of MeHg exists. The lack of evidence might be due to the inadequate study of those interactions in human populations and animal models. Therefore, the extent to which interindividual variability in effects at similar doses is attributable to genetic differences in susceptibility remains unknown (Tamashiro et al. 1986).

MECHANISMS OF NUTRITIONAL INFLUENCE ON MeHg HEALTH EFFECTS

Overall, nutritional status and dietary interactions can affect the outcomes of MeHg studies, either by influencing the toxicity of Hg or by having effects on the end points measured. The main source of exposure to MeHg is through the food chain, largely through consumption of nonherbivorous fish and marine mammals, with smaller amounts contributed by intake of other fish and seafood. The nature of dietary exposures is such that consumption of one food group is generally related to a reduction or avoidance of other food groups. Establishing causality becomes particularly complex under those circumstances. Pathways through which diet and nutrients might affect the results of MeHg toxicity studies include the potential for attenuating a MeHg effect, exacerbating a MeHg effect, or acting as a confounder by causing toxicity due to other common food components or contaminants. Those three pathways are outlined in Figure 3-1.

Potentially harmful effects of MeHg might be attenuated by protective effects of such nutrients as selenium and omega-3 fatty acids. At the other extreme, malnourishment could affect study results either by directly reducing the sensitivity of an end point tested or by exacerbating the effects of MeHg, thereby increasing the sensitivity to MeHg toxicity. Nutritional factors that disrupt neuronal development, such as iron or folate deficiencies, might increase the impact of MeHg on neural development. Conversely, adequate levels of iron and folate in the diet might reduce the impact of MeHg. Such nutrient deficiencies could arise from an inadequate diet or, secondarily, from repeated infection, intestinal parasites, or excessive alcohol consumption.

The available data for the birth weight, gestation, and weight of the children studied in the Faroe Islands, Seychelles, and New Zealand do not suggest that there are energy or macronutrient (protein, carbohydrate, and fat) deficiencies in these populations. However, micronutrient deficiencies, such as iron and zinc, due to low intake of fortified or unrefined grains, fruits, and vegetables are possible. There

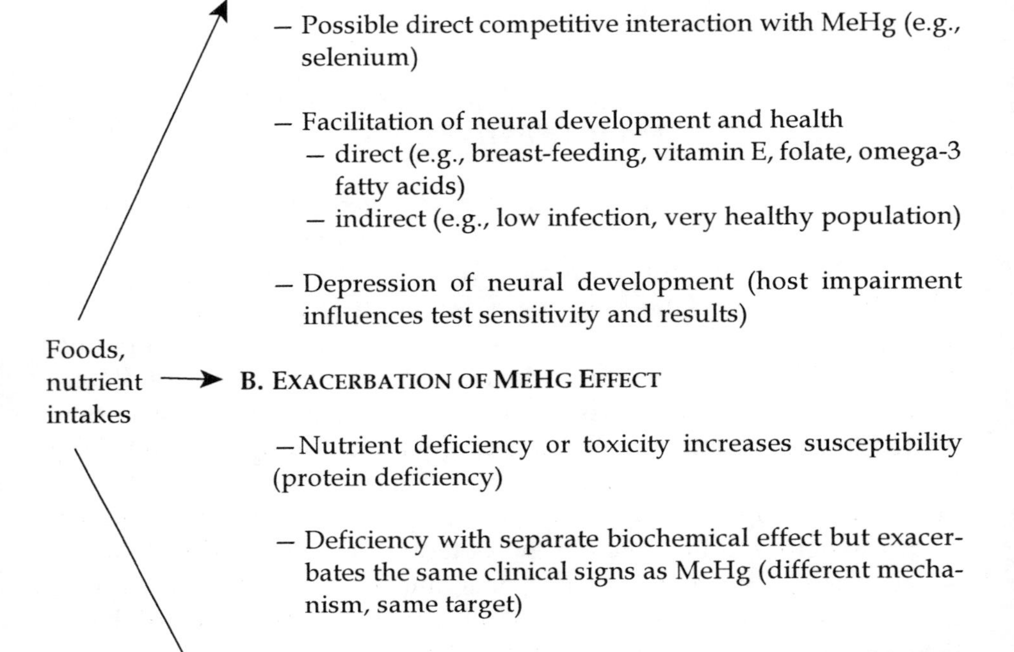

FIGURE 3-1 Hg nutrient interactions. Source: Modified from NIEHS 1998.

is insufficient information on the extent of breast-feeding of the infants to determine whether the use of other sources of milk and milk substitutes affected the outcome because of inadequate levels of iron, vitamins, and other minerals in those sources. The effects of such deficiencies on neurobehavioral end points might be evident long before any clinical signs of deficiency are present, and such deficiencies might not have been obvious in the study populations. Although those deficiencies could affect the risk estimate, an artifactual response will be seen only if those deficiencies are disproportionally distributed among the individuals exposed to MeHg at different doses.

Dietary Interactions and Confounding

Dietary factors can also confound studies of the effects of MeHg when consumption of a food contributes to the measured outcome through more than one component in that food. If a factor is associated with both MeHg exposure and outcome measures but is not part of the pathway by which MeHg effects neurological or other responses, it can be considered a confounding factor and must be controlled for in the analyses. Because the primary source of exposure is from fish consumption, it is difficult to determine whether the Hg in that source is the only cause of the fish-related effects. Other contaminants that could be present in fish, such as polychlorinated biphenyls (PCBs) or dichlorodiphenyltrichloroethane (DDT), could confound a study. High fish consumption could also result in the absence of another important food or nutrient from the diet. Conversely, fish consumption might be associated with the intake of protective substances, such as selenium and omega-3 fatty acids. Such an association was seen by Osman et al. (1998), who examined blood MeHg and selenium concentrations in Polish children from Katowice.

The understanding of the causal relationship between MeHg and adverse effects, therefore, would be greatly enhanced by information on the intake of all dietary constituents and adjustment for other toxicants. Availability of quantitative dietary information and incorporation of that information into assessments of MeHg effects could improve the analysis of the studies.

It is important to remember that fish and shellfish are high-quality food sources of protein and nutrients and that they are low in saturated fats. They contain nutrients that are essential for proper central-nervous-system development and function, and they might have potential health benefits in the prevention of cardiovascular disease and cancer. A reduction in the consumption of fish and shellfish might result in dietary patterns that are generally more harmful.

Selenium

Although the effect of selenium on MeHg toxicity has not been well documented in humans, it has been known for over two decades that organic and inorganic selenium can influence the deposition of MeHg in the body and protect against its toxicity in animals (Ganther et al. 1972). In animals, selenium deficiency has been associated with enhanced fetotoxicity following MeHg exposure (Potter and Matrone 1974; Nishikido et al. 1987). At the other extreme, the toxicity of MeHg can be enhanced in the presence of very high selenium supplementation (as in its absence) (Nobunaga et al. 1979). Over 40 studies have examined the interaction of selenium and mercury in various systems. These have recently been reviewed by Chapman and Chan (2000).

Selenium also influences tissue deposition in a form- and dose-dependent manner. Administration of seleno-methionine increased MeHg and total Hg content in the blood of rats exposed to MeHg through fish consumption. Administration of selenium dioxide lowered Hg concentrations by 24-29% in the blood and liver of rats in the same model system. Selenite supplementation in the diet of female rats before mating, during gestation, and during lactation antagonized the central-nervous-system effects following in utero exposure to MeHg (Fredriksson et al. 1993). Selenium injection during gestation has been shown to increase Hg concentrations in the neonatal brain (Satoh and Suzuki 1979), whereas ingestion has been shown to reduce brain levels (Fredriksson et al. 1993).

Therefore, animal experiments show that selenium might be protective in terms of neurodevelopmental responses but this is not clear. The selenium dose, form, and exposure route (injection vs ingestion) might affect the tissue deposition profiles. Although selenium appears to have a protective effect in animals, no association has been confirmed in humans. The mechanism by which selenium influences the deposition of Hg is not established. Proposed mechanisms include the formation of seleno-MeHg complexes, a selenium-induced release of MeHg from sulfhydryl bonds in the blood, and tissue-specific mechanisms that influence intracellular uptake (Glynn and Lind 1995).

Garlic

Garlic might be an important effect modifier in MeHg studies. Many compounds (or their metabolites) in garlic could act as metal chelating or complexing agents and increase MeHg excretion. Such chemicals can be converted to thiols or include thiols (diallyldisulfide, diallyltrisulfide, propylallyldisulfide, and diallylsulfide), glutathione, vitamin C, and thiol amino acids (see review by Block 1985). Garlic also contains selenium (0.72-1.52 μg of selenium per gram of garlic), which, as previously discussed, might influence Hg toxicokinetics.

Animal studies support a protective role of garlic against Hg toxicity (Cha 1987; Rhee et al. 1985). Male Sprague-Dawley rats were simultaneously administered MeHg (4 ppm), cadmium, and phenylmercury in their drinking water as well as 8% peeled, crushed raw garlic (*Allium satirum*) in the feed (200 ppm allicin) for 12 weeks. Results indicate a statistically significant reduction in Hg tissue concentrations compared with rats that did not receive garlic (Cha 1987). It is not clear from that study whether the garlic removed Hg that had been deposited in the tissues or whether it prevented its accumulation before deposition. Severe pathology was noted in the kidneys of rats receiving the MeHg in the absence of garlic, and only mild or no damage was noted in MeHg-exposed rats receiving 6.7% or 8% garlic, respectively. Interpretation of that study must be done cautiously, however, because the effects might be due to cadmium and not Hg toxicity. It should also be noted that the protection is against the renal effects of MeHg, not the neurotoxicity.

In an earlier paper, Rhee et al. (1985) exposed rats (40 per group) intraperitoneally to MeHg (5 mg/kg of body weight per day) for 8 days. One group also received garlic. Tissue Hg concentrations were lower in the garlic-exposed animals than in the rats that did not receive garlic.

It should be noted that the doses of garlic used in those studies (6-8% by animal weight) are well above the expected garlic content in the human diet, even in those cultures that use relatively high amounts of garlic in their cooking. More extensive study of the interactions between garlic and MeHg is needed.

Omega-3 Fatty Acids

Polyunsaturated fatty acids are essential for brain development. During perinatal development, docosahexaenoic acid (DHA), an omega-3 fatty acid, accumulates in membrane phospholipids of the nervous system. Deficiency in DHA impairs learning and memory in rats (Greiner et al. 1999).

The ratio of omega-3 to omega-6 fatty acids might also be important. The largest source of omega-3 fatty acids, in particular eicosapentanoic acid and its metabolite DHA, in the human diet is oily fish, such as salmon, herring, and other cold- water fish. Chalon et al. (1998) demonstrated that fish oil affects monoaminergic neruotransmission and behavior in rats. Omega-3 fatty acids might enhance neurotoxicological function and their deficiency might contribute to lower test results, which would confound MeHg toxicological studies in human populations. Individuals consuming less fish might perform more poorly. Individuals on a diet high in fish might demonstrate the competing effects of enhanced function from these fatty acids and reduced function because of the presence of MeHg in the same food source. A case-control study in Greece concluded that low fish intake is associated with an increased risk of cerebral palsy (Petridou et al. 1998). Populations eating diets rich in fish might have enhanced neural development that could mask adverse effects on development caused by MeHg. Therefore, controlling for intake of essential fatty acids in MeHg studies is important. That can be done by measuring biomarkers of long-term exposure to fatty acids, such as adipose tissue (Kohlmeier and Kohlmeier 1995). However, there is no evidence to date that supplementation of omega-3 fatty acid to the diet of a well-nourished term infant further enhances neurological development or attenuates the toxic effects of Hg.

Protein

The type and amount of protein consumed might affect the uptake and distribution of Hg in the body. Low protein intakes have been associated with increased Hg in the brain of the mouse (Adachi et al. 1994; Adachi et al. 1992). Sulfur amino acid ingestion might also increase blood, renal, and hepatic concentrations of Hg. Cysteine appeared to enhance transport of MeHg to the brains of rodents (Aschner and Clarkson 1987; 1988; Aschner and Aschner 1990; Hirayama 1985) when the amino acid was injected into the animals at the time of oral dosing or injection of MeHg chloride. There is some indication that leucine might inhibit MeHg uptake (Mokrzan et al. 1995). In contrast, in vitro studies indicate that methionine might stimulate MeHg uptake (Alexander and Aaseth 1982; Wu 1995).

Alcohol

Ethanol has been shown to potentiate MeHg toxicity in mice and rats (Takahashi et al. 1978; Turner et al. 1981). Five studies conducted in rodents examined the potential for alcohol interactions with MeHg. These studies indicate that the coconsumption of alcohol with MeHg can potentiate toxicity, particularly in the kidney (Takahashi et al. 1978 as cited in Chapman and Chan 2000; Rumbeiha et al. 1992; Tamashiro et al. 1986b; Turner et al. 1981; McNeil et al. 1988). Ethanol administered to male rats in conjunction with daily injections of MeHg chloride has resulted in a dose-dependent increase in tissue concentrations of both total Hg and MeHg in the brain and kidneys and in the morbidity and mortality of these animals. The applicability of these findings to human alcohol consumption and MeHg exposure patterns is unknown.

Other Foods That Might Influence Hg Uptake

Two studies indicated that the addition of milk to rodent diets increases the total body burden of Hg as well as Hg concentrations in the brain (Landry et al. 1979; Rowland et al. 1984). Landry et al. (1979) showed a 56% increase in the whole-body retention of Hg in female BALB/c mice fed liquid diets of evaporated whole milk as compared with their standard diet. That increase was attributed to the binding of heavy metals to the milk triglycerides, enhancing gut absorption. Those findings of an increased retention of MeHg with a diet containing evaporated milk were confirmed by Rowland et al. (1984).

There are strong indications that wheat bran, but neither cellulose nor pectin, when consumed concurrently with MeHg administration, might reduce Hg concentrations in the brain. In a study of male BALB/c mice, a dose-response relationship between brain Hg concentrations and the percentage of wheat bran was seen across 0%, 5%, 15%, and 30% wheat bran in the diet. The highest dose of wheat bran decreased the half-time of Hg elimination by 43%, and decreased the brain Hg concentrations by 24%. Corresponding reductions were seen in the Hg concentrations in the blood of the bran-fed animals. Reductions of that magnitude have been associated with a lower incidence and severity of symptoms of neurotoxicity in rats. The effect has been attributed partially to binding of the Hg to bran, reducing its absorption from the gut and decreasing intestinal transit time. Using evidence of an increase in mercuric Hg in the large intestines of the bran-fed mice, it has also been hypothesized that wheat bran increased the rate of demethylation of organic Hg in the gut (Rowland et al. 1986).

Vitamin E

The protective effect of coconsumption of α-tocopherol supplementation in the diet has been shown to be protective against Hg toxicity in tissue cultures and animals models. For example, in studies of male golden hamsters, the injection of 2 ppm α-tocopherol acetate completely prevented the neurotoxic effects and histological changes associated with injection of 2 ppm MeHg (Chang et al. 1978). The hypothesized mechanism is an antioxidant effect related to lipid peroxidation and to the prevention of neuronal degeneration (Kling and Soares 1982; Kling et al. 1987; Chang et al. 1978; Kasuya 1975; Park et al. 1996; Prasad et al. 1980).

Nutrient Enhancement of Toxicity

In addition to the effects of protein, milk, and alcohol discussed earlier, four other nutrients have been implicated in the enhancement of MeHg toxicity: vitamin A, vitamin C, iron, and β carotene. Welsh (1977) reported in an abstract that vitamin A (10,000 IU/kg) decreased the time of onset of Hg toxicity in Fischer 344 rats given methylmercuric chloride at 10-15 ppm in drinking water. Vitamin C was shown to increase the absorption of Hg from the gastrointestinal tract, shortening the survival time of guinea pigs exposed to methylmercuric iodide at 8 mg/kg per day (Murray and Hughes 1976). The iron chelator, deferoxamine, was shown to inhibit the formation of reactive oxygen species in the cerebellum of rats treated with MeHg at 5 mg/kg (LeBel et al. 1992). Finally, Andersen and Andersen (1993) reported that β-carotene at 1,000 to 10,000 IU/kg enhanced Hg-induced lipid peroxidation in the liver, kidney, and brain of CBA mice exposed to methylmercuric chloride. β-carotene did not affect the activity of total glutathione peroxidase or selenium-dependent glutathione peroxidase.

Beneficial Effects of Fish Consumption

The committee is aware of the other nutritional advantages of diets rich in fish, including fish being a rich source of vitamin D, omega-3 fatty acids, protein, and other nutrients that might be marginal in some diets. Cardiovascular disease, osteoporosis, and cancer might be partially prevented by the regular consumption of fish. Those are major chronic diseases that afflict large proportions of the U.S. population. For that reason, the long-term goal needs to be a reduction in the concentrations of MeHg in fish, rather than a replacement of fish in the diet by other foods, such as saturated fat rich sources of protein like red meat.

In the interim, the best methods of maintaining fish consumption and minimizing Hg exposure is the consumption of smaller fish within a species and the selection of species of fish known to have lower MeHg concentrations. Ikarashi et al. (1996) reported that, within a species, the MeHg content of the fish relates to the size of the fish, presumably because larger fish have had a longer life span and more time to accumulate MeHg.

TOXICOKINETIC VARIABILITY

Typically, biomarkers of MeHg exposure (i.e., hair or blood total Hg concentrations) are used as a surrogate for dose in the derivation of a reference dose (RfD) for MeHg. As discussed in Chapter 4, hair Hg is approximately 90% MeHg. Total Hg concentration in blood can reflect exposures to both MeHg and inorganic Hg. One goal of a dose-response assessment is thus to identify a biomarker concentration that is associated with a no-observed-adverse-effect level (NOAEL), low-observed-adverse-effect level (LOAEL), or a benchmark dose. That biomarker concentration is then translated into an RfD. An RfD is a dose of MeHg that is considered safe to ingest and is expressed in micrograms of MeHg per kilogram of body weight per day. Therefore, to derive the RfD, it is necessary to determine the ingested dose that resulted in the measured Hg concentration in the biomarker. That is referred to as dose reconstruction. That determination requires the back-calculation of dose using a toxicokinetic model. There are several toxicokinetic parameters that determine the tissue (or biomarker) MeHg concentration after ingestion of a given dose of MeHg. Those parameters include the uptake of MeHg from the gastrointestinal tract, the distribution of MeHg to the various body tissues (including the biomarker tissues), and the elimination of MeHg or Hg from those tissues. The uptake, distribution, metabolism, and elimination have been described by a physiologically based pharmacokinetic (PBPK) model (Clewell et al. 1999) and by a one-compartment pharmacokinetic model (IPCS 1990; EPA 1997). Both models require a quantitative description of several physiological and toxicokinetic inputs (e.g., body weight, blood volume, and hair-blood partition coefficients).

The PBPK model of Clewell et al. (1999), illustrated in Figure 3-2, attempts to characterize the distribution and redistribution of MeHg among several body compartments, including maternal hair and fetal cord blood. Although the PBPK model is conceptually more accurate and flexible than the one-compartment model, it is also considerably more complex, and thus, more difficult to evaluate.

In contrast, the one-compartment model, illustrated in Figure 3-3, collapses the maternal-body compartments to a single maternal-blood compartment. The blood concentration of MeHg (and Hg^{2+} resulting from MeHg metabolism) is assumed to be at steady state, and the model permits the estimation of the blood Hg concentration resulting from a given ingested dose. The corresponding hair Hg concentration can then be estimated by using an empirically derived hair-to-blood Hg concentration ratio.

The rate of MeHg entry into the blood, I (micrograms per day), is calculated by

$$I = D \times W \times A \times F, \qquad (1)$$

where D is the ingested dose (micrograms per kilogram of body weight

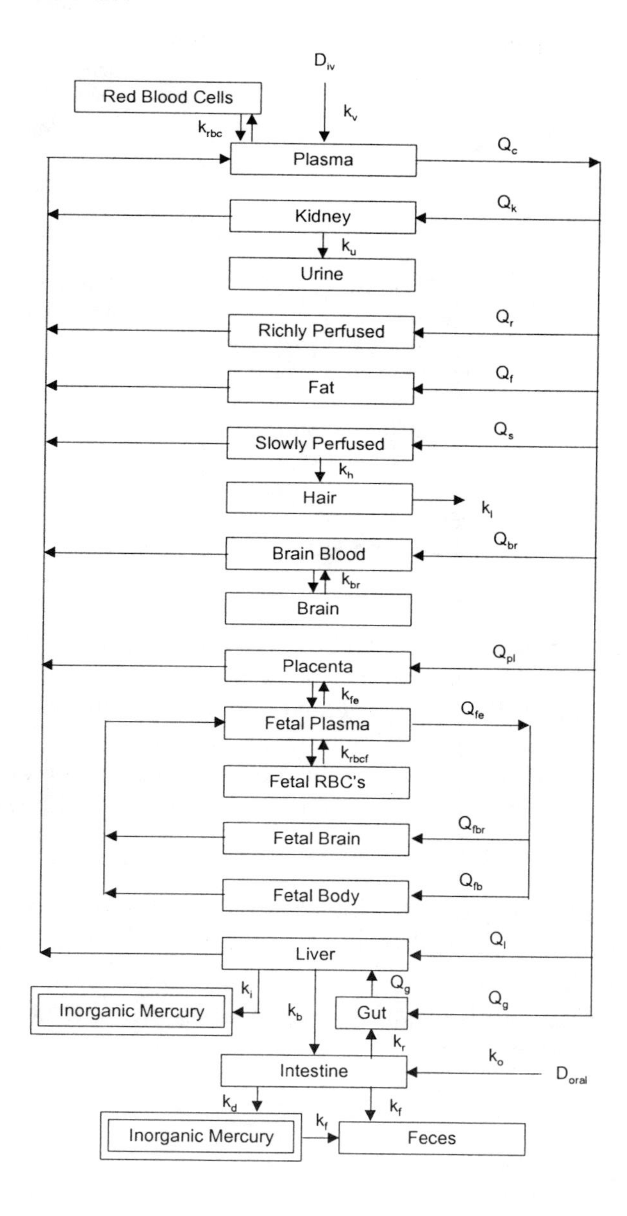

FIGURE 3-2 PBPK model for MeHg. Model parameters denoted by k represent rate constants for MeHg. Model parameters denoted by Q represent plasma flow rates. D represents the dose of MeHg. Source: Clewell et al. 1999. Reprinted with permission from *Risk Analysis*; copyright 1999, Plenum Publishing Corporation.

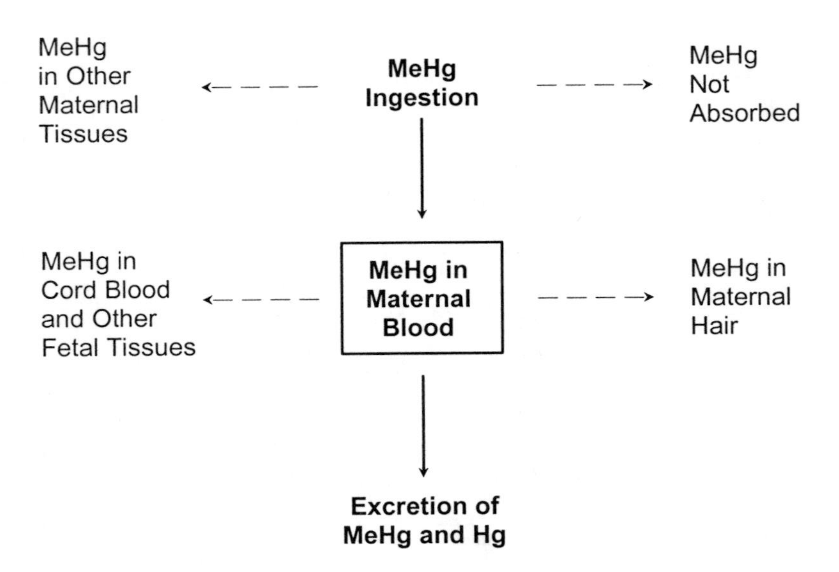

FIGURE 3-3 The one-compartment model relating ingestion of MeHg to MeHg in maternal blood. The one-compartment model predicts the steady-state MeHg concentration in the maternal-blood compartment under the assumption that the daily mass of MeHg entering the compartment from ingestion is equal to the daily mass leaving the compartment by excretion. Dotted lines indicate other toxicokinetic compartments that are not directly considered in the model.

per day); W is the body weight (kilograms); A is the fraction of ingested MeHg that is absorbed; and F is the fraction of absorbed MeHg that is distributed to the blood.

The rate of MeHg elimination from the blood, E (micrograms per day), is calculated by

$$E = C \times \mathrm{b} \times V, \qquad (2)$$

where C is the concentration of MeHg in the blood (micrograms per liter); b is the elimination-rate constant, expressed as the fraction of the concentration eliminated per day (day^{-1}); and V is the blood volume (liters).

By definition of steady state, the rate of MeHg entry into the blood is equal to the rate of MeHg elimination from the blood. Therefore, at steady state, $I = E$ and

$$D \times W \times A \times F = C \times b \times V. \qquad (3)$$

Equation 3 can be solved for C to calculate the blood MeHg concentration resulting from a given steady-state dose:

$$C = \frac{D \times W \times A \times F}{b \times V}. \qquad (4)$$

Equation 3 can also be solved for D to calculate the steady-state dose corresponding to a given blood concentration:

$$D = \frac{C \times b \times V}{W \times A \times F}. \qquad (5)$$

The MeHg concentration in hair, H, is related to the concentration in blood, C, through the empirically derived hair-to-blood Hg concentration ratio (μg/g)/(μg/L), R, as follows:

$$C = (1/R) \times H. \qquad (6)$$

The inverse form, $(1/R)$, is used to maintain the ratio in the form in which it is traditionally expressed in the scientific literature.

To calculate the ingested dose that gave rise to a measured hair concentration, Equation 6 can be combined with Equation 5 as follows:

$$D = \frac{(1/R)\, H \times b \times V}{W \times A \times F} \qquad (7)$$

Equation 5 and 7 can thus be used to estimate the ingested dose of MeHg that gave rise to a maternal-blood and maternal-hair MeHg concentration, respectively, which are associated with a given level of adverse effects.

When fetal-cord-blood MeHg concentration is the biomarker measured, the corresponding maternal-blood concentration can be estimated using an empirically derived ratio of cord-blood concentration to maternal-blood concentration.

Due to interindividual variability in physiology and kinetics, there is no single correct value that can be assigned to any of the parameters in either toxicokinetic model. Each of the model parameters is a random variable whose possible values in a population can be described by a probability distribution. The ingested dose of MeHg corresponding to a measured biomarker concentration, therefore, is also described by a probability distribution. That distribution is determined by the combination of the distributions of the individual model parameters according to the mathematic form of the model. The central-tendency value of the ingested dose corresponding to a given biomarker concentration could be estimated using the central-tendency value for each parameter of the model. However, no single value, including the central tendency, can capture the range of possible values for a parameter in a heterogenous population. Furthermore, no combination of single-number (point-estimate[1]) parameter values in a model can estimate the range of possible ingested doses. Because the RfD is intended to protect the most sensitive individuals in a population, an estimate based solely on the central tendency of the distribution (without uncertainty adjustment) would not provide a protective RfD. To be protective of the sensitive population, the ingested dose used as the basis for the RfD should be at the lower range of doses that could result in a given MeHg biomarker concentration. Figure 3-4 presents an estimate, using the one-compartment model, of the percentage of women of childbearing age having a hair Hg concentration of 11 ppm with a given MeHg ingested dose. The use of central-tendency values for each of the model parameters would approximate the ingested dose at which 50% of the population would achieve a hair MeHg concentration of 11 ppm. In that example, the dose is approximately 0.4 µg/kg per day. However, 50% of the population is predicted to achieve a hair Hg concentration of a 11 ppm with an ingested dose of MeHg of less than 0.4 µg/kg per day. An RfD based on an ingested dose of 0.4 µg/kg per day without further adjustment would, therefore, be protective of only 50% of the population. In that example, to protect 95% of the population from having a hair Hg

---

[1]A point estimate is a single value, selected from a distribution of values, that is intended to represent the entire distribution.

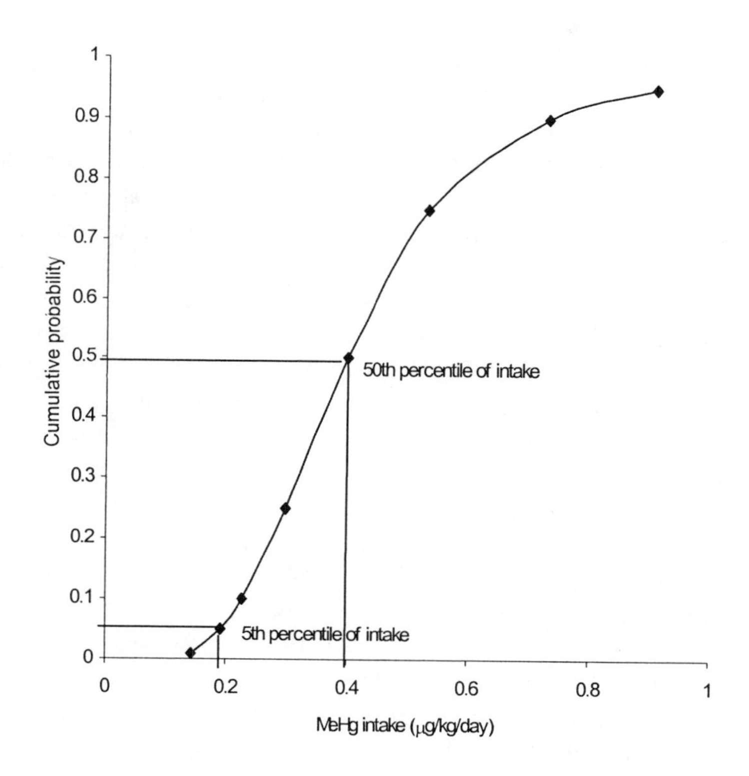

FIGURE 3-4 Predicted mean probability of MeHg intake corresponding to 11 ppm MeHg in hair. Source: Data from Stern 1997.

concentration above 11 ppm, the ingested dose would be about 0.2 μg/kg per day, or half the dose predicted using central-tendency values for the model parameters.

In estimating the range of ingested doses which could have resulted in a given biomarker concentration, there are three main sources of variability errors in model selection, errors in estimation of model parameters, and true population variability (i.e., heterogeneity). The variability due to the first two sources can be reduced by the collection of more or better data and through development of more accurate models. However, heterogeneity in toxicokinetics is inherent in human populations, and the variability in estimates due to population variability cannot be decreased.

In its derivation of an RfD for MeHg, the U.S. Environmental Protection Agency (EPA) used point estimates to calculate an ingested dose from its benchmark hair concentration (EPA 1997). EPA did not address the distribution of the model parameters or of the predicted dose, nor did it characterize how inclusive its point estimate of dose was of the range of ingested doses that could have given rise to the benchmark hair concentration. An uncertainty factor of 3, however, was used to account for "variability in the human population, in particular, variation in the biological half-life of MeHg, and the variation that occurs in the hair/blood ratio for Hg." Appendix D of volume 5 of the *EPA Mercury Study Report to Congress* (EPA 1997; see also Swartout and Rice 2000), presents an ad hoc probabilistic assessment of interindividual toxicokinetic variability, using the one-compartment model, in the calculation of the ingested dose. The results of that analysis, however, were not used by EPA in the derivation of its RfD.

In the derivation of an MRL for MeHg, ATSDR (1999) also used point-estimate values in the one-compartment model to calculate the ingested dose without addressing the distribution of the model parameters or of the predicted dose. Parameter estimates in the model were selected to reflect the central tendency of the range of values in the population. Two uncertainty factors of 1.5 each (summed to give an overall uncertainty factor of 3) were applied to the NOAEL to address "variability in hair-to-blood ratios among women and fetuses in the U.S. population, as determined by pharmacokinetic modeling of actual data by Clewell et al. (1998)," and "to address the remainder of any inter-individual variability (i.e., pharmacodynamics) in the U.S. population." The appropriateness of such an overall adjustment that addresses interindividual variability in only one parameter of the toxicokinetic model (i.e., the hair-to-blood ratio) and the poorly characterized adjustment to account for the "remainder" of variability are difficult to assess.

Three analyses have been carried out to characterize the interindividual toxicokinetic variability in the estimates of the MeHg-ingested dose corresponding to a given concentration of Hg in a biomarker (Stern 1997; Swartout and Rice 2000 (see also EPA 1997); Clewell et al. 1999). Each of those analyses used a Monte Carlo simulation to combine the probability distributions for the individual model parameters to generate a probability distribution of the corresponding ingested dose. That probability distribution estimates the fraction of the maternal population who could achieve a specific hair Hg concentration from a given MeHg ingestion. The analyses by Stern (1997) and Swartout and Rice (2000) are based on the one-compartment model, and that by Clewell et al. (1999) is based on the PBPK model.

Stern (1997) identified data on the distribution of parameters in the one-compartment model from the published literature. Blood volume and body weight were assumed to be correlated. A similar approach was used by Swartout and Rice (2000). In that analysis, however, some of the parameters are described by different distributional shapes or by distributions from different data sources than those used by Stern (1997). Swartout and Rice (2000) assumed correlations between several pairs of parameters: the hair-to-blood ratio and the elimination-rate constant; body weight and blood volume; and the fraction of the absorbed dose in the blood and body weight.

Clewell et al. (1999) likewise identified data on the distribution of parameters from the literature, but because the PBPK model contains many parameters that are not used in the one-compartment model, the distributions used in this analysis are not directly comparable to those used in one-compartment-model analyses. In addition, given the large number of parameters and the inconsistent availability of distributional data for those parameters, Clewell et al. (1999) tended to use default distributions for their model parameters.

The variability in the relationship between the concentration of Hg in maternal hair or cord blood and the ingested dose of MeHg predicted by the three analyses is summarized in Table 3-1. If the ingested dose is calculated from the Hg concentration in hair or blood from the central-tendency estimates of model parameters in either the one-compartment or the PBPK model, then the resulting ingested dose should approximate the 50th percentile of the population distribution. The ratio of the ingested dose corresponding to the 50th percentile of the distribution to the dose at the 5th percentile of the distribution, therefore, is an estimate of the factor by which the central-tendency estimate of the ingested dose should be divided to make the dose estimate inclusive of the variability in 95% of the population. Likewise, the ratio of the ingested dose

TABLE 3-1 Comparison of Results from Three Analyses of the Interindividual Variability in the Ingested Dose of MeHg Corresponding to a Given Maternal-Hair or Blood Hg Concentration

Study	Maternal Medium	50th percentile[a] (μg/kg-d)	50th percentile/5th[b] percentile	50th percentile/1st percentile[c]
Stern (1997)	hair	0.03-0.05[d] (mean = 0.04)	1.8-2.4 (mean = 2.1)	2.3-3.3 (mean = 2.7)
	blood	0.01	1.5-2.2 (mean = 1.8)	1.7-3.0 (mean = 2.4)
Swartout and Rice (2000)	hair	0.08	2.2	Data not reported
	blood[e]	0.02	2.1	2.8
Clewell et al. (1999)	hair	0.08	1.5	1.8
	blood[f]	0.07	1.4	1.7

[a]Predicted 50th percentile of the ingested dose of methylmercury that corresponds to 1 ppm Hg in hair or 1 ppb in blood.

[b]Ratio of 50th percentile of ingested dose of methylmercury that corresponds to 1 ppm Hg in hair or 1 ppb in blood to the 5th percentile.

[c]Ratio of 50th percentile of ingested dose of methylmercury that corresponds to 1 ppm Hg in hair or 1 ppb in blood to the 1st percentile.

[d]Range reflects minimum and maximum values among eight alternative analyses.

[e]Data from J. Swartout, U.S. Environmental Protection Agency, personal commun.; June 9, 2000.

[f]Data from H.J. Clewell, ICF Consulting, personal commun.; April 19, 2000.

corresponding to the 50th percentile of the distribution to the dose at the 1st percentile is an estimate of the factor by which the central tendency estimate should be divided to make the dose estimate inclusive of the variability in 99% of the population. In general, Stern (1997) and Swartout and Rice (2000) predicted similar variability in the relationship between hair Hg concentration and ingested dose, but both predicted a somewhat larger variability than did Clewell et al. (1999). Nonetheless, the three studies differ only by a factor of 1.5 in their predictions of both the 50th:5th and the 50th:1st percentile ratios. Although the three studies predict similar relative variability, examination of the median (i.e., 50th percentile) ingested dose predicted to correspond to 1 ppm Hg in maternal hair (Table 3-1) indicates that the absolute value of the central tendency of the distribution of ingested doses predicted by Stern

(1997) is lower than that predicted by the other two studies. In other words, the distribution predicted by Stern appears to be shifted toward lower-ingested dose values compared with the other two analyses. That difference appears to be due to selection of different data sets for several key model parameters. Given the existence of several valid data sets for those parameters, it is not clear which central-tendency estimate is more appropriate.

Each of those analyses (Stern 1997; Swartout and Rice 2000 (see also EPA 1997); Clewell et al. 1999) estimated the ingested dose corresponding to maternal-hair Hg concentration. For studies in which the biomarker measured is the Hg concentration in cord blood (e.g., the Faroe Islands studies), the estimation of the ingested dose would follow the same approach, except that the hair-to-blood ratio would be omitted from the model. The analyses of variability of the ingested dose can be recalculated for the variability of the ingested dose corresponding to a given blood Hg concentration. The results of that calculation for the three analyses are presented in Table 3-1. Comparison of the relative variability in the ingested dose based on the ratio of hair Hg concentration to blood Hg concentration indicates that the variability for maternal-hair Hg concentration is 1.1 to 1.2 times greater than that for the maternal-blood Hg concentration. That result is consistent with the results of sensitivity analyses conducted in each of the three studies, which identified the hair-to-blood ratio as a major contributor to the variability in the predicted ingested dose. Nonetheless, the Table 3-1 data, which are intended to describe the distribution of ingested doses that corresponds to a given cord-blood Hg concentration, actually describe the relationship between the ingested dose and the maternal-blood Hg concentration. The application of estimates based on the ratio of maternal-blood Hg concentration to estimates of cord-blood Hg concentration assumes that those two concentrations are equal. Some observations suggest that Hg concentrations in cord blood are larger than in maternal blood by at least 20-30 % (Dennis and Fehr 1975; Pitkin et al. 1976; Kuhnert et al. 1981), however, such differences are not seen consistently (Fujita and Takabatake 1977; Sikorski et al. 1989). Therefore, the data in Table 3-1 might underestimate the variability in ingested doses calculated from cord-blood Hg concentrations, however, that is not entirely clear.

On the basis of the data in Table 3-1, if an estimate of the ingested dose from a benchmark hair Hg concentration is generated using point estimates of the central tendency for each model parameter, then dividing this initial estimate by an uncertainty factor of 2 would result in a dose that includes approximately 95% of the interindividual toxicokinetic variability in the population. Dividing the initial estimate by an uncertainty factor of 2-3 would include approximately 99% of the interindividual toxicokinetic variability. Similarly, for estimates of the ingested dose based on a benchmark blood Hg concentration, the data in Table 3-1 indicate that adjustment of a central-tendency estimate of the ingested dose by an uncertainty-factor adjustment of about 2 takes into account 95-99% of the interindividual toxicokinetic variability.

The use of uncertainty factors to adjust a central-tendency estimate of the ingested dose for interindividual variability is an indirect, or "back-end," approach to accounting for such variability in the RfD. A direct, or "front-end," approach would be to select as the starting point for the derivation of the RfD the ingested dose that corresponds to a given (e.g., benchmark) hair or blood Hg concentration for the percentile of the population variability that is to be accounted for. In that case, no uncertainty-factor adjustments would be necessary to account for toxicokinetic variability in the dose conversion. For example, with reference to Figure 3-4, if the benchmark (or NOAEL) hair concentration is 11 ppm and the RfD is intended to include the toxicokinetic variability in 95% of the population, then the corresponding ingested dose would be approximately 0.2 μg/kg per day. The difficulty with using such an approach is that, in the direct approach the estimate of the absolute value of the ingested dose is the critical determination. Whereas in the uncertainty factor approach the estimate of the relative variability in the ingested dose is critical. As discussed previously, for a given hair concentration the absolute value of the ingested dose for any given percentile of the population is not consistent in the analyses of Stern (1997), Swartout and Rice (2000), and Clewell et al. (1999). The analysis of Stern (1997) predicts lower absolute values of the ingested dose for a given percentile of the population than the other two analyses. Therefore, the use of the direct approach requires that a choice be made among the probability distributions predicted by those analyses. The differences in the analyses are due to the use of different data sets for parameter estimates, and there is no clear basis for choosing one data set over another. Even when central-tendency estimates and uncertainty factors are used, the most appropriate value for each model parameter must be selected. Selection of different values for model parameters could underlie differences in the modeling results. The advantage of the uncertainty-factor approach, however, is that the choice for each model parameter is explicit. That allows for a more reasoned and detailed discussion of those choices. The analyses of Stern (1997), Swartout and Rice (2000), and Clewell et al. (1999) all discuss their choices of parameter estimates. The information presented in those discussions should be considered in the selection of the central-tendency estimates of the individual parameters.

CONCLUSIONS

- Sensitivity to the toxic effects of MeHg is related to the age at which exposure occurs. Because of that, the fetus and young infants exposed during periods of rapid brain development are particularly vulnerable.
- Sex differences appear to affect the metabolism, tissue uptake, excretion and toxicity of Hg.
  —Gender specific effects due to developmental exposure to MeHg typically indicate a greater sensitivity for male offspring.
  —Gender sensitivity in toxicity appears to be dependent on the species used and outcome studied.
- Dietary nutrients and supplements might protect against the toxicity of MeHg. Data regarding the relative presence or absence of such nutrients and supplements either in the populations studied or in the United States are not available. The lack of that information contributes to overall data-base uncertainty, but it does not detract from the suitability of those studies for determining the risk associated with MeHg.
- In addition to the above factors, intraindividual differences are clearly noted in responses to similar exposures. Those are explained, in part, by nutritional factors that might exacerbate or attenuate the effects of Hg toxicity in the host. Currently unknown genetic susceptibilities could be expected to play a role in response variability.
- In any MeHg risk assessment in which the exposure metric is a Hg

biomarker concentration, it is necessary to use a toxicokinetic model to estimate the ingested dose that gave rise to the critical biomarker concentration (e.g., benchmark or NOAEL concentration).

  - The simpler and more easily manipulated one-compartment model and the more complex but more realistic PBPK model have been used for that purpose.
  - The parameters in those models are variables whose possible values are described by probability distributions reflecting the interindividual variability in the population.
  - The ingested doses predicted by the one-compartment and PBPK models, therefore, are also probability distributions that reflect the likelihood that any given ingested dose could give rise to the critical biomarker concentration.

- Failure to consider interindividual toxicokinetic variability can result in an RfD that is not protective of a substantial portion of the population.
  - Interindividual toxicokinetic variability can be addressed in the derivation of the RfD by application of an uncertainty factor to a central- tendency estimate of the ingested dose.
  - It is uncertain which values are most appropriate for the model parameters used to derive the central-tendency estimates. The basis for each choice should be carefully considered with reference to discussions already presented in the published analyses of toxicokinetic variability.

RECOMMENDATIONS

- Future studies of MeHg exposures in humans should include a thorough assessment of the diet during the periods of vulnerability and exposure. They should involve assessment of the nutritional adequacy of the group, including the assessment of nutritional and environmental factors that might attenuate or exacerbate the effect of MeHg on the health end points measured.
  - Dietary assessment should be conducted concurrently with the exposures, because retrospective assessment is influenced by many factors, including memory, changes in eating behavior,

food fortification, and use of prenatal and postnatal vitamin and mineral supplementation. Dietary assessment should be conducted on a person-specific basis, with particular effort to estimate quantitatively individual consumption and consumption patterns of fish and pilot whale.

—For all the studies, the estimates of consumption of fish (and whale meat as appropriate) should be used with information on MeHg concentrations in the food to estimate possible MeHg intake by pregnant women, young children, and adults. Attempts should be made to validate estimates of intake by using experimental data on the relationship between hair Hg concentration and diet intake.

—Future studies should include a standardized measure of the duration of breast-feeding and the quantity of breast milk ingested by infants. The dose of MeHg is dependant on the amount of milk ingested and the MeHg content of the milk. Historical recording of duration of breast-feeding is likely to be biased; therefore, a prospective diary of breast-feeding and weaning should be considered.

- Studies using animal models should examine changes in the dose response characteristics of Hg effects associated with nutritional or genetic factors.
- Any biomarker-based RfD for MeHg should specifically address interindividual toxicokinetic variability in the estimation of dose corresponding to a given biomarker concentration.

  —The starting point for addressing interindividual toxicokinetic variability should be a central-tendency estimate of the ingested dose corresponding to a critical biomarker concentration (e.g., a benchmark hair concentration).

  —The central-tendency estimate of the ingested dose should be based on careful consideration of the several possible and sometimes contradictory data sets for each parameter. A starting point for such consideration is the discussion of parameter distributions presented in the analyses of Stern (1997), Swartout and Rice (2000), and Clewell et al. (1999).
- An uncertainty-factor adjustment should be applied to any central-tendency estimate of the ingested dose corresponding to the critical biomarker concentration.

- For an RfD based on maternal-hair Hg concentration, an uncertainty-factor adjustment of 2 should be applied to the central-tendency estimate of dose to be inclusive of 95% of the toxicokinetic variability in the population. An uncertainty-factor adjustment of 2-3 should be applied to be inclusive of 99% of the toxicokinetic variability.
- For an RfD based on blood Hg concentration, an uncertainty factor adjustment of about 2 should be applied to the central-tendency estimate of dose to be inclusive of 95-99% of the toxicokinetic variability in the population.
- Because of the recognized nutritional benefits of diets rich in fish, the best method of maintaining fish consumption and minimizing Hg exposure is the consumption of fish known to have lower MeHg concentrations.

REFERENCES

Adachi, T., A. Yasutake, and K. Hirayama. 1992. Influence of dietary protein levels on the fate of methylmercury and glutathione metabolism in mice. Toxicology 72(1):17-26.

Adachi, T., A. Yasutake, and K. Hirayama. 1994. Influence of dietary levels of protein and sulfur amino acids on the fate of methylmercury in mice. Toxicology 93(2-3):225-23.

Alexander, J., and J. Aaseth. 1982. Organ distribution and cellular uptake of methyl mercury in the rat as influenced by the intra- and extracellular glutathione concentration. Biochem. Pharmacol. 31(5):685-690.

Andersen, H.R., and O. Andersen. 1993. Effects of dietary alpha-tocopherol and beta- carotene on lipid peroxidation induced by methyl mercuric chloride in mice. Pharmacol. Toxicol. 73(4):192-201.

Aschner, M., and J.L. Aschner. 1990. Mercury neurotoxicity: Mechanisms of blood-brain barrier transport. Neurosci. Biobehav. Rev. 14(2):169-176.

Aschner, M., and T.W. Clarkson. 1987. Mercury 203 distribution in pregnant and nonpregnant rats following systemic infusions with thiol-containing amino acids. Teratology 36(3):321-328.

Aschner, M., and T.W. Clarkson. 1988. Uptake of methylmercury in the rat brain: Effects of amino acids. Brain Res. 462(1):31-39.

ATSDR (Agency for Toxic Substances and Disease Registry). 1999. Toxicological Profile for Mercury (Update). U.S. Department of Health and Human Services, Public Health Service. Agency for Toxic Substances and Disease Registry Atlanta, GA. March.

Bakir, F., S.F. Damluji, L. Amin-Zaki, M. Murthadha, A. Khalidi, N.Y. al-Rawi, S. Tikriti, S H.I. Dahahir, T.W. Clarkson, J.C. Smith, and R.A. Doherty. 1973. Methylmercury poisoning in Iraq. Science 181:230-241.

Block, E. 1985. The chemistry of garlic and onions. Sci. Am. 252(3):114-119.

Burbacher, T.M., M.K. Mohamed, and N.K. Mottett. 1988. Methylmercury effects on reproduction and offspring size at birth. Reprod. Toxicol 1(4):267-278.

Cha, C.W. 1987. A study on the effect of garlic to the heavy metal poisoning of rat. J. Korean Med. Sci. 2(4):213-224.

Chalon, S., S. Delion-Vancassel, C. Belzung, D. Guilloteau, A.M. Leguisquet, J.C. Besnard, and G. Durand. 1998. Dietary fish oil affects monoaminergic neurotransmission and behavior in rats. J. Nutr. 128(12):2512-2519.

Chang, L.W., M. Gilbert, and J. Sprecher. 1978. Modification of methylmercury neurotoxicity by vitamin E. Environ. Res. 17(3):356-66.

Chapman, L., and H.M. Chan. 2000. The influence of nutrition on methyl mercury intoxication. Environ. Health Perspect. 108(Suppl.1):29-56.

Clewell, H.J., P.R. Gentry, A.M. Shipp, and K.S. Crump. 1998. Determination of a Site-Specific Reference Dose for Methylmercury for Fish-Eating Populations. ICF Kaiser International, KS Crump Group, Ruston, Louisiana.

Clewell, H.J., J.M. Gearhart, P.R. Gentry, T.R. Covington, C.B. VanLandingham, K.S. Crump, and A.M. Shipp. 1999. Evaluation of the uncertainty in an oral reference dose for methylmercury due to inter-individual variability in pharmacokinetics. Risk Anal. 19(4):547-558.

Dennis, C.A., and F. Fehr. 1975. The relationship between mercury levels in maternal and cord blood. Sci. Total Environ. 3(3):275-277.

EPA (U.S. Environmental Protection Agency). 1997. Mercury Study Report for Congress. Vol. V: Health Effects of Mercury and Mercury Compounds. EPA-452/R-97-007. U.S. Environmental Protection Agency, Office of Air Quality Planning and Standards, and Office of Research and Development.

Fowler, B.A. 1972. Ultrastructural evidence for nephropathy induced by long-term exposure to small amounts of methyl mercury. Science 175(23):780-781.

Fredriksson, A., A.T. Gardlund, K. Bergman, A. Oskarsson, B. Ohlin, B. Danielsson, and T. Archer. 1993. Effects of maternal dietary supplementation with selenite on the postnatal development of rat offspring exposed to methylmercury in utero. Pharmacol. Toxicol. 72(6):377-382.

Fujita, M., and E. Takabatake. 1977. Mercury levels in human maternal and neonatal blood, hair and milk. Bull. Environ. Contam. Toxicol. 18(2):205-209.

Ganther, H.E., C. Goudie, M.L. Sunde, M.J. Kopecky, and P. Wagner. 1972. Selenium: Relation to decreased toxicity of methylmercury added to diets containing tuna. Science 175(26):1122-1124.

Glynn, A.W., and Y. Lind. 1995. Effect of long-term sodium selenite supplementation on levels and distribution of mercury in blood, brain and kidneys of methyl mercury-exposed female mice. Pharmacol. Toxicol. 77(1):41-47.

Greiner, R.S, T. Moriguchi, A. Hutton, B.M. Slotnick, and N. Salem, Jr. 1999. Rats with low levels of brain docosahexaenoic acid show impaired performance in olfactory-based and spatial learning tasks. Lipids 34(Suppl.): S239-S243.

Harada, Y. 1968. Congenital (or fetal) Minamata disease. Pp. 93-118 in Minamata Disease. Study group of Minamata Disease. Japan: Kumamoto University.

Hirayama, K. 1985. Effects of combined administration of thiol compounds and methylmercury chloride on mercury distribution in rats. Biochem. Pharmacol. 34(11):2030-2032.

Hirayama, K., and A. Yasutake. 1986. Sex and age differences in mercury distribution and excretion in methylmercury-administered mice. J. Toxicol. Environ. Health 18(1):49-60.

Ikarashi, A., K. Sasaki, M. Toyoda, and Y. Saito. 1996. Annual daily intakes of Hg, PCB and arsenic from fish and shellfish and comparative survey of their residue levels in fish by body weight. [in Japanese]. Eisei Shikenjo Hokoku (114):43-47.

IPCS (International Programme on Chemical Safety). 1990. Environmental Health Criteria Document 101—Methylmercury. Geneva: World Health Organization.

Kasuya, M. 1975. The effect of vitamin E on the toxicity of alkyl mercurials on nervous tissue in culture. Toxicol. Appl. Pharmacol. 32(2):347-54.

Kling, L.J., and J.H. Soares, Jr. 1982. Effect of mercury and vitamin E on tissue glutathione peroxidase activity and thiobarbituric acid values. Poult. Sci. 61:1762-1765.

Kling, L.J., J.H. Soares, Jr., and W.A. Haltman. 1987. Effect of vitamin E and synthetic antioxidants on the survival rate of mercury-poisoned Japanese quail. Poult. Sci. 66:325-331.

Kohlmeier, L., and M. Kohlmeier. 1995. Adipose tissue as a medium for epidemiologic exposure assessment. Environ. Health Perspect. 103(Suppl.3): 99-106.

Kuhnert, P.M., B.R. Kuhnert, and P. Erhard. 1981. Comparison of mercury levels in maternal blood, fetal cord blood, and placental tissues. Am. J. Obstet. Gynecol. 139(2):209-213.

Landry, T.D., R.A. Doherty, and A.H. Gates. 1979. Effects of three diets on mercury excretion after methylmercury administration. Bull. Environ. Contam. Toxicol. 22(1-2):151-158.

LeBel, C.P., S.F. Ali, and S.C. Bondy. 1992. Deferoxamine inhibits methyl mercury induced increases in reactive oxygen species formation in rat brain. Toxicol. Appl. Pharmacol. 112(1):161-165.

Magos, L., F. Bakir, T.W. Clarkson, A.M. Al-Jawad, and M.H. Al-Soffi. 1976. Tissue levels of mercury in autopsy specimens of liver and kidney. Bull. WHO 53(Suppl.):93-97.

Marsh, D.O., G.J. Myers, T.W. Clarkson, L. Amin-Zaki, S. Tikriti, and M.A. Majeed. 1980. Fetal methylmercury poisoning: Clinical and toxicological data on 29 cases. Ann. Neurol. 7(4):348-353.

McKeown-Eyssen, G.E., J. Ruedy, and A. Neims. 1983. Methylmercury exposure in northern Quebec. II. Neurologic findings in children. Am. J. Epidemiol. 118(4):470-479.

McNeil, S.I., M.K. Bhatnager, and C.J. Turner. 1988. Combined toxicity of ethanol and methylmercury in rat. Toxicology 53(2-3):345-363.

Miettinen, J.K. 1973. Absorption and elimination of dietary (Hg^{++}) and methylmercury in man. Pp. 233-246 in Mercury, Mercurial, and Mercaptans, M.W. Miller, and T.W. Clarkson, eds. Springfield, IL: C.C. Thomas.

Mokrzan, E.M., L.E. Kerper, N. Ballatori, and T.W. Clarkson. 1995. Methylmercury-thiol uptake into cultured brain capillary endothelial cells on amino acid system L. J. Pharmacol. Exp. Ther. 272(3):1277-1284.

Mottet, N.K., C.M. Shaw, and T.M. Burbacher. 1987. The pathological lesions of methyl-mercury intoxication in monkeys. Pp. 73-103 in The Toxicity of Methyl Mercury, C.U. Eccles, and Z. Annau, eds. Baltimore: Johns Hopkins.

Murray, D.R., and R.E. Hughes. 1976. The influence of dietary ascorbic acid on the concentration of mercury in guinea-pig tissues. Proc. Nutr. Soc. 35(3):118A-119A.

NIEHS (National Institute of Environmental Health Sciences). 1998. Scientific Issues Relevant to Assessment of Health Effects from Exposure to Methylmercury. Workshop organized by Committee on Environmental and Natural Resources(CENR) Office of Science and Technology Policy (OSTP) The White House. November 18-20, 1998. Raleigh, NC.

Nielsen, J.B., and O. Andersen. 1991. Methyl mercuric chloride toxicokinetics in mice. II: Sexual differences in whole-body retention and deposition in blood, hair, skin, muscles and fat. Pharmacol. Toxicol. 68(3):208-211.

Nishikido, N., K. Furuyashiki, A. Naganuma, T. Suzuki, and N. Imura. 1987. Maternal selenium deficiency enhances the fetolethal toxicity of methyl mercury. Toxicol. Appl. Pharmacol. 88(3):322-328.

Nobunaga, T., H. Satoh, and T. Suzuki. 1979. Effects of sodium selenite on methyl mercury embryotoxicity and teratogenicity in mice. Toxicol. Appl. Pharmacol. 47:79-88.

Osman, K., A. Schutz, B. Akesson, A. Maciag, and M. Vahter. 1998. Interactions between essential and toxic elements in lead exposed children in Katowice, Poland. Clin. Biochem. 31(8):657-665.

Park, S.T., K.T. Lim, Y.T. Chung, and S.U. Kim. 1996. Methylmercury-induced neurotoxicity in cerebral neuron culture is blocked by antioxidants and NMDA receptor antagonists. Neurotoxicology 17(1):37-45.

Pitkin, R.M., J.A. Bahns, L.J. Filer, Jr., and W.A. Reynolds. 1976. Mercury in human maternal and cord blood, placenta, and milk. Proc. Soc. Exp. Biol. Med. 151(3):565-567.

Petridou, E., M. Koussouri, N. Toupadaki, S. Youroukos, A. Papavassiliou, S. Pantelakis, J. Olsen, and D. Trichopoulos. 1998. Diet during pregnancy and the risk of cerebral palsy. Br. J. Nutr. 79(5):407-412.

Potter, S., and G. Matrone. 1974. Effect of selenite on the toxicity of dietary methylmercury and mercuric chloride in the rat. J. Nutr. 104(5):638-647.

Prasad, K.N., and S. Ramanujam. 1980. Vitamin E and vitamin C alter the effect of methylmercuric chloride on neuroblastoma and glioma cells in culture. Environ. Res. 21(2):343-349.

Rhee, M.G., C.W. Cha, and E.S. Bae. 1985. The chronological changes of rat tissue and the effect of garlic in acute methylmercury poisoning. Kor. Univ. Med. J. 22(1):153-164.

Rice, D.C., and S.G. Gilbert. 1990. Effects of developmental exposure to methyl mercury on spatial and temporal visual function in monkeys. Toxicol. Appl. Pharmacol. 102(1):151-163.

Rowland, I.R., R.D. Robinson, and R.A. Doherty. 1984. Effects of diet on mercury metabolism and excretion in mice given methylmercury: Role of gut flora. Arch. Environ. Health 39(6):401-408.

Rowland, I.R., A.K. Mallett, J. Flynn, and R.J. Hargreaves. 1986. The effect of various dietary fibres on tissue concentration and chemical form of mercury after methyl mercury exposure in mice. Arch. Toxicol. 59(2):94-98.

Rumbeiha, W.K., P.A. Gentry, and M.K. Bhatnagar. 1992. The effects of administering methylmercury in combination with ethanol in the rat. Vet. Hum. Toxicol. 34(1):21-25.

Satoh, H., and T. Suzuki. 1979. Effects of sodium selenite on methylmercury distribution in mice of late gestational period. Arch. Toxicol. 42(4):275-279.

Sikorski, R., T. Paszkowski, P. Slawinski, J. Szkoda, J. Zmudzki, and S. Skawinski. 1989. The intrapartum content of toxic metals in maternal blood and umbilical cord blood. Ginekol. Pol. 60(3):151-155.

Spyker, J.M., S.B. Sparber, and A.M. Goldberg. 1972. Subtle consequences of methyl mercury exposure: Behavioral deviations in offspring of treated mothers. Science 177(49):621-623.

Stern, A.H. 1997. Estimation of the inter-individual variability in the one-compartment pharmacokinetic model for methylmercury: Implications for the derivation of a reference dose. Reg. Toxicol. Pharmacol. 25(3):277-288.

Strange, R.C., and A.A. Fryer. 1999. Chapter 19. The glutathione S-transferases: Influence of polymorphism on cancer susceptibility. IARC Sci. Publ. (148):231-249.

Swartout, J., and G. Rice. 2000. Uncertainty analysis of the estimated ingestion rates used to derive the methylmercury reference dose. Drug Clin. Toxicol. 23(1):293-306.

Tagashira, E., T. Urano, and S. Yanaura. 1980. Methylmercury toxicosis. I. Relationship between the onset of motor incoordination and mercury contents in the brain. [in Japanese]. Nippon Yakurigaku Zasshi 76(2):169-177.

Takahashi, H., K. Shibuya, and Y. Fukushima. 1978. A study of the factors influencing toxicity of methylmercury. [in Japanese]. Kumamota University Medical School Toxicol. Rep. 11:15-16.

Takeuchi, T. 1968. Pathology of Minamate disease. Pp. 141-228 in Minamata Disease. Study Group of Minanata Disease, ed. Kumamoto, Japan: Kumamoto University.

Tamashiro, H., M. Arakaki, H. Akagi, K. Hirayama, K. Murao, and M.H. Smolensky. 1986a. Sex differential of methylmercury toxicity in spontaneously hypertensive rats (SHR). Bull. Environ. Contam. Toxicol. 37(6):916-924.

Tamashiro, H., M. Arakaki, H. Akagi, K. Murao, K. Hirayama, and M.H. Smolensky. 1986b. Effects of ethanol on methyl mercury toxicity in rats. J. Toxicol. Environ. Health 18(4):595-605.

Thomas, D.J., H.L. Fisher, M.R. Sumler, A.H. Marcus, P. Mushak, and L.L. Hall. 1986. Sexual differences in the distribution and retention of organic and inorganic mercury in methyl mercury-treated rats. Environ. Res. 41(1):219-234.

Turner, C.J., M.K. Bhatnagar, and S. Yamashiro. 1981. Ethanol potentiation of methylmercury toxicity: A preliminary report. J. Toxicol. Environ. Health 7(3-4):665-668.

Vorhees, C.V. 1985. Behavioral effects of prenatal methylmercury in rats: A parallel trial to the Collaborative Behavioral Teratology Study. Neurobehav. Toxicol. Teratol. 7(6):717-725.

Welsh, S.O. 1977. Contrasting effects of vitamins A and E on mercury poisoning. [Abstract]. Fed. Proc. 36(1146):4627.

Wu, G. 1995. Screening of potential transport systems for methyl mercury uptake in rat erythrocytes at 5 degrees by use of inhibitors and substrates. Pharmacol. Toxicol. 77(3):169-176.

Yasutake, A., and K. Hirayama. 1988. Sex and strain differences of susceptibility to methylmercury toxicity in mice. Toxicology 51(1):47-55.

Yasutake, A., K. Hirayama, and M. Inouye. 1990. Sex difference in acute renal dysfunction induced by methylmercury in mice. Ren. Fail. 12(4):233-240.

4

DOSE ESTIMATION

IN assessing the risks of exposure to MeHg, quantitative exposure assessments are required to derive dose-response relationships from epidemiological data. A quantitative exposure assessment also allows risk assessment of an exposed population by comparing actual exposures to a reference dose (or similar benchmark) derived from critical studies. In contrast to experimental animal studies, in which the dose can be closely controlled, the dose in population-based epidemiological studies is not controlled and is therefore viewed as a random variable distributed across the study population. Three metrics for retrospective dose estimation and reconstruction are available for MeHg: dietary assessment, hair analysis, and blood analysis. Each metric has advantages and disadvantages. Ponce et al. (1998) proposed an approach for examining the relative uncertainties of those metrics.

DIETARY ASSESSMENT

With the exception of intakes through breast milk, which is less well characterized, exposure to MeHg occurs almost entirely from a single dietary category—fish (IPCS 1990, 1991). For that reason, the task of assessing dietary intake or assessing ongoing intake in populations with uncontrolled exposures is relatively straight forward compared to assessment of multiple types of food. There are several basic approaches to the estimation of MeHg exposure from dietary intake: collection of duplicate portions of foods consumed; food-consumption diaries, in which daily fish intakes are recorded quantitatively; recall methods, such as 24-hr recall of fish consumption; diet histories of usual consumption at various meals; and food-frequency measures of usual frequency of consumption of fish and shellfish. Duplicate-diet collections and food-consumption diaries are prospective approaches, and the others are retrospective approaches.

General considerations for duplicate-diet studies were recently discussed by Berry (1997) and Thomas et al. (1997). In duplicate-diet studies, participants collect an identical portion of the food they consume and provide it to the investigator for laboratory analyses. In theory, duplicate-diet studies have the potential to provide the most accurate information on the ingested dose of MeHg, because the mass of fish and other nutrients and contaminants, in addition to MeHg, can be measured directly. The fact that only the fish portion of any given meal will contain MeHg simplifies the burden of duplicate-diet collection. In practice, however, this approach is limited by the demands it makes on the participants, the difficulty in identifying individuals who are willing to carry out such a study, the influence exerted by investigator observation, and the potential change in diet resulting in response to the burden of food collection. Thomas et al. (1997), working with nine highly motivated households, was able to collect duplicate samples for 97% of meals and 94% of snacks over a 7-day period. The number of uncollected meals, however, tripled after the first 3 days, and participants strongly recommended that future studies be limited to a maximum of 3-4 days. When such studies are confined to fish consumption, 3-4 days of collection might be useful only for populations with very frequent and highly regular patterns of fish consumption. Because of the practical limits on the length of the collection period, the authors recommended that duplicate-diet studies for risk-assessment purposes should be done over multiple intervals of time. Moreover, when the calorie content of collected food was compared with the estimated energy requirements of participants, duplicate portions were found to be underestimated.

Duplicate-diet studies have been specifically applied to the estimation of MeHg exposure by Sherlock et al. (1982) and Haxton et al. (1979). Sherlock et al. (1982) carried out a 1-week duplicate-diet study with 98 participants selected on the basis of frequent fish consumption. In addition, a 1-month dietary diary was kept by the participants; the last week of the diary corresponded to the duplicate-diet collection. No indication is provided of the completeness of the duplicate-diet collection, but the weight of fish calculated from the diary during the week of duplicate-diet collection corresponded closely to the weight of fish measured from the duplicate samples. The authors noted however, indirect evidence of undercollection of duplicate-diet portions relative to consumed portions. It should also be noted that the preselection of subjects with frequent fish consumption increased the likelihood of collecting a meaningful number of samples over a 1-week period. A similar study with a randomly selected study sample would be less likely to provide adequate representation of infrequent consumers.

Haxton et al. (1979) conducted a 1-week duplicate-diet study with 174 subjects selected from fishermen and their families in coastal communities to obtain a population with high fish-consumption rates. No simultaneous diaries were kept, but the characteristic intake for each individual was identified from pre-collection interviews. No estimate of the completeness of the duplicate-diet collection was provided. However, the authors noted that the measured weight of weekly fish intake from the duplicate-diet samples was lower than that calculated from the interviews, and all measured intakes were below the calculated mean intake. The authors suggested that the discrepancy resulted from misidentification of characteristic intake in the interviews rather than from undercollection of dietary samples. No data are provided to support that assertion. As with the Sherlock et al. (1982) study, the preselection of subjects with frequent fish consumption made the relatively short collection period feasible.

Multiple-day food records (food-consumption diaries) are often used in conjunction with duplicate-diet studies (Sherlock et al. 1982, Thomas et al. 1997). This method, if conducted appropriately, has the advantage of recording information prospectively with little reliance on recall. It also requires less effort from participants than the duplicate-diet approach. However, daily recording of foods eaten at each meal requires a continuous and significant time commitment. Because fish are consumed relatively infrequently, the duration of the recording period might require many weeks to adequately capture infrequent consumers as well as variability in consumption among more frequent consumers. Furthermore, the design must be such that possible seasonal patterns of consumption are observable. The determination of the mass of food consumed when using food-consumption diaries can be made by weighing samples or by participants' estimating portion size. The former is preferable but more invasive, especially when foods are consumed away from home. Participants' estimation of portion size introduces a degree of measurement error not seen with duplicate-diet methods. Furthermore, if the diary approach is used without duplicate-diet collection, analysis of Hg concentration in each fish meal consumed cannot be made directly but must be based on the characteristic Hg concentration in each reported species. Such studies must, therefore, rely on participants for correct identification of species. Incorrect species identification can lead to errors in estimation of MeHg intake. Consumers, as well as the markets from which they purchase fish, might not know or correctly identify the species that was bought and consumed.

The data from the Continuing Survey of Food Intake by Individuals (CSFII) generated by the U.S. Department of Agriculture from 1989 to 1995 rely on self-administered food consumption diaries for the second and third days of its 3 days of reporting (discussed in EPA 1997). The CSFII data have been used by the U.S. EPA to estimate fish consumption in the U.S. population (Jacobs et al. 1998) and to estimate MeHg intake (EPA 1997). The National Purchase Diary conducted by the Market Research Corporation used dietary diaries over 1-month periods between 1973 and 1974 (discussed in EPA 1997). The fish-consumption portions of these diary data were used to estimate MeHg exposure in the U.S. population (Stern 1993, EPA 1997).

Retrospective dietary-assessment methods are simpler and less expensive than prospective and duplicate-diet methods, and therefore are used more often as the basis of dietary exposure assessments. Food-frequency studies take the form of participants identifying their typical fish consumption (e.g., "How many times per week/month do you usually eat fish A?"). Diet histories involve recollection of specific meals over a specific time (e.g. 24-hr or 1-week periods). In the studies mentioned above, Sherlock et al. (1982) and Haxton et al. (1979) used retrospective assessment of typical consumption to preselect subjects. In a recent study of MeHg exposure among pregnant women in New Jersey (Stern et al. 2000), participants were asked to identify their typical consumption frequency and typical portion size of 17 species of fish and fish dishes (e.g., fish sticks). MeHg intake was estimated as the product of the characteristic MeHg concentration for each fish species, the self-reported yearly frequency of consumption, and the self-reported average portion size. The yearly MeHg intake estimated in that manner was poorly correlated with the Hg concentration in hair from the same individuals. The authors attributed the discrepancy to the relatively infrequent consumption of fish in general. Therefore, the hair segments might have been too short to provide an adequate sample of the yearly intake. Uncertainty in the reporting of characteristic consumption frequency and portion size was also suspected as a contributing factor to the poor correlation.

The usefulness of studies using dietary recollection over a specific period depends on the participants' ability and willingness to recall information about fish meals over the target period. Recall of fish consumption seems to be much better than recall of other dietary items or of food intake in general. However, short-term recall methods of dietary assessment will tend to underrepresent the consumption characteristics of infrequent consumers (Whipple et al. 1996; Stern et al. 1996). If the species of fish (and thus the characteristic Hg concentration in the fish) consumed by frequent and infrequent consumers differ, or if the average portion size consumed by each group differs, the estimate of MeHg intake in the overall population will not be accurate.

The CSFII data used by EPA to estimate fish consumption (Jacobs et al. 1998) and MeHg exposure (EPA 1997) nationally are, as noted above, based on 1 day of recall and 2 succeeding days of diary entries. The National Health and Nutrition Examination Surveys (NHANES III) dietary data, generated from 1-day recall, were also used by the EPA to generate estimates of MeHg in the U.S. population (EPA 1997). Stern et al. (1996) used data from a fish-consumption-specific telephone survey of New Jersey residents. The survey elicited a 7-day recall. Relatively short-term recall studies can miss long-term patterns of variability in consumption and might not adequately capture consumption patterns of infrequent consumers. To address those issues, information on respondents' usual frequency of fish consumption was also elicited. That information allowed identification of infrequent consumers of fish in the sample. The information was used to investigate reweighting of the data to estimate the distribution of consumption frequency represented in a hypothetical 1-year recall study. Interestingly, the reweighting of the data using several different approaches resulted in only minor differences in estimates of fish consumption and MeHg exposure.

Retrospective dietary data and diary data on fish consumption have frequently been used to stratify a study population into broad classes of MeHg intake before more quantitative estimation of exposure by measurement of Hg in biomarkers. Such data have also been used to provide a rough validation of biomarker analyses (e.g., Dennis and Fehr 1975; Skerfving 1991; Grandjean et al. 1992; Holsbeek et al. 1996; Vural and Ünlü 1996; Mahaffey and Mergler 1998). Less frequently, retrospective and diary data on fish consumption are used directly in quantitative estimations of MeHg exposure (Buzina et al. 1995; Stern et al. 1996; Chan et al. 1997). Such estimates, however, generally require species-specific Hg concentrations (microgram of Hg per gram of fish), which are combined with the reported consumption frequency (grams of fish per day), to yield an Hg intake rate (micrograms Hg per day). The assignment of species-specific concentration data is a potential source of error in such studies for several reasons. First, the identity of the species on the part of the retailer or the consumer is often ambiguous. Second, Hg concentrations characteristic of a given species in local and regional markets or waters might differ from the characteristic concentrations identified on the basis of nationwide sampling. Finally, characteristic Hg concentrations derived from data that are often decades old might not be valid today.

In the United States, data on Hg concentration in commercial fish are largely available from two sources: (1) the National Marine Fisheries Service (NMFS) study, which sampled fish that were intended for human consumption and which were landed in the United States in the early to mid-1970s (Hall et al. 1978); and (2) the U.S. Food and Drug Administration (FDA) sampling conducted in the early 1990s (FDA 1992). Both data bases represent samples of fish collected from landings and markets in various parts of the United States but do not identify the locations at which samples were obtained or sold. The NMFS data were collected more systematically, represent more species, and generally contain considerably more samples for each species than the FDA data. However, the FDA data are about 20 years more recent than the NMFS data. Analysis of species represented in both data bases by at least three samples (n=15) indicates that, in almost all cases, the Hg concentration reported by FDA in a particular species is significantly lower than the concentration reported by NMFS. The most likely explanation for the discrepancy is the decreased availability of large fish due to overfishing (Stern et al. 1996). Those data cannot be used, therefore, to reflect potentially important local or regional differences in the characteristic concentrations of Hg by species. Studies addressing smaller populations with fewer varieties of fish (e.g., Buzina et al. 1995) can generate population-specific estimates of Hg concentrations by species. Given the variability in concentration within species, the assignment of a single representative value of Hg concentration is another potential source of error in such studies. The NMFS data base provides no estimate of such variability in U.S. commercial supplies. The FDA data base provides concentration ranges as well as average concentrations by species, but the ability to assess intraspecies variability in Hg concentration is limited by the generally small sample sizes. A study of Hg concentration in canned tuna (Yess 1993) indicates coefficients of variation of 55-120% across the several types of tuna commonly sold in cans. In general, sparse data on fish from commercial sources in the United States (FDA 1992) and data on food fish from noncommercial sources (e.g., Schuhmacher et al. 1994; Castilhos et al. 1998) often show a 2-3-fold difference between the mean and the maximum concentrations of Hg. Interspecies variability can be considerably larger. When data are available on intraspecies variability in MeHg concentrations, the variability can be integrated into estimates of intake through Monte Carlo probabilistic analysis (Chan et al. 1997). Intraspecies variability in Hg concentration might be less of a source of error in studies of frequent consumers of that species. With repeated consumption of a species of fish, total MeHg intake by consumers will approach the average concentration in that species. However, for populations with infrequent or sporadic consumption of a species, the effect of ignoring intraspecies variability in Hg concentrations could be significant.

BIOMARKERS OF EXPOSURE

MeHg lends itself to assessment of exposure through direct measurement in blood and hair. Assessment of Hg exposure through analysis of nail clippings has also been done (Pallotti et al. 1979; MacIntosh et al.

1997), but its correlation with fish consumption has yet to be clearly established. Hg exposure through breast milk has also been investigated (Pitkin et al. 1976; Fujita and Takabatake 1977; Skerfving 1988; Grandjean et al. 1995; Oskarsson et al. 1996). Compared with whole blood, breast milk (which is derived from maternal plasma) contains a much higher proportion of inorganic Hg (Skerfving 1988; Oskarsson et al. 1996). Therefore conclusions regarding the exposure of infants to MeHg from breast milk should use MeHg-specific analysis. Finally, there have been no reports of measurement of Hg in the fetal brain, the ultimate target of MeHg developmental neurotoxicity, although Cernichiari et al. (1995a) reported on a small set of measurements of Hg in infant brain.

The relationship among the several possible indicators of exposure is shown in Figure 4-1. Some of the indicators, such as the biomarkers of hair and blood Hg concentration, are commonly measured directly, whereas others, particularly fetal brain Hg concentration, are assumed to be correlated with the directly measured quantities. For the purposes of risk assessment, biomarker concentrations of MeHg serve two functions. First, a biomarker concentration is used as a surrogate for the unknown biologically relevant dose of MeHg in the developing fetal brain. That permits the development of a "dose"-response relationship in which the dose is represented by the biomarker concentration. Second, once such a dose-response relationship has been established, the biomarker concentration identified as the critical (e.g., benchmark) concentration must be translated into an estimate of the ingested dose. At that point, public-health interventions and regulatory measures can be guided by that estimate. The translation of the biomarker concentration to the ingested dose involves the use of toxicokinetic modeling to recapitulate the steps that precede the measured biomarker compartment in Figure 4-1 (see Chapter 3).

Methylmercury in Blood

The detection limit for total Hg in blood is generally in the range of 0.1 to 0.3 μg/L (ppb) (Grandjean et al. 1992; Girard and Dumont 1995; Oskarsson et al. 1996; Mahaffey and Mergler 1998). The mean concentration reported in U.S. studies in which high-fish-consuming populations were not specifically selected appears to be in the range of 1 to 5

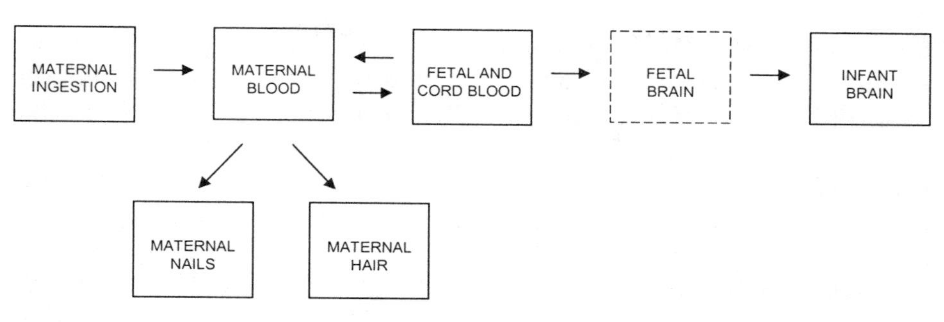

FIGURE 4-1 Relationship among the various indicators of MeHg exposure. Maternal ingestion of MeHg refers to the ingested dose, the magnitude of which depends on the amount of fish consumed and the concentration of MeHg in the fish. The concentration of Hg measured in the maternal blood, fetal blood, cord blood, maternal nails, and maternal hair are all biomarkers of exposure. Concentrations of Hg in the fetal brain, if available, would be considered the effective dose. (Note: The Fetal Brain box is shown with a dotted line because no direct data are available on Hg concentrations in the fetal brain.)

μg/L (Humphrey 1975; Brune et al. 1991; Nixon et al. 1996; EPA 1997; Kingman et al. 1998; Stern et al. 2000). Thus, current methods for blood MeHg determination appear to be adequate for fully characterizing population distributions of MeHg exposure and are, in practice, limited only by the volume of blood that can be obtained. Fish and other seafood, including marine mammals, are the only significant source of MeHg exposure (IPCS 1990). Therefore, the blood Hg concentrations in populations with little or no fish consumption should reflect exposure to inorganic Hg. The mean blood Hg concentration in such populations was reported to be about 2 μg/L (standard deviation (SD) = 1.8 μg/L) (Brune et al. 1991). Blood Hg concentrations in populations with high fish consumption are usually considerably higher than that value. For example, median cord-blood concentration in a cohort with high fish consumption in the Faroe Islands was 24 μg/L (Grandjean et al. 1992). Therefore, the measurement of the concentration of total Hg in blood is generally a good surrogate for the concentration of MeHg in blood in populations with high fish consumption. In populations with relatively low fish consumption, inorganic Hg concentration might constitute a larger fraction of total Hg concentrations. Therefore, for such populations, estimates of MeHg exposure from cord blood might be unreliable. Adult blood Hg concentration has frequently been used as a biomarker of adult MeHg exposure, and it has been used to assess dose-response relationships in adult neurotoxicity (e.g., Hecker et al. 1974; Dennis and Fehr 1975; Gowdy et al. 1977; Palotti et al. 1979; Skerfving 1991; Mahaffey and Mergler 1998). Cord-blood or maternal-blood Hg concentrations have also been used with some frequency in assessing exposure to the developing fetus (Dennis and Fehr 1975; Pitkin et al. 1976; Fujita and Takabatake 1977; Kuhnert et al. 1981; Kuntz et al. 1982; Sikorski et al. 1989; Grandjean et al. 1992; Girard and Dumont 1995; Oskarsson et al. 1996).

In assessing the appropriateness of a particular biomarker of exposure, it is important to consider three factors: (1) how well the biomarker of exposure (i.e., the concentration of Hg in hair or blood) correlates with the ingested dose of MeHg; (2) how well the biomarker of exposure correlates with the Hg concentration in the target tissue; and (3) how well the variability over time in the biomarker of exposure correlates with changes in the effective dose at the target tissue over time.

For developmental neurotoxicity, the target organ is the developing fetal brain. The kinetics of MeHg transport among compartments is subject to interindividual variability at each step, and therefore, the more closely a compartment is kinetically related to the target tissue, the more closely the concentration measured in that compartment is likely to correlate with the concentration in the target tissue. As shown in Figure 4-1, the fetal and cord-blood compartment is one compartment removed from the fetal-brain compartment. Thus, the cord-blood Hg concentration might be a reasonable surrogate for the biologically relevant dose to the fetal brain. Having determined a critical concentration of Hg in the blood, it is then necessary to back-calculate the ingested dose (micrograms of Hg per kilogram of body weight per day) corresponding to the critical concentration in blood (Stern 1997). Just as the kinetic proximity of the biomarker compartment to the target tissue increases the correlation between biomarker concentration and dose to the target tissue, the kinetic closeness of the biomarker compartment to the ingested dose will increase the correlation between the critical biomarker concentration and the estimated intake. The cord-blood Hg concentration is more closely linked to the fetal-brain compartment than to the ingested dose. Maternal-blood Hg concentration is more closely linked to the ingested dose than to the fetal-brain compartment. Thus, with the use of blood as a biomarker of MeHg exposure, there is a trade-off between the precision in the derivation of the dose-response relationship and the precision in the estimate of the corresponding ingested dose.

The mean half-life of total MeHg in blood in humans is about 50 days (Stern 1997; EPA 1997), but much longer half-lives (more than 100 days) are observed. Blood Hg concentration, therefore, reflects relatively short-term exposures relative to the total period of gestation. However, the Hg blood concentration at any given time reflects both the decreasing concentration from earlier exposures and the increase in concentration from recent exposures. Individuals with frequent and regular patterns of fish consumption achieve, or approximate, steady-state blood Hg concentrations (IPCS 1990). At steady state, the daily removal of Hg from the blood equals the daily addition to the blood from intake. Under such conditions, an individual's blood Hg concentration at any given time provides a good approximation of the mean blood Hg concentration over time. For individuals with infrequent or irregular fish consumption, however, recent fish consumption will result in peaks in blood Hg concentration. A single blood sample showing an elevated concentration, without additional exposure information, does not provide a temporal perspective and does not permit differentiation between increasing peak concentrations, decreasing peak concentrations, and steady-state exposure. Conversely, a single blood sample obtained between peak exposures and showing a low blood Hg concentration provides no evidence of peak exposures. That result can introduce error into dose-response and risk assessment in adult populations in whom short-term peak exposures might be relevant to chronic toxicity. The blood Hg concentration can correlate well with the dose presented to the brain at the time of sampling, but such information cannot necessarily be extrapolated to dose at the target tissue at other times. A blood Hg measurement that might be adequate to reflect exposure over time can be determined to some extent by obtaining dietary intake data that corresponds to several half-lives preceding the Hg measurement. In assessing exposure and dose-response relationship in utero, the temporal considerations associated with the use of blood Hg concentration as a biomarker are further complicated by two additional factors: (1) the fetal brain is developing during much of gestation and might not be equally sensitive to the MeHg during all periods; and (2) the half-life of MeHg in cord blood might not be the same as that in maternal blood.

As summarized by IPCS (1990) and Gilbert and Grant-Webster (1995), there are clear differences between fetal and adult MeHg neurotoxicity. However, few data provide specific information on the differences in sensitivity to MeHg developmental neurotoxicity across the fetal period. The existence of such differences can be inferred, however, from summaries of experimental animal data presented by Gilbert and Grant-Webster (1995) and ATSDR (1999). The timing of maternal dosing in these studies was generally not chosen to relate differences in effect to specific stages of neurological development; therefore, it is difficult to infer specific information about developmental periods of specific sensitivity in the human fetus. It is assumed that there are windows of vulnerability to MeHg during neurological development (Choi 1989), and specific types of developmental effects (e.g., motor and cognitive) might have separate windows of vulnerability. In general, the embryonic period of development (fewer than 4 weeks of gestation), when there is no brain per se, might show little sensitivity to MeHg developmental neurotoxicity. Fetal stages during which the structure of the brain is forming are the periods in which the broad abnormalities in brain architecture, most characteristic of MeHg developmental neurotoxicity, are likely to occur. MeHg exposure during late fetal development, when brain structure is basically established, is likely to produce more function-specific effects on brain architecture. Even within early-to-middle fetal developmental stages, there might be discrete windows of sensitivity. As discussed above, the existence of such windows of sensitivity might have little practical significance if maternal MeHg intake does not vary substantially during pregnancy. However, individuals and populations with irregular patterns of MeHg intake will have peaks of exposure that might or might not occur during a window of vulnerability. If the half-life of MeHg in fetal blood is the same as in maternal blood (~50 days), the cord-blood MeHg concentration would be expected to reflect to some extent fetal exposures over about three half-lives (150 days) prior to delivery. That time (calculating backwards from birth) corresponds approximately to the second half of the second trimester and the third trimester. However, the cord-blood concentration would be most heavily influenced by exposures during the most recent half-life, which corresponds to the last half of the third trimester. If that period is not critical for MeHg neurological developmental toxicity, dose-response assessments conducted using cord-blood Hg might lead to misclassification of exposure.

Because the fetus (and presumably the infant) has no independent mechanism for excreting or metabolizing MeHg to mercuric mercury (Grandjean and Weihe 1993), any elimination of Hg by the fetus will be by passage across the placenta to the maternal blood. Therefore, if the fetus does not have a specific affinity for MeHg, the half-life of MeHg in the fetal blood will be the same as that in the maternal blood, and the ratio of MeHg in the fetal blood to the maternal blood will be 1.0. However, Dennis and Fehr (1975), Pitkin et al. (1976), and Kuhnert et al. (1981), as well as additional studies cited by the latter two studies, found the concentration of Hg to be about 20-30% higher in cord blood than in maternal blood. On the other hand, Kuntz et al. (1982) and Sikorski et al. (1989) found a ratio close to 1.0. If the Hg concentration in cord blood is 20-30% higher than that in maternal blood (because of a longer half-life in fetal blood), the cord-blood Hg concentration would be more influenced than the maternal-blood concentration by exposures during the latter portion of the second trimester and the first half of the third trimester.

Methylmercury in Hair

In contrast to adult blood sampling, hair sampling is noninvasive and can be done without medical supervision. Although cord-blood collection is also essentially noninvasive, the logistics of its collection can be difficult. Hg concentration in hair is often used to estimate exposure to MeHg. In some studies, hair and blood Hg are measured for comparison. More often, hair is used as the sole biomarker of exposure. Using the standard cold-vapor analytical techniques, the detection limit for total Hg in human hair is generally reported to be in the range of 0.01 to 0.04 μg/g hair (e.g., Airey 1983; Bruhn et al. 1994; Lópes-Artiguez et al. 1994; Holsbeek et al. 1996; Gaggi et al.1996; Stern et al. 2000). Few studies have reported on hair Hg concentrations in U.S. populations that were not specifically selected for high fish consumption. Among these, the mean hair Hg concentration appears to be in the range of 0.3 to 1.0

μg/g (Smith et al. 1997; EPA 1997; Stern et al. 2000). Thus, the sensitivity of the standard methodology should be adequate to characterize population distributions of MeHg exposure in the United States. Among individuals, whose hair Hg concentrations are presumed to reflect inorganic Hg exposure because of little or no fish consumption, hair Hg concentrations are reported to be in the range of 0.2 to 0.8 μg/g (Pallotti et al. 1979; Grandjean et al. 1992; Oskarsson et al. 1994; Bruhn et al. 1994; Batista et al. 1996; Smith et al. 1997). Populations selected for dose-response analysis on the basis of high fish consumption generally have considerably higher hair Hg concentrations (e.g., the mean maternal-hair Hg concentration in the Seychelles main study cohort was 6.8 μg/g (Cernichiari et al. 1995), and the median maternal-hair Hg concentration in the Faroes cohort was 4.8 μg/g (Grandjean et al. 1992)). Hair Hg concentrations that exceed those attributable to inorganic Hg exposure in fish-consuming populations must arise from MeHg exposure. Thus, the use of total hair Hg concentration in fish-consuming populations as a surrogate for hair MeHg concentration in fish-consuming populations should not lead to significant exposure misclassification.

Blood Hg concentration, unless supplemented by additional temporal exposure data, provides no clear information about the magnitude or timing of the exposures that yield the total Hg concentration observed in a given sample. In contrast, hair Hg concentration as a biomarker of MeHg exposure has the advantages of being able to integrate exposure over a known and limited time and recapitulate the magnitude and the timing of exposure. The ability to obtain such information from hair is predicated on two assumptions: that growing hair shafts incorporate Hg from the circulating blood in proportion to the concentration of Hg in the blood, and that hair shafts grow at a constant rate that does not vary significantly among individuals. The first of these assumptions is necessary to establish a quantitative relationship between hair Hg concentration and MeHg intake, the blood Hg concentration being an intermediate kinetic compartment. The second assumption is necessary to establish a relationship between location along the hair strand and time of exposure.

Although the proximal portion of the growing hair shaft is exposed to circulating blood for several days, that exposure appears to be indirect, as the shaft grows from a group of matrix cells located in the dermis, and these matrix cells are in direct contact with the capillaries

(Hopps 1977). Furthermore, the growing portion of the shaft is also in direct contact with sweat and sebum, both of which can contribute to the incorporation of trace elements into the shaft (Hopps 1977; Katz and Chatt 1988). Therefore, measured concentrations of Hg in blood and hair can be separated, at least in part, by one or more kinetic compartments. Interindividual pharmacokinetic variability in these compartmental transfers could explain some of the scatter seen in plots of hair and blood Hg concentrations (e.g., Sherlock et al. 1982; Grandjean et al. 1992).

The growth rate of hair varies both within and among individuals. Among individuals, variations in hair-growth rate occurs because individual hair follicles experience a cycle of growth, transition, and terminal resting (Katz and Chatt 1988). Direct incorporation of trace elements, including MeHg, into the hair occurs only during the growth phase. The growth phase is the longest phase, although for scalp hair (the hair commonly used as an MeHg biomarker), estimates of the proportion of the total cycle during which growth occurs vary from 70% to 90% (Katz and Chatt 1988). In humans, individual hair follicles have independent growth cycles (Hopps 1977), and given the predominance of the growth portion of the follicle's life cycle, a sample of multiple hairs largely reflects hair that was recently incorporating MeHg. However, such a sample potentially has 10% to 30% of its follicles in the terminal resting phase. The Hg concentration in such follicles reflects less-recent exposure than that reflected by follicles in the growth phase. That difference can lead to exposure misclassification for the period of interest. The potential for exposure misclassification due to collection of follicles in the terminal resting phase is a particular concern in single-strand hair analysis for Hg. That analysis implicitly assumes that a point on a hair strand at a given distance from the scalp corresponds to the same point in time on all other strands from that individual.

There appears to be significant interindividual variability in hair-growth rate. An average growth rate of 1.1 centimeters (cm) per month for scalp hair is commonly assumed (Grandjean et al. 1992; Cernichiari et al. 1995; Boischio and Cernichiari 1998). However, Katz and Chatt (1988) characterize hair-growth rates as highly variable and dependent on age, race, gender, and season. They provide a summary of studies of hair-growth rates expressed as ranges. Interstudy values typically range from 0.6 to 1.5 cm per month, but ranges of 2.3 to 3.4 and 3.3 ± 0.6 cm per month are also reported. Thus, a 9-cm length of maternal hair intended to correspond to approximately 8 months of gestation (assuming a hair-growth rate of 1.1 cm per month) could correspond to a period of 6-15 months (assuming a growth rate of 0.6 to 1.5 cm per month). There is also evidence of intraindividual variability in hair-growth rates (Giovanoli-Jakubczak and Berg 1974). Furthermore, during pregnancy, the rate of hair growth slows slightly (approximately 7% during the second trimester), and the interindividual variability in growth rate appears to increase (Pecoraro et al. 1967), thus adding to the temporal uncertainty inherent in assessing MeHg exposure from hair analysis. In addition, the physical characteristics of the hair alter somewhat, and the percentage of thick hairs increases (Pecoraro et al. 1967) . Those physical changes suggest that the uptake and binding of MeHg might be altered. Attempts to identify segments of hair corresponding to all or part of the period of gestation (Grandjean et al. 1992; Cernichiari et al. 1995) by using the average growth rate of 1.1 cm per month might include exposure data from unintended time periods or exclude exposure data from a portion of the intended period. The use of such misidentified segments can result in exposure misclassification in dose-response analysis.

An additional difficulty in identifying the segment of hair corresponding to the entire period of gestation or to any specific period of gestation is the location of the most recently formed portion of the hair shaft, which is below the scalp until pushed out by subsequent growth. To assign an exposure period to a segment of hair, a chronological benchmark on the hair strand is needed to relate measurements of length and time. The proximal end of the shaft, as it emerges from the scalp, is generally taken as such a benchmark, even though the hair is not cut exactly at the scalp level (Hislop et al. 1983). Because the hair below the scalp represents the hair formed at the time of sampling, the proximal end of the cut hair must be assigned a time of formation that accounts for the lag time between formation and sampling (Cernichiari et al. 1995).

Hislop et al. (1983) related the time course of MeHg elimination from blood and hair cut at the scalp with the assumption that the hair Hg concentration is proportional to the blood Hg concentration. The blood was sampled at regular intervals. The hair was sampled once and divided into 8-mm segments. The measurement of the hair-growth rate was 8 mm of growth per 20 days. The presence of a distinct maximal concentration in the serial blood samples and the segmental hair analysis allowed calculation of the lag between equivalent concentration points in the blood and hair samples. The hair segment with the maximum Hg concentration was found to be offset from the appearance of the maximum concentration in the blood by 20 days. It should be noted that the 20-day estimate is based on a measurement of hair-growth rate specific to this study. Different characteristic hair-growth rates in different populations and variability in the growth rates among individuals in a population would yield different estimates of the time difference in hair and blood measurements. Because the concentration of Hg in the blood represented a precise time point but was compared with the average concentration in the 8-mm segment representing 20 days of exposure, the 20-day estimate is somewhat uncertain.

Cernichiari et al. (1995) attempted to further refine this estimate by assuming that the average concentration in the 8-mm segment is the concentration in the mid-point of the segment and by estimating the time at which that point on the strand appeared just above the scalp. The validity of that assumption is not clear, as there does not appear to be any justification for assuming that the average concentration in a hair segment necessarily represents the concentration at any specific point along that segment. Furthermore, it is not clear that an estimation of the time necessary for a given point along a hair strand to appear just above the scalp is particularly useful unless one is analyzing segments shorter than 8 mm. Grandjean et al. (1998) reported that the appearance of Hg in a hair strand above the scalp is delayed by about 6 weeks. That is more than twice the delay reported by Hislop et al. (1983). However, Grandjean and coworkers provide no specific data to evaluate their assertion.

As discussed for blood Hg analysis, temporal uncertainties might not be critical for individuals with steady-state MeHg concentrations. However, for individuals with variable or peak exposures that might occur at critical periods during development, the uncertainties in assigning a specific time during pregnancy to specific hair segments might result in significant misclassification of exposure.

Despite the potential for temporal misclassification of exposure, the potential for identifying the segment of hair corresponding to a specific period of gestation (and neurological development) has a distinct advantage over cord-blood analysis. However, analysis of Hg concentration in any given segment of a hair sample will yield only the average exposure over the corresponding time period. Details of exposure within that time period, including peak exposures, will not be elucidated except as they influence the overall average concentration within a segment. As an illustration, consider a 3-cm-long segment of maternal hair (the hair samples analyzed in the Faroe Islands study were generally 3 cm in length (Grandjean et al. 1998) corresponding to approximately 3 months of exposure and intended to correspond to a given trimester of pregnancy. Assume that during that time period, the individual contributing the hair consumed several fish meals high in Hg in close succession and achieved a peak hair concentration that was double the steady-state concentration before consumption of the fish. Assume further, that MeHg is removed from her blood following first-order decay kinetics (IPCS 1990) with a half-life of 50 days (Stern 1997). It can be calculated that (even if blood and hair Hg concentrations are perfectly correlated) the Hg concentration detected in such a hair segment would be only 20% higher than the concentration in the segment before the high-Hg fish consumption. That small observed increase occurs because the rise and return to background of the Hg-concentration peak occurs over a shorter time period than the exposure period represented by the entire 3-cm hair segment. Therefore, the average segment concentration reflects the dilution of the peak concentration by the adjacent stretches of background concentration in the segment. The true peak concentration, representing a doubling in exposure, would likely be identified as a significant increase in exposure if the concentration at that point could be measured, but it is not clear whether the observed 20% increase would be identified as a significant increase in exposure. Such an approach to segmental hair analysis would not give an accurate indication of the magnitude or duration of the peak concentration in the maternal or fetal blood. To some extent, the sensitivity to peak exposures can be increased by analyzing smaller-length segments of hair corresponding to narrower periods of exposure. Following the example above, a 1-cm (approximately 30-day) segment of hair containing the record of a peak doubling of exposure would be seen as an average increase of 50% above the steady-state background concentration. Such an increase is more likely to be recognized as significant but still does not provide a clear indication of the true peak concentration. If peak concentrations of fetal exposure are important to the elucidation of a dose-response relationship, even the accurate identification and analysis of the segment of hair corresponding to a putative window of developmental sensitivity might result in exposure misclassification. Furthermore, practical considerations might prohibit increasing the number of analyses that would be required for the analysis of shorter hair segments. Despite those limitations, hair samples have the potential to provide temporal information on Hg exposures.

An alternative to segmental hair analysis is continuous single-strand hair analysis using x-ray fluorescence (XRF) (Marsh et al. 1987; Cox et al. 1989). This nondestructive method involves measurement along the length of the strand. It is not truly continuous because determinations are made on consecutive 2-mm segments. Assuming a mean hair-growth rate of approximately 1.1 cm per month, 2 mm corresponds to about 6 days of growth. Assuming first-order decay kinetics, a peak concentration on a single day would decrease by only 8% during this 6-day averaging period. Thus, single-strand analysis will give a much finer picture of exposure peaks than individual segmental hair analysis. In addition, as illustrated in Cox et al. (1989), single-strand analysis avoids errors in the alignment of multiple strands, which will tend to flatten and broaden peaks. Localization of portions of a hair strand corresponding to a given period of gestation, however, is still subject to uncertainty arising from variability in hair-growth rate. In addition, as discussed above, analysis of individual strands in terminal resting phase will give misleading estimates of the exposures corresponding to any time period. For 45 individuals in the Iraqi poisoning, Cox et al. (1989) compared the maximal concentration in two hair strands from the same individual. The overall correlation was good, and the peaks in the Iraqi poisoning episode were distinct and easily identifiable, thus reducing the error in comparing corresponding points in each analysis. Furthermore, it appears that the correlation was based on matching the *value* rather than the *location* of the peak in each strand. Thus, this determination does not necessarily address the errors inherent in the temporal calibration of hair strands or in the selection of hair strands in the terminal resting phase. In the Cox et al. (1989) analysis, a few of the residual errors in the comparison between concentrations on alternate strands appear to be on the order of 25%. Such observations might reflect errors in temporal calibration. Continuous single-strand analysis allows the investigation of multiple plausible dose metrics in dose-response analysis. Those metrics include peak concentration in a specific trimester, peak concentration at any time during gestation, average peak concentration, average concentration during a specific trimester, and average concentration during the entire gestation. Overall, single-strand hair analysis by XRF appears to be a powerful tool with the distinct advantage of being able to determine short-term changes in exposure, including peak exposures.

As is the case for blood Hg, the use of maternal hair Hg as a dose metric in the derivation of a reference dose for effects of MeHg on neurological development requires that the hair Hg concentration be used in two separate determinations. The first determination is the derivation of a dose-response relationship between hair Hg concentration and effects. The second determination is the estimation of the MeHg ingested dose that corresponds to the critical Hg concentration in hair identified in the dose-response relationship. In the first determination, the maternal-hair Hg concentration is a surrogate for the unknown dose to the fetal brain. In the second, the critical Hg concentration is used in a pharmacokinetic model to back-calculate the ingested dose.

Comparison of Biomarkers of Exposure

As shown in Figure 4-1, the fetal brain is one kinetic compartment further removed from hair Hg than from cord-blood Hg. Therefore, for the somewhat uncertain period of gestation represented by the cord-blood Hg concentration, the fetal-brain Hg concentration would be expected to correlate more closely with the cord-blood Hg concentration than with the maternal-hair Hg concentration. Cernichiari et al. (1995a), however, presented data comparing the correlations of maternal-hair and infant-brain Hg concentrations, and infant-blood and infant-brain Hg concentrations measured from autopsy samples. The hair samples were collected at delivery, and represent a period of approximately 20 days before delivery. The correlation of maternal-hair and infant-brain Hg concentrations (r = 0.6-0.8, depending on the specific brain region) was generally comparable to the correlation of infant-blood and infant-brain Hg concentrations (0.4-0.8). That finding suggests that, as predictors of Hg concentration in the infant brain, maternal hair and infant blood might have equal validity. However, the error of the regression slope of infant-brain Hg to maternal-hair Hg is about 3-6 times the error of the slope of infant-brain Hg to infant-blood Hg (although the coefficient of variation for the brain-hair relationship is smaller than that for the brain-blood relationship) (Stern and Gochfeld 1999; Davidson et al. 1999). Perhaps more important in considering the relevance of those comparisons to the choice of dose metric for reference-dose development is the fact that Cernichiari et al. (1995a) examined the correlation between maternal hair and infant brain rather than between maternal hair and fetal brain. Likewise, infant blood rather than cord blood was compared with infant brain. Cernichiari et al. (1995a) do not give the age of the infants in this study, but postnatal infant brain cannot be considered identical to fetal brain, especially since the fetal brain changes substantially during development. Although the vulnerable periods for MeHg effects on neurological development are unknown, they might occur much earlier in gestation than the perinatal period. Furthermore, infant blood is not necessarily comparable to fetal blood due to the ongoing replacement of fetal hemoglobin with adult hemoglobin. At birth, fetal hemoglobin constitutes about 75% of total hemoglobin, but after about 50 days, it constitutes only about 50% of the total (Lubin 1987). Therefore, it is not clear to what extent these observations elucidate the relationship of fetal-brain to either cord-blood, or maternal-hair Hg concentrations.

For the back-calculation of the average ingested dose corresponding to a given biomarker critical concentration, the maternal-hair Hg compartment and the cord-blood Hg compartment are equally distant kinetically from ingestion (see Figure 4-1). The estimation of the ingested dose corresponding to a critical biomarker concentration requires the intermediate estimation of the corresponding maternal-blood Hg concentration. Although no study was found that specifically supplies data on the variance inherent in the ratio of cord-blood to maternal-blood Hg concentrations, the ability of MeHg to pass freely through the placenta (IPCS 1990) suggests that there might be interindividual variability in the extent of transfer of MeHg between the cord-blood and maternal-blood compartments. As discussed above, the mean cord-blood/maternal-blood Hg ratios reported for several populations differed by 20-30% at most. For maternal-hair Hg, the few studies reporting data on the variance in the maternal hair /maternal-blood Hg ratio within a given study population give widely differing coefficients of variation (Stern 1997). The mean maternal-hair/maternal-blood Hg ratios reported for different population groups can differ at most by a factor of about 2, although nearly all observations fall within approximately 20% of the overall mean of the various observations (Stern 1997; ATSDR 1999). Nonetheless, when the estimation of the ingested dose from a critical concentration in hair is carried out probabilistically and interindividual variability in the various pharmacokinetic inputs is taken into account, sensitivity analysis reveals that the maternal-hair/maternal-blood ratio is one of the key contributors to the variability in the predicated ingested dose (Stern 1997; Clewell et al. 1999).

Overall, in comparing maternal hair and cord blood as possible biomarkers of in utero MeHg exposure, each has significant advantages and disadvantages. At least conceptually, cord blood is kinetically more closely linked to the fetal brain-target and could, therefore, yield a more precise dose-response relationship if the critical period for toxicity coincides with the time period reflected in the cord-blood Hg measurement. However, the cord-blood Hg measurement is not capable of providing information about the specific patterns of exposure during gestation and does not reflect exposure over a clearly delineated period of gestation. In addition, cord blood is not capable of providing information about variability in exposure, even for the time period it most directly reflects. Simple maternal-hair analysis can provide information about average exposure over the entire period of gestation but provides no information about variability in exposure during that period. Identification of the specific portion of a hair strand corresponding to all of gestation is uncertain and is a potential source of exposure misclassification. In addition, maternal-hair Hg concentration is kinetically more distant from the fetal brain than is cord-blood Hg. Segmental hair analysis has the potential to provide information about exposure during specific portions (e.g., trimesters) of gestation, but uncertainties related to hair-growth rate make the identification of segments corresponding to periods as short as a single trimester uncertain. Although segmental hair analysis can provide some information about variability in exposure during different periods of gestation, it is of limited use in identifying either the magnitude or the duration of peak exposures. Continuous single-strand hair analysis, on the other hand, can provide precise information on peak exposures and thus permits the investigation of several different dose metrics in dose-response assessment. This approach is potentially the most powerful for investigation of dose-response relationships. However, single-strand analysis is still hampered by uncertainty in assigning specific periods of gestation to a given section of hair strand. The utility of cord blood and hair as biomarkers of MeHg exposure can be substantially improved by linking them to accurate dietary intake information. Data on frequency, amount, and type of fish consumption in the period during and immediately preceding pregnancy can provide information on the overall variability in exposure as well as on peak exposures. Furthermore, accurate dietary information can provide benchmarks for the temporal calibration of both cord blood and hair Hg data. Recognizing that each of the available metrics provides different and complementary information, the most useful and powerful approach to exposure and dose assessment for MeHg is the collection of comparable dietary, cord-blood and single-strand hair data.

ANALYTICAL ERROR IN BIOMARKER MEASUREMENTS

In comparing the outcomes of the Faroe Islands, Seychelles, and New Zealand studies, it is important to consider the relative analytical errors in the measurement of the biomarker of exposure among those studies. Unfortunately, the reporting of such data in those studies is inconsistent and incomplete. The Seychelles study analyzed only maternal hair. Several analytical methods were used for various purposes in that study, the dose-response analysis used hair Hg concentration determined by cold vapor atomic absorption (CVAA). Although determinations were carried out to compare CVAA results with those from a reference method (counts of exogenously applied ^{203}Hg) (Cernichiari et al. 1995), no summary statistics of the comparison are provided. Results of an interlaboratory comparison of CVAA determinations of hair Hg concentration are reported (Cernichiari et al. 1995), and 100% of all samples analyzed were less than or equal to ± 2 standard deviation (SD) of the target value. The nature of the target value is not, however, discussed. The reporting of analytical quality-control data in the Faroe Islands study is somewhat confusing, because different analytical methods were used at different times for hair and blood, and various analyses were carried out in different laboratories (Grandjean et al. 1992). It appears, however, that all hair Hg analyses used in the dose-response analyses were carried out in the laboratory of the Seychelles study group (i.e., University of Rochester) using CVAA. Presumably then, the analytical errors in the analyses of hair Hg concentrations in both studies are highly comparable. It appears that all cord-blood Hg analyses used for dose response were carried out in the laboratory of the Faroe Islands group (i.e., Odense University) using ultraviolet (UV) absorption spectrometry. The accuracy of this method was determined relative to four reference samples of trace metals in blood. For the sample with a reference Hg concentration of 9.9 μg/L, the mean reported value was 9.9 μg/L (0% difference). For the three reference samples with much larger Hg concentrations (98, 103, and 103 μg/L), the percent difference between the mean reported values and the reference value was +13.0%, +11.2%, and +10.0%, respectively. The authors report that all the reported values were within the "acceptable range," although this range is not further defined. Analytical imprecision (coefficient of variation for repeated analyses of the same sample) for the first three of those four reference samples ranged from 7.0% to 14.1%, the lowest concentration having the largest imprecision. The New Zealand study generated data only on hair Hg concentration (Kjellstrom et al. 1986, 1989). The reporting of analytical quality control data from the New Zealand study is somewhat complicated because the study was carried out in two stages. During the first stage, no reference samples were available. Reference samples were available for the second stage of the study, and samples from additional mothers were analyzed for that stage. For the most part, however, it appears that samples analyzed during the first stage of the study were not re-analyzed during the second stage. Given the lack of reference samples during the first stage of the study, analytical quality during that stage was based on interlaboratory comparison. Sixteen samples analyzed by CVAA in the laboratory of the New Zealand study group (i.e., University of Auckland) (reporting Hg concentrations of more than 10 ppm) were re-analyzed by CVAA at the University of Rochester. The percent difference (sum of absolute values) for 13 of those samples was 22.8%; 62% of University-of-Auckland values were smaller than the corresponding University-of-Rochester values. The values for the remaining three samples were grossly different. When those three samples were re-analyzed by the University of Auckland, a much closer agreement with the University-of-Rochester values was achieved. This raises some concern. In the second stage of the New Zealand study, 12 reference samples of hair were analyzed and the results compared with the maximum acceptable deviation (MAD) which was defined as (true $value_{ppm}$ ± (10% of true value ± 1 ppm)). The basis for that metric is not entirely clear. The regression line for the reported values versus the true values was found to lie completely within the MAD lines. Some individual values, however, were marginally outside the MAD lines, and the regression line was biased toward reported values underestimating reference values.

In summary, given the nature of the reporting of the quality-control data in these three studies, it is difficult to assess the analytical error inherent in the biomarker concentrations and ultimately in the dose-response relationships. It is also difficult to quantitatively or qualitatively compare the extent of analytical error in these studies. Some concern, however, is warranted with respect to the analytical error inherent in the New Zealand study. The extent to which that error might affect the interpretation of the dose-response relationship based on the New Zealand study is not clear.

EXPOSURE AND DOSE ASSESSMENT IN THE SEYCHELLES, FAROE ISLANDS, AND NEW ZEALAND STUDIES

Exposure in the Seychelles studies was measured as total Hg in maternal-hair samples. Other biomarkers of exposure were not investigated. Cernichiari et al. (1995) reported that the majority of mothers in the study provided at least two hair samples. The first sample was obtained at delivery, and the second, 6 months after delivery. In the first sample, the hair was cut at the scalp, and the proximal 9 cm was used for Hg analysis. A growth rate of 1.1 cm per month was assumed, and therefore a 9-cm segment of hair corresponded to 8.2 months of gestation. The hair representing the last month of gestation was assumed to lie beneath the scalp at the time of sampling. In the second sample, a segment of hair intended to correspond to the same time period as the first sample of hair was identified by assuming a 1.1 cm per month growth rate. Cernichiari et al. (1995) presented a scatterplot of the correspondence of the mean Hg concentration in those two samples. Regression statistics were not supplied, but the regression slope was reported to be insignificantly different from unity. There is, however, significant scatter around the regression line, and the data are not symmetrical at the line of equality. The most likely explanation for the scatter and asymmetry appears to be intraindividual variability in hair-growth rates. It is not clear whether the Hg concentration from one of the samples or the average concentration of both samples was used as the actual dose metric in the dose-response assessment. Hair samples from 86% of the main cohort were also divided into segments intended to represent the three trimesters of pregnancy. The average Hg concentration in each of the segments was compared with the average concentration in the complete 9-cm segment. The correlations (r) were all similar and ranged from 0.85 to 0.91. In addition, intercepts were each close to zero. The general comparability of each of the segmental average Hg concentrations to the total average Hg concentration suggests that intake did not vary greatly by approximate trimester for the cohort as a whole. Potential seasonal variations in Hg exposure would not likely be detectable in such an analysis because the cohort was not in synchrony with respect to the onset of gestation. In considering individual variability, the overall strong correlations notwithstanding, a considerable number of outliers can be seen in the scatterplots of these trimester comparisons, particularly in the assumed third-trimester segment. Those outliers suggest that some individuals might have had significant variability in exposure over the course of gestation. As discussed previously however, such analyses are relatively insensitive to short-term peaks in exposure. More specific information about intraindividual exposure variability or peak exposures cannot be deduced from these data. Data from this segmental analysis was not used in dose-response assessment.

Dietary information was obtained from the Seychelles cohort 6 months after delivery (Shamlaye et al. 1995). The extent to which this survey included the entire cohort is not reported. A median fish consumption of 12 meals per week, as well as some additional population percentiles of fish consumption, was reported. Those data, however, reflect self-reported average intake and do not provide information on variability in fish intake during pregnancy. In addition, although data on characteristic Hg concentrations for commonly consumed species of fish were generated (and included some species with relatively high characteristic Hg concentrations) (Davidson et al. 1998), data on consumption by species does not appear to have been collected. Thus, dietary data in the Seychelles studies cannot be used to suggest the extent to which individual exposure was variable or peaked during pregnancy.

In the Faroe Islands studies, exposure was measured by the concentration of Hg in maternal hair obtained at delivery and by cord-blood Hg concentration (Grandjean et al. 1992). The collection and analysis of the cord blood appears to be standard. However, the hair samples analyzed were not of uniform length and varied from 3 to 9 cm (Grandjean et al. 1999), thus reflecting exposure over variable times during gestation. As discussed above, cord-blood Hg concentration is influenced by exposure over an indeterminate time period, possibly including the latter part of the second trimester but weighted most heavily toward the latter part of the third trimester. Assuming a delay of about 20 days between incorporation of Hg into a growing hair strand and its appearance above the scalp, a 3-cm hair sample proximal to the scalp would reflect average exposure from the end of the second trimester to the second-third of the third trimester. A 9-cm hair sample would reflect average exposure beginning before conception. If, as asserted by Grandjean et al. (1998), the delay in appearance above the scalp of a section of a hair strand containing a given Hg concentration is 6 weeks rather than 20 days, the 3-cm hair segment would reflect exposure starting before the middle of the second trimester. Taking into account the apparent inconsistency in the length of the hair segments, as well as the inherent variability in hair-growth rates, the extent to which the hair and cord-blood MeHg concentrations reflect common exposures is uncertain. The correlation of hair and blood MeHg concentrations following log transformation ($r = 0.78$) was reasonably strong (Grandjean et al. 1998). However, it is not clear whether that correlation indicates consistency of the hair and blood measurements or it reflects little or no intraindividual variability in exposure during gestation.

Dietary exposure to MeHg in the Faroe population is complicated. The cohort generally consumed fish frequently; 48% ate fish dinners three or more times per week (Grandjean et al. 1992). However, the species of fish generally consumed were coalfish, ling, turbot, and salmon, which characteristically have relatively low concentrations of Hg (Grandjean et al. 1998). On the other hand, pilot whale is a traditional Faroese food that has considerably higher characteristic MeHg concentrations (less than 1 to more than 3 ppm) (Grandjean et al. 1998). The availability of pilot whale meat is somewhat irregular, as the catch is opportunistic rather than systematic. In the Faroese cohort, 79% of the mothers reported in prenatal interviews that they ate at least one whale dinner per month, but only 27% reported eating three or more whale dinners per month (Grandjean et al. 1992). It should be noted, however, that these data do not provide information on portion size, and refer to dinners only. Grandjean et al. (1998) suggested that whale meat is also eaten at other meals and as snacks (dried). This incomplete dietary intake information makes assessment of variability in exposure difficult. Nonetheless, sporadic consumption of meals high in Hg is expected to result in temporal variability in exposure and possibly in peak exposure. Grandjean et al. (1992) reported on the results of analysis of Hg in multiple segments, each 1.1 cm long, from each of the six women in the Faroese cohort. Coefficients of variation (i.e., comparison among segments from the same individual) ranged from 8.1 to 23.8%. Although that suggests low-to-moderate intraindividual variability in MeHg intake over time, generalization to the entire cohort is not warranted because of the small sample size.

The relative magnitude of potential peak exposures from sporadic consumption of whale meat is possible to estimate. Assume that a pregnant woman consumes whale meals of 113 g each on 3 consecutive days when whale meat is available, and that her maternal-hair Hg concentration is 4.5 ppm (the median maternal-hair concentration in the Faroese cohort (Grandjean et al. 1992). Assume that pilot whale contains Hg at 3.3 ppm (Grandjean and Weihe 1993), and assume that absorption of MeHg from the gastrointestinal tract is 95% complete, that 5% of the ingested dose is distributed to the blood (IPCS 1990), and that the blood volume for a woman of child-bearing age is 3.6 L (Stern 1997). Ginsberg and Toal (2000) have shown that the one-compartment pharmacokinetic model for MeHg provides a reasonable approximation of the accumulation of Hg in hair for single exposures. Using the one-compartment model, therefore, assume that the rate constant for elimination of MeHg from the blood is 0.014 per day (equivalent to a half-life of 50 days) (Stern 1997). Finally, assume that the ratio of maternal-hair Hg concentration to maternal-blood concentration is 0.250 (μg/g)/(mg/L) (IPCS 1990). Based on those assumptions, the one-compartment model predicts that the hair Hg concentration will increase by about 3.6 ppm to a total concentration of 8.1 ppm and decrease to the baseline concentration (assuming no additional exposures above background) in about 6 weeks. Thus, this scenario of a fairly large intake over a short time period is predicted to result in a hair concentration that less than doubles the original concentration and returns to the original concentration all within a length of hair slightly greater than 1 cm.

The hair samples analyzed in the Faroe Islands study were generally 3 cm long (Grandjean et al. 1998) and thus represent a longer time period than that incorporating the entire rise and decline of such peaks. As discussed previously, the average Hg concentration in the 3-cm segment is a dilution of the peak value. With the moderate increases that might be represented by such peaks, it is unclear to what extent such peaks would have been discernable when averaged over the length of the longer segment.

In the New Zealand study (Kjellström et al. 1986, 1989), hair Hg concentration was the only biomarker of MeHg exposure used. Cord-blood samples were collected but were analyzed only for lead. Hair samples were obtained from all mothers in the original cohort shortly after delivery. The proximal 9 cm of the sample were analyzed for total Hg to give an average Hg concentration over that entire length. Those 9-cm average values were the dose metric used in the dose-response analyses. As discussed previously, the length of hair approximately corresponding to the last 20 days of gestation remained beneath the scalp, and (assuming a hair-growth rate of approximately 1 cm per month) the distal 1 cm of the 9-cm segment analyzed corresponded to the period preceding conception. In addition to the 9-cm sample of hair, when the mothers provided a sufficient quantity of hair, the sample was split, and another bundle of 9-cm length hair was sectioned into nine 1-cm segments. Analyses of the segments were carried out on samples from 47 of the 237 (19.8%) mothers in the second stage of the study (children at 6 years of age, (Kjellström et al. 1989)). A 1-cm segment of hair represents about 30 days of exposure. As discussed previously, a rapid doubling in Hg exposure during that period, such as that resulting from a few successive high-Hg fish meals, for a 1-cm segment would be reflected as a 50% increase compared with neighboring segments with no such peak exposures. Analysis of these segments would likely detect significant peaks in exposure but would not necessarily provide accurate information on the absolute magnitude of those peaks. Peak concentration was defined as the single largest excursion above the overall 9-cm average concentration. On average, the ratios of individual peak concentrations to the average 9-cm concentrations ranged between 1.4 and 1.6, the highest ratio (1.64) being in the group with hair Hg concentrations in the 6-10-ppm range. The group with the highest average hair concentration (at or above 10 ppm) had a ratio of 1.44. The largest individual ratio of peak-to-average concentration was 3.61, and the next largest value was 1.94. Those data do not permit an assessment of the number of peak exposures during gestation, but the range of average ratios is consistent with actual doublings in exposure at least once during gestation. Generalization to the entire cohort is difficult given the relatively small fraction for which segmental data were obtained. However, those data suggests that MeHg exposure in the New Zealand cohort might have been relatively spiky as opposed to constant and regular. It is interesting that the peak exposures were not regularly distributed across the period of gestation. The largest fraction of peak exposures (30%) occurred in the 9-cm segment most distal to the scalp, and 57% of the peak exposures occurred in the three distal-most segments. Only 19% of peak exposures occurred in the three segments most proximal to the scalp. The reason for the disparity is not clear, but it suggests that, at least for this subsample of the cohort, peak exposures might have been less common during the third trimester.

Information on fish consumption was obtained at about the same time as the hair sample through the administration of a questionnaire. The questionnaire requested information on the overall frequency of consumption of fish and shellfish. In addition, more detailed information on consumption frequency and portion size was obtained for specific fresh fish (lemon fish, snapper, gurnard, and "all other fresh fish"), canned fish (tuna, salmon, smoked, and "all other canned fish"), fish products (fish cakes and fish "fingers"), shellfish (oysters, scallops, mussels, and "all other shellfish"), and fried "takeaway" (i.e., fast food, fish-and-chips) fish. Although consumption of shark was not specifically queried, shark was stated to be a common source of takeaway fish. No data were provided on the characteristic Hg concentration in the species identified by the mothers. That lack precludes quantitative estimates of the contribution of individual species and eating patterns to possible peaks in exposure. In terms of overall fish consumption, 1.5% of mothers claimed daily fish consumption during pregnancy, and consumption of fish "a few times a week" during pregnancy was identified by 19% of mothers. The most frequently identified fish-consumption category (32% of mothers) was "once a week." Thus, about 53% of the mothers in the original cohort ate fish at least once per week. Therefore, although this population cannot be considered subsistence fish consumers, it is clear that fish constituted a significant fraction of the overall diet. Furthermore, such an overall consumption pattern, in which fish is eaten frequently, but not continuously, is consistent with the possibility of peak or spiky MeHg exposures. Among the possible choices of fish type, consumption of snapper was most closely correlated with hair Hg concentration. Information on the correlation between consumption and hair Hg concentration is not provided for any other species. It appears that additional information on those mothers who were likely to have experienced short-term peak exposures can be recovered from the questionnaire data, particularly from more detailed consideration of the frequency of takeaway fish. Further analysis of these data, therefore, might provide some indication of the influence of peak exposures on MeHg dose-response relationships for neurodevelopmental effects.

The frequency of overall fish consumption was used in the first stage of the study to screen for both the high Hg-dose group (consumption of fish more than three times per week), and the reference group (one or less than one fish meal per week). Ultimately, however, the high Hg-dose group for the first and second stages of the study was selected from among frequent consumers on the basis of hair Hg concentrations of more than 6 ppm. In the second stage of the study, each child in the high Hg-dose group was matched with three control children on the basis of low hair Hg concentration. One of these control children was additionally selected on the basis of frequent (more than three times per week) maternal fish consumption during pregnancy.

Given the differences, uncertainties, and limitations of the exposure-assessment approaches used in the Seychelles, Faroe Islands, and New Zealand studies, none of the approaches can be identified as better or more relevant. It is clear, however, that each of the approaches supplied different, and not necessarily comparable pictures of exposure and dose. Grandjean et al. (1999) noted in the Faroe Islands study that cord-blood MeHg appeared to better predict deficits in cognitive functions (language, attention, and memory), and maternal-hair MeHg appeared to better predict deficits in fine-motor function. The authors attributed those qualitative differences to the different periods of development reflected by each of the measurements. That conclusion is consistent with the idea that discrete windows of vulnerability in the developmental toxicity of MeHg are differentially represented by hair- and cord-blood Hg measurements. However, the lack of uniformity in the lengths of the hair segments analyzed in the Faroe Islands study (Grandjean et al. 1999) make a clear interpretation of such differences somewhat problematic. Therefore, the uncertainties and limitations in the various biomarkers that are used for MeHg exposure assessment could result in exposure misclassification in the dose-response assessment.

Misclassification of exposure in these studies could take several forms. Those include incorrectly considering exposures that occurred during developmental periods during which there is little or no vulnerability of the observed developmental endpoints to MeHg; failing to identify peak concentrations that might be more toxicologically relevant than the measured average concentrations; and using portions of hair with Hg concentrations that accumulated before or after pregnancy. Generally, exposure misclassification biases to the null—that is, use of an incorrect exposure level in a regression analyses of outcome data leads to decreased power to detect a real effect. Thus, the likely implication of the uncertainties and limitations in the dose metrics used in the Seychelles and the Faroe Islands studies is that the probability of observing true associations of dose and response will be reduced. In addition, the magnitude of those observed associations may be underestimated. Therefore, the existence of uncertainties and limitations notwithstanding, those statistically significant dose-response associations observed with any of the dose metrics are likely to reflect (perhaps indirectly) true associations (if other sources of bias have been adequately addressed). Failure to observe statistically significant dose-response associations could well be due to exposure misclassification resulting from one or more of the uncertainties and limitations discussed above.

SUMMARY AND CONCLUSIONS

- Duplicate diet data can potentially provide accurate data on MeHg intake, although interindividual pharmacokinetic variability creates uncertainty in the use of such data to estimate the dose to the fetal brain. The collection of duplicate dietary data places demands on study participants. This approach is, therefore, generally limited to short periods of observation that might not capture critical intake variability in populations with high intraindividual variability in intake of fish.

- Retrospective dietary data (diary and recall) are relatively simple to collect, but diary-based data are subject to participant errors in species identification, portion estimation, and assignment of MeHg concentration by species. The number of days of dietary-intake data collected needs to be long enough to characterize adequately the frequent fish consumer and to differentiate the levels of less frequent consumption. Recall-based data are additionally subject to recall errors. Such data might be useful in stratifying exposure and in temporal calibration of hair strands.
- On the other hand, prospective data on all sources of Hg exposure, such as vaccines and dental amalgams and, in particular, dietary intakes of MeHg are essential to understanding the effects of environmental Hg exposures on any outcomes. Quantitative dietary intake data on intakes of all marine food sources can and should be collected in any serious study of this contaminant. Such data are essential for quantifying exposures, separating out the effect modifiers that account for the differences between exposures and target tissue concentrations. Intake data are also essential for identifying possible confounding factors, such as other contaminants or nutrients that are abundant in some of these food sources but not in others.
- Cord-blood Hg concentration is closely linked kinetically to the fetal-brain compartment and should correlate closely with the concentrations at the target organ near the time of delivery. Cord-blood Hg is less closely linked to the ingested dose. That separation can introduce uncertainty into the back-calculation of a reference dose from cord-blood-based dose-response data.
- Cord-blood Hg measurement cannot show temporal variability in exposure. It can provide data on a limited portion of gestation whose duration is somewhat uncertain but which occurs late in gestation. That portion of gestation might not correspond to the periods of greatest fetal sensitivity to MeHg neurotoxicity.

- Maternal-hair Hg concentration is less closely linked kinetically to the fetal-brain compartment than is cord-blood Hg concentration, and the kinetic distance between the maternal-hair and fetal-brain compartments might be a significant source of statistical error in dose-response assessment.
- Hair Hg measurement can potentially provide a range of dose metrics. Analysis of longer strands corresponding to all or part of gestation will provide average exposure data but no information on temporal variability in exposure. Segmental analysis can isolate specific periods of gestation, but peak exposures might be inadequately represented. Continuous single-strand analysis is a powerful technique that can recapitulate MeHg exposure during the entire period of gestation with accurate representation of peak exposures. This approach presents a range of dose metrics that can be investigated in assessing dose response.
- Because of intraindividual and interindividual variability in hair-growth rates, attempts to identify hair Hg concentrations corresponding to specific time periods during gestation might be subject to significant error which can result in exposure misclassification in dose-response assessment. The temporal calibration of Hg measurements along a hair strand can be aided by consideration of corresponding dietary intake data for Hg.
- Each of the dose metrics—dietary records, cord blood, and hair—provides different exposure information. Use of data from two or more of these metrics will increase the likelihood of uncovering a true dose-response relationship.
- In the Seychelles studies, dose was estimated from the average Hg concentration in a length of hair assumed to represent the first 8 months of pregnancy. That approach precluded observing any intraindividual variability in exposure over the course of gestation.
- Fish-consumption data for the Seychelles cohort established the generally high level of fish consumption but could not provide any data on intraindividual variability in exposure.
- In the Faroe Islands studies, dose was estimated from cord-blood and single-sample maternal-hair Hg concentration. The cord-blood Hg data cover exposures over an indeterminate period late in gestation. The hair Hg samples appear not to have been of uniform length and therefore do not necessarily reflect comparable periods

of gestation. Those differences in length are relevant only if there is significant variability in MeHg exposure during gestation. Neither of these metrics has the ability to show intraindividual variability over the course of gestation.

- Fish-consumption data for the Faroe cohort indicated a high rate of consumption of fish with low Hg concentrations, and less frequent consumption of pilot whale containing high concentrations of MeHg. Such a diet suggests a pattern of peaking exposures. Exposure modeling suggests that as reflected by accumulation in hair such peaks might represent a moderate increase above baseline concentrations.
- The uncertainties and limitations in exposure assessment in these studies can result in exposure misclassification, which will lessen the ability to detect significant dose-response associations and might result in inaccuracies in the derivation of dose-response relationships.
- If exposure misclassification occurred in the studies of MeHg, such misclassification would tend to obscure any true effect. Therefore, statistically significant dose-response associations are likely to reflect true dose-response relationships, assuming that other sources of bias are adequately addressed.
- Dose-response assessments using either cord-blood or maternal-hair Hg concentrations are adequate to support the derivation of an RfD.

RECOMMENDATIONS

- Quantitative dietary intake data on patterns of consumption of the primary sources of MeHg including all marine food sources, should be collected in all prospective studies of MeHg exposure. Estimates of exposures will improve dose-response analyses that have implications for regulatory purposes.
- In future studies, data on maternal fish intake by species and by meal should be collected along with Hg biomarker data. Those data should be used to provide estimates of temporal variability in MeHg intake during pregnancy.
- Future studies should collect data on maternal-hair, blood, and

cord-blood Hg concentrations. All three dose metrics should be considered in attempting to identify dose-response relationships.

- Data are needed that reliably measure both Hg intake and biomarkers of Hg exposure to clarify the relationship between the different dose metrics. NHANES IV data should be examined when it becomes available to determine if it satisfies those needs.
- To detect exposure variability, archived hair strands from both the Seychelles and the Faroe Islands studies should be analyzed by continuous single-strand XRF analysis. The possible dose metrics that can be derived from XRF analysis should be examined in the dose-response assessment. Such considerations should also be addressed in future studies.

REFERENCES

Airey, D. 1983. Total mercury concentrations in human hair from 13 countries in relation to fish consumption and location. Sci. Total Environ. 31(2):157-180.

ATSDR (Agency for Toxic Substances and Disease Registry). 1999. Toxicological Profile for Mercury (Update). U.S. Department of Health and Human Services, Agency for Toxic Substances and Disease Registry. Atlanta, GA.

Batista, J., M. Schuhmacher, J.L. Domingo, and J. Corbella. 1996. Mercury in hair for a child population from Tarragona Province, Spain. Sci. Total Environ. 193(2):143-148.

Berry, M.R. 1997. Advances in dietary exposure research at the United States Environmental Protection Agency-National Exposure Research Laboratory. J. Expo. Anal. Environ. Epidemiol. 7(1):3-16.

Boischio, A.A.P., and E. Cernichiari. 1998. Longitudinal hair mercury concentration in riverside mothers along the Upper Madeira river (Brazil). Environ. Res. 77(2):79-83.

Bruhn, C.G., A.A. Rodríguez, C. Barrios, V.H. Jaramillo, J. Becerra, U. Gonzáles, N.T. Gras, O. Reyes, and Seremi-Salud. 1994. Determination of total mercury in scalp hair of pregnant women resident in fishing villages in the Eighth Region of Chile. J. Trace Elem. Electrolytes Health Dis. 8(2):79-86.

Brune, D., G.F. Nordberg, O. Vesterberg, L. Gerhardsson, and P.O. Wester. 1991. A review of normal concentration of mercury in human blood. Sci. Total Environ. 100(spec No):235-282.

Buzina, R., P. Stegnar, S.K. Buzina-Suboticanec, M. Horvat, I. Petric, and T.M.

Farley. 1995. Dietary mercury intake and human exposure in an Adriatic population. Sci. Total Environ. 170(3):199-208.

Castilhos, Z.C., E.D. Bidone, and L.D. Lacerda. 1998. Increase of the background human exposure to mercury through fish consumption due to gold mining at the Tapajos River region, Para State, Amazon. Bull. Environ. Contam. Toxicol. 61(2):202-209.

Cernichiari, E., T.Y. Toribara, L. Liang, D.O. Marsh, M.W. Berlin, G.J. Myers, C. Cox, C.F. Shamlaye, O. Choisy, P. Davidson, et.al. 1995. The biological monitoring of mercury in the Seychelles study. Neurotoxicology 16(4):613-628.

Cernichiari, E., R. Brewer, G.J. Myers, D.O. Marsh, L.W. Lapham, C. Cox, C.F. Shamlaye, M. Berlin, P.W. Davidson, and T.W. Clarkson. 1995a. Monitoring methylmercury during pregnancy: Maternal hair predicts fetal brain exposure. Neurotoxicology 16(4):705-710.

Chan, H.M., P.R. Berti, O. Receveur, and H.V. Kuhnlein. 1997. Evaluation of the population distribution of dietary contaminant exposure in an Arctic population using Monte Carlo statistics. Environ. Health Perspect. 105(3):316-321.

Choi, B.H. 1989. The effects of methylmercury on the developing brain. Prog. Neurobiol. 32(6):447-470.

Clewell, H.J., J.M. Gearhart, R. Gentry, T.R. Covington, C.B. VanLandingham, K.S. Crump, and A.M. Shipp. 1999. Evaluation of the uncertainty in an oral Reference Dose for methylmercury due to interindividual variability in pharmacokinetics. Risk Anal. 19(4):547-558.

Cox, C., T.W. Clarkson, D.O. Marsh, L. Amin-Zaki, S. Tikriti, and G.G. Myers. 1989. Dose-response analysis of infants prenatally exposed to methyl mercury: An application of a single compartment model to single-strand hair analysis. Environ. Res. 49(2):318-322.

Davidson, P.W., G.J. Myers, C. Cox, C. Axtell, C. Shamlaye, J. Sloane-Reeves, E. Cernichiari, L. Needham, A. Choi, Y. Wang, M. Berlin, and T.W. Clarkson. 1998. Effects of prenatal and postnatal methylmercury exposure from fish consumption on neurodevelopment: Outcomes at 66 months of age in the Seychelles Child Development Study. JAMA 280(8):701-707.

Davidson, P.W., G.J. Myers, C. Cox, E. Cernichiari, T.W. Clarkson, and C. Shamlaye. 1999. Effects of methylmercury exposure on neurodevelopment. [Letter]. JAMA 281(10):896-897.

Dennis, C.A., and F. Fehr. 1975. Mercury levels in whole blood of Saskatchewan residents. Sci. Total Environ. 3(3):267-274.

EPA (U.S. Environmental Protection Agency). 1997. Mercury Study Report to Congress. Vol. IV: An Assessment of Exposure to Mercury in the United

States. EPA-452/R-97-006. U.S. Environmental Protection Agency, Office of Air Quality Planning and Standards, and Office of Research and Development.

FDA (U.S. Food and Drug Administration). 1992. Compilation of methylmercury results by species for FY=91 and FY=92. Office of Seafood, Washington, DC.

Fujita, M., and E. Takabatake. 1977. Mercury levels in human maternal and neonatal blood, hair and milk. Bull. Environ. Contam. Toxicol. 18(2):205-209.

Gaggi, C., F. Zino, M. Duccini, and A. Renzoni. 1996. Levels of mercury in scalp hair of fishermen and their families from Camara de Lobos-Madeira (Portugal): A preliminary study. Bull. Environ. Contam. Toxicol. 56(6):860-865.

Gilbert, S.G., and K.S. Grant-Webster. 1995. Neurobehavioral effects of developmental methylmercury exposure. Environ. Health Perspect. 103(Suppl. 6):135-142.

Ginsberg, G.L., and B.F. Toal. 2000. Development of a single meal fish consumption advisory for methyl mercury. Risk Anal. 20(1):41-48.

Giovanoli-Jakubczak, T., and G.G. Berg. 1974. Measurement of mercury in human hair. Arch. Environ. Health 28(3):139-144.

Girard, M., and C. Dumont. 1995. Exposure of James Bay Cree to methylmercury during pregnancy for the years 1983-91. Water Air Soil Pollut. 80:13-19.

Gowdy, J.M., R. Yates, F.X. Demers, and S.C. Woodward. 1977. Blood mercury concentration in an urban population. Sci. Total Environ. 8(3):247-251.

Grandjean, P., and P. Weihe. 1993. Neurobehavioral effects of interuterine mercury exposure; potential sources of bias. Environ. Res. 61(1):176-183.

Grandjean, P., E. Budtz-Jørgensen, R.F. White, P. Weihe, F. Debes, and N. Keiding. 1999. Methylmercury exposure biomarkers as indicators of neurotoxicity in children aged 7 years. Am. J. Epidemiol. 150(3):301-305.

Grandjean, P., P. Weihe, P.J. Jørgensen, T. Clarkson, E. Cernichiari, and T. Viderø. 1992. Impact of maternal seafood diet on fetal exposure to mercury, selenium, and lead. Arch. Environ. Health 47(3):185-195.

Grandjean, P., P. Weihe, L.R. Needham, V. W. Burse, D.G. Patterson, Jr., E.J. Sampson, P.J. Jørgensen, and M. Vahter. 1995. Relation of a seafood diet to mercury, selenium, arsenic, and polychlorinated biphenyl and other organochlorine concentrations in human milk. Environ. Res. 71(1):29-38.

Grandjean, P., P. Weihe, R.F. White, N. Keiding, E., Budtz-Jørgensen, K. Murato, and L. Needham. 1998. Prenatal exposure to methylmercury in the Faroe Islands and neurobehavioral performance at age seven years. Response to workgroup questions for presentation on 18-20 November, 1998. In: Scientific Issues Relevant to Assessment of Health Effects from Exposure to Methylmercury. Appendix II-B.- Faroe Islands Studies. National Institute for Environmental Health Sciences. [Online]. Available: http://ntp-server.niehs.nih.gov/Main_Pages/PUBS/MethMercWkshpRpt. html

Hall, R.A., E.G. Zook, and G.M. Meaburn. 1978. Survey of Trace Elements in the Fisheries Resource. Technical Report NMFS SSRF-721. National Oceanic and Atmospheric Administration. National Marine Fisheries Service. 313pp.

Haxton, J., D.G. Lindsay, J.S. Hislop, L. Salmon, E.J. Dixon, W.H. Evans, J.R. Reid, C.J. Hewitt, and D.F. Jeffries. 1979. Duplicate diet study on fishing communities in the United Kingdom: Mercury exposure in a "critical group". Environ. Res. 18(2):351-368.

Hecker, L.H., H.E. Allen, and B.D. Dinman. 1974. Heavy metals in acculturated and unacculturated population. Arch. Environ. Health 29(4):181-185.

Hislop, J.S., T.R. Collier, G.P. White, D.T. Khathing, and E. French. 1983. The use of keratinised tissues to monitor the detailed exposure of man to methylmercury from fish. Pp. 145-148 in Clinical Toxicology and Clinical Chemistry of Metals, S.S. Brown, ed. New York: Academic Press.

Holsbeek, I., H.K. Das, and C.R. Joiris. 1996. Mercury in human hair and relation to fish consumption in Bangladesh. Sci. Total Environ. 186(3):181-188.

Humphrey, H.E.B. 1975. Mercury concentrations in humans and consumption of fish containing methylmercury. Pp. 33 in Heavy Metals in the Aquatic Environment, Proceedings of the International Conference held in Nashville, TN, December, 1973, P.A. Krenkel, ed. Oxford: Pergamon Press.

Hopps, H.C. 1977. The biologic basis for using hair and nail for analyses of trace elements. Sci. Total Environ. 7(1):71-89.

IPCS (International Programme on Chemical Safety). 1990. Environmental Health Criteria Document 101 - Methylmercury. Geneva: World Health Organization.

IPCS (International Programme on Chemical Safety). 1991. Environmental Health Criteria Document 118 -Inorganic Mercury. Geneva: World Health Organization.

Jacobs, H.L., H.D. Kahn, K.S. Stralka, and D.B. Phan. 1998. Estimates of per capita fish consumption in the U.S. based on the continuing survey of food intake by individuals (CSFII). Risk Anal. 18(3):283-291.

Katz, S., and A. Chatt. 1988. Pp. 6-12 in Hair Analysis: Application in the Biomedical and Environmental Sciences. New York: VCH Publishers.

Kingman, A., T. Albertini, and L.J. Brown. 1998. Mercury concentrations in urine and whole blood associated with amalgam exposure in a U.S. military population. J. Dent. Res. 77(3):461-471.

Kjellström, T., P. Kennedy, S. Wallis, and C. Mantell. 1986. Physical and

Mental Development of Children with Prenatal Exposure to Mercury from Fish. Stage I: Preliminary Tests at Age 4. National Swedish Environmental Protection Board Report 3080. Solna, Sweden.

Kjellström, T., P. Kennedy, S. Wallis, A. Stewart, L. Friberg, B. Lind, T. Wutherspoon, and C. Mantell. 1989. Physical and Mental Development of Children with Prenatal Exposure to Mercury from Fish. National Swedish Environmental Protection Board Report No. 3642.

Kuhnert, P.M, B.R. Kuhnert, and P. Erhard. 1981. Comparison of mercury levels in maternal blood, fetal cord blood, and placental tissues. Am. J. Obstet. Gynecol. 139(2):209-213.

Kuntz, W.D., R.M. Pitkin, A.W. Bostrom, and M.S. Hughes. 1982. Maternal and cord blood background mercury levels: A longitudinal surveillance. Am. J. Obstet. Gynecol. 143(4):440-443.

López-Artiguez, M., A. Grilo, D. Martinez, M.L. Soria, L. Nuñez, A. Ruano, E. Moreno, F. García-Fuente, and M. Repetto. 1994. Mercury and methylmercury in population risk groups on the Atlantic coast of southern Spain. Arch. Environ. Contam. Toxicol. 27(3):415-419.

Lubin, B.H. 1987. Reference values in infancy and childhood. Pp. 1683 in Hematology of Infancy and Childhood, 3rd Ed., D.G. Nathan, and F.A. Oski, eds. Philadelphia: W.B. Saunders.

MacIntosh, D.L, P.L. Williams, D.J. Hunter, L.A. Sampson, S.C. Morris, W.C. Willett, and E.B. Rimm. 1997. Evaluation of a food frequency questionnaire-food composition approach for estimating dietary intake of inorganic arsenic and methylmercury. Cancer Epidemiol. Biomarkers Prev. 6(12):1043-1050.

Mahaffey, K.R., and D. Mergler. 1998. Blood levels of total and organic mercury of residents of the upper St. Lawrence River basin, Quebec: Association with age, gender, and fish consumption. Environ. Res. 77(2):104-114.

Marsh, D.O., T.W. Clarkson, C. Cox, G.J. Myers, L. Amin-Zaki, and S. Al-Tikriti. 1987. Fetal methylmercury poisoning. Relationship between concentration in single strands of maternal hair and child effects. Arch. Neurol. 44(10):1017-1022.

Nixon, D.E., G.V. Mussmann, and T.P. Moyer. 1996. Inorganic, organic and total mercury in blood and urine: Cold vapor analysis with automated flow injection sample delivery. J. Anal. Toxicol. 20(1):17-22.

Oskarsson, A., B.J. Lagerkvist, B. Ohlin, and K Lundberg. 1994. Mercury levels in the hair of pregnant women in a polluted area in Sweden. Sci. Total Environ. 151 (1):29-35.

Oskarsson, A., A. Schütz, S. Skerfving, I.P. Hallén, B. Ohlin, and B.J. Lagerkvist. 1996. Total and inorganic mercury in breast milk and blood in relation to fish consumption and amalgam fillings in lactating women. Arch. Environ. Health 51(3):234-241.

Pallotti, G., B. Bencivenga, and T. Simonetti. 1979. Total mercury levels in whole blood, hair and fingernails for a population group from Rome and its surroundings. Sci. Total Environ. 11(1):69-72.

Pecoraro, V., J.M. Barman, and I. Astore. 1967. The normal trichogram of pregnant women. Pp. 203-210 in Advances in Biology of Skin, Vol. IX, Hair Growth. Proceedings of the University of Oregon Medical School Symposium on the Biology of Skin. Pub. No. 277, Oregon Regional Primate Research Center, W. Montagna, and R.L. Dobson, eds. Oxford: Pergamon Press.

Pitkin, R.M, J.A Bahns, L.J. Filer, Jr., and W.A. Reynolds. 1976. Mercury in human maternal and cord blood, placenta and milk. Proc. Soc. Exp. Biol. Med. 151(3): 565-567.

Ponce, R.A., S.M. Bartell, T.J. Kavanagh, J.S. Woods, W.C. Griffith, R.C. Lee, T.K. Takaro, and E.M. Faustman. 1998. Uncertainty analysis methods for comparing predictive models and biomarkers: A case study of dietary methyl mercury exposure. [Review]. Regul. Toxicol. Pharmacol. 28(2):96-105.

Schuhmacher, M., J. Batiste, M.A. Bosque, K.L. Domingo, and J. Corbella. 1994. Mercury concentrations in marine species from the coastal area of Tarragona Province, Spain. Dietary intake of mercury through fish and seafood consumption. Sci. Total Environ. 156(3):269-273.

Shamlaye, C.F., D.O. Marsh, G.J. Myers, C. Cox, P.W. Davidson, O. Choisy, E. Cernichiari, A. Choi, M.A. Tanner, and T.W. Clarkson. 1995. The Seychelles child development study on neurodevelopmental outcomes in children following in utero exposure to methylmercury from a maternal fish diet: background and demographics. Neurotoxicology 16(4):597-612.

Sherlock, J.C., D.G. Lindsay, J.E. Hislop, W.H. Evans, and T.R. Collier. 1982. Duplicate diet study on mercury intake by fish consumers in the United Kingdom. Arch. Environ. Health 37(5):271-278.

Sikorski, R., T. Poszkowski, P. Slawinski, J. Szkoda, J. Zmudzki, and S. Skawinski. 1989. The intrapartum content of toxic metals in maternal blood and umbilical cord blood. Ginekol. Pol. 60(3):151-155.

Skerfving, S. 1988. Mercury in women exposed to methylmercury through fish consumption and in their newborn babies and breast milk. Bull. Environ. Contam. Toxicol. 41(4):475-482.

Skerfving, S. 1991. Exposure to mercury in the population. Pp. 411-425 in Advances in Mercury Toxicology, T. Suzuki, N. Imura, and T.W. Clarkson, eds. New York: Plenum Press.

Smith, J.C., P.V. Allen, and R. Von Burg. 1997. Hair methylmercury levels in U.S. women. Arch. Environ. Health 52(6):476-480.

Stern, A.H. 1993. Re-evaluation of the reference dose for methylmercury and assessment of current exposure levels. Risk Anal. 13(3):355-364.

Stern, A.H. 1997. Estimation of the interindividual variability in the one-compartment pharmacokinetic model for methylmercury: implications for the derivation of a reference dose. Regul. Toxicol. Pharmacol. 25(3):277-288.

Stern, A.H., L.R. Korn, and B.E. Ruppel. 1996. Estimation of fish consumption and methylmercury intake in the New Jersey population. J. Expo. Anal. Environ. Epidemiol. 6(4):503-525.

Stern, A.H., and M. Gochfeld. 1999. Effects of methylmercury exposure on neurodevelopment. [Letter]. JAMA 281(10):896-897.

Stern, A.H., M. Gochfeld, C. Weisel, and J. Burger. 2000. Mercury and methylmercury exposure in the New Jersey pregnant population. Arch. Environ. Health. In press.

Thomas, K.W., L.S. Sheldon, E.D. Pellizzari, R.W. Handy, J.M. Roberds, and M.R. Berry. 1997. Testing duplicate diet sample collection methods for measuring personal dietary exposures to chemical contaminants. J. Expo. Anal. Environ. Epidemiol. 7(1):17-36.

Vural, N., and H. Ünlü. 1996. Methylmercury in hair of fisherman from Turkish coasts. Bull. Environ. Contam. Toxicol. 57:315-320.

Whipple, C., L. Levin, and C. Seigneur. 1996. Sensitivity of mercury exposure calculations to physical and biological parameters of fish consumption. Presented at Fourth International Conference on Mercury as a Global Pollutant, Hamburg, August 4-8.

Yess, N.J. 1993. U.S. Food and Drug Administration survey of methyl mercury in canned tuna. J. AOAC Int. 76(1):36-38.

5

HEALTH EFFECTS OF METHYLMERCURY

THIS chapter begins with a brief review of the carcinogenicity of MeHg and its immunological, reproductive, renal, cardiovascular and hematopoietic toxicity. Because the central nervous system is widely viewed as the organ system most sensitive to MeHg, the remainder of this chapter focuses on the adverse effects of MeHg on neurological function. Neurological effects in infants, children, and adults are discussed. Studies carried out in populations exposed to high concentrations of MeHg are described, followed by a discussion of epidemiological data on populations exposed chronically to low concentrations of MeHg. Animal data following in utero, early postnatal, and adult exposure are also discussed.

The information available on the human health effects of MeHg are derived from studies of various designs. Each type of design has strengths and weaknesses and might be the most appropriate choice for a given set of circumstances. The methodology, strengths, and weaknesses of environmental epidemiological studies have been discussed in previous NRC reports (NRC 1991, 1997). The data on the Minamata and Iraqi episodes, the collection of which were initiated in response to the occurrence of recognizable illness in the population, are derived from case reports, descriptive studies of convenience samples, and ecological studies of rates. A major advantage of such studies is that the end points assessed are often of clear clinical significance. The inferences permitted from such studies, as described in greater detail in the following sec-tions, can be limited by methodological weaknesses, such as the absence of detailed information on the sampling frame or referral patterns that generated the study sample, the degree to which the study sample is representative of the population from which it was drawn, exposure histories of the subjects, detailed assessments of health status, and the nature of severity of possible confounding biases.

Case-control studies, in which the exposure status (or history) of individuals with a certain health outcome (case) is compared with the exposure status of individuals without the health outcome (controls), can provide a much stronger basis for drawing inferences about exposure-disease associations. Among the challenges of such studies, however, are assembling a representative group of cases and a comparable group of controls, collecting adequate information on critical aspects of exposure history (which, in the case of long-latency diseases, might mean exposures that occurred decades before), and identifying the critical potential confounding biases. A case-control design, however, might be the only efficient way to study rare health outcomes.

Cohort designs (e.g., cross-sectional, retrospective, and prospective) provide a number of advantages. Instead of being selected on the basis of outcome status, as in case-control studies, study subjects are either randomly selected from the target population or selected on the basis of particular exposure characteristics (e.g., over-sampling of extremes of exposure distribution). The former strategy might be used if the goal is to enhance the generalizability of the study inferences to the target population, and the latter might be used if the goal is to estimate, with the greatest precision, the nature of the dose-response relationship within a certain region of the dose distribution. Another advantage of a cohort design is that multiple health outcomes can be measured and related to the index of exposure. A cohort study that incorporates prospective assessments of the study sample generally provides opportunities to assemble more-comprehensive exposure histories of the study subjects and to examine the natural history of a dose-response relationship, including factors that modify risk. As with all epidemiological studies, the methodological challenges of cohort studies include accurate classification of exposure and outcome status and the assessment and control of confounding bias.

CARCINOGENICITY

None of the epidemiological studies found an association between Hg exposure and overall cancer rates; however, two studies found an association between exposure to Hg and acute leukemia. The interpretation of those results is difficult due to the small study populations, the problem of assessing historical exposures to Hg, and the inability of investigators to control for other risk factors. In animals, chronic exposure to MeHg increased the incidence of renal tumors in male mice in some of the studies; however, the increase was observed only at doses that were toxic to the kidneys. Therefore, the tumorigenic effect is thought to be secondary to cell damage and repair. MeHg did not cause tumors in female mice or in rats of either sex. Therefore, in the absence of a tumor initiator, long-term exposure to subtoxic doses of MeHg does not appear to increase tumor formation.

On the basis of the available human and animal data, the International Agency for Research on Cancer (IARC) and the U.S. Environmental Protection Agency (EPA) have classified MeHg as a "possible" human carcinogen.

Human Studies

Four epidemiological studies examined the effect of Hg exposure on cancer incidence or cancer death rate. Those studies are summarized in Table 5-1. Tamashiro et al. (1984) carried out a cohort study that evaluated the causes of death of 334 individuals who had survived Minamata disease (MD) and died between 1970 and 1980. Control cases were selected from deaths that occurred in the same city or town as the MD cases and were matched for sex, age and year of death. No significant difference in cancer death rates was observed between the subjects and the controls, suggesting that the risk of dying from cancer was not correlated with patient history of MeHg poisoning. Specific types of cancer, however, were not evaluated.

Tamashiro et al. (1986) compared the death rates among residents of the Fukuro and Tsukinoura districts with those of age-matched residents of Minamata City. Residents of the two districts were assumed to have a higher intake of local seafood and higher Hg exposure than residents

TABLE 5-1 Summary of Cancer Studies in Humans

Type of Study	Size of Study	Finding	Reference
Retrospective cohort	334 deaths in high-exposure cohort; 668 in low-exposure cohort	No increase in cancer death rate; site-specific rates not analyzed	Tamashiro et al. 1984
Retrospective cohort	416 deaths in high-exposure cohort; 2,325 deaths in low-exposure cohort	Increased liver-cancer death rate among males in high-exposure cohort	Tamashiro et al. 1986
Case-control study of hair Hg concentrations in leukemia patients	47 cases; 79 controls	Increased hair Hg concentrations in acute leukemia patients	Janicki et al. 1987
Retrospective cohort study of Minamata-disease (MD) survivors	1,351 MD survivors; 5,667 referents	Increased leukemia death rate among MD survivors; relative risk, 8.35	Kinjo et al. 1996

of Minimata City. No statistically significant increase in the overall cancer mortality was observed. However, an increase in liver- cancer death rates was observed among males who resided in the areas thought to have high Hg exposure (standardized mortality ratio (SMR[1]), 250.5; 95% confidence interval (CI), 133.4-428.4). Males also had significantly higher mortality due to chronic liver disease and cirrhosis in those areas than in Minamata City. The investigators indicated that the increases could not be attributed solely to MeHg, because the alcohol consumption rates and the prevalence of hepatitis B infection were higher in the Fukuro and Tsukinoura districts than in Minamata City. The study is also limited by its failure to fully characterize Hg concentrations in subjects in each cohort.

In a case-control study in Poland, Janicki et al. (1987) found a statisti-

[1]The SMR is the ratio of the number of deaths observed in a study group divided by the number expected (based on age- or sex-specific rates in the general population) and multiplied by 100. An SMR greater than 100 indicates that the death rate was higher than would be expected.

cally significant increase in the Hg content in hair collected from 47 patients with leukemia compared with 52 healthy unrelated subjects (mean 1.24 versus 0.49 ppm). The Hg content in hair from a subgroup of 19 leukemia patients was also significantly greater than that from 52 healthy relatives who had shared the same home for at least 3 years (0.69 versus 0.43 ppm). When those data were analyzed for specific types of leukemia, only patients with acute leukemia had significantly higher hair Hg concentrations. No significant difference was seen in the Hg content in hair collected from nine patients with chronic granulocytic leukemia or from 15 patients with chronic lymphocytic leukemia compared with the healthy unrelated subjects. The study is limited by the small study population, inadequate description of case and control populations, uncertainty about the source of Hg exposure, and lack of adjustment for other leukemia risk factors. In addition, all the hair Hg concentrations were within normal limits.

Kinjo et al. (1996) compared cancer death rates for a cohort (1,351 cases) of MD survivors with those of a referent population (5,667 subjects) who lived in the same region of Japan and consumed fish daily. After adjusting for age, gender, and length of follow-up period, they found no excess relative risk (RR) for overall mortality, all cancer deaths combined, or all noncancer deaths combined. Analysis of site-specific cancers found that Minamata survivors were less likely to die of stomach cancer than the referent population (RR, 0.49; 95% confidence interval (CI), 0.26-0.94). However, on the basis of five observed deaths, survivors were eight times more likely than the referent population to have died from leukemia (RR, 8.35; 95% CI, 1.61-43.3).

Animal Studies

The carcinogenic potential of MeHg was examined in several chronic exposure animal studies. Those studies are summarized in Table 5-2.

Newberne et al. (1972) carried out a 2-year multigeneration study in which Sprague-Dawley rats (30 per sex) were fed diets with MeHg doses of 0 or 0.008 mg/kg per day. Tumor incidence was similar in both groups; however, the maximum tolerated dose (MTD) was not achieved.

A 2-year feeding study conducted by Verschuuren et al. (1976) also failed to provide evidence of carcinogenic effects. Rats (25 per sex per

TABLE 5-2 Summary of Cancer Studies in Animals

Animal	Dose (mg/kg/d)	Tumor response	Study Duration (wk)	Reference
Sprague-Dawley rat	0, 0.008	None	104	Newberne et al. 1972
Rats, unspecified strain	0, 0.004, 0.02, 0.1	None	104	Verschuuren et al. 1976
Sprague-Dawley rats			130	Mitsumori et al. 1983, 1984
Males	0, 0.01, 0.05, 0.28	None		
Females	0, 0.01, 0.06, 0.34	None		
Swiss Albino mice	0, 0.19, 0.95[a]	None	Weaning to death	Schroeder and Mitchener 1975
ICR mice			78	Mitsumori et al. 1981
Males	0, 1.6, 3.1	0/37, 11/16, NA		
Females	0, 1.6, 3.1	0, 0, NA		
Swiss mice	0, 0.03, 0.07, 0.27	Increased tumor response to urethane	15	Blakley 1984
ICR mice			104	Hirano et al. 1986
Male	0, 0.03, 0.15, 0.73	1/32, 0/25,		
Female	0, 0.02, 0.11, 0.60	0/29, 13/26 None in any group		
$B6C3F_1$ mice	0, 0.03, 0.14, 0.69	0/60, 0/60,	104	Mitsumori et al. 1990
Male	0, 0.03, 0.13, 0.60	0/60, 13/60		
Female		0/60, 0/60, 0/60, 1/60		

[a] 0.95 mg/kg per day for 70 days and then 0.19 mg/kg per day thereafter due to high mortality at 0.95 mg/kg per day.

Abbreviation: NA, not available.

group) were exposed to MeHg chloride at 0, 0.004, 0.020, or 0.10 mg/kg per day for 2 years. Survival decreased in the mid- and high-dose groups, and kidney weights increased in the high-dose group. However, tumors occurred at similar rates in all the groups.

Mitsumori et al. (1983, 1984) also exposed Sprague-Dawley rats to MeHg chloride in feed (males, 0, 0.011, 0.05, or 0.28 mg/kg per day;

females, 0, 0.014, 0.064, or 0.34 mg/kg per day) for up to 130 weeks. Effects were seen in the central nervous system, kidney, arterial wall, and spleen. The MTD was achieved in males in the mid-dose group and exceeded in males and females in the high-dose group. No increase in tumor incidence was observed.

A lifetime study conducted in Swiss albino mice failed to detect a tumorigenic response (Schroeder and Mitchener 1975). Groups of mice (54 per sex per group) were exposed from weaning until death to methylmercuric acetate in drinking water at two doses. The low-dose group received 1 ppm (0.19 mg/kg per day) and the high dose group received 5 ppm (0.95 mg/kg/day) for the first 70 days and then 1 ppm thereafter due to high mortality at the higher dose. Although no increase in tumors was noted, interpretation of the study is limited because of cessation of the high-dose exposure and failure to conduct complete histological examinations.

The incidence of renal tumors was increased in males in a study of ICR mice (60 per sex) fed diets containing MeHg chloride (0, 1.6, or 3.1 mg/kg per day) for 78 weeks (Mitsumori et al. 1981). The majority of mice in the high-dose group died by week 26 of the study. Males in the low-dose group had significantly higher numbers of renal epithelial adenocarcinomas (0 of 37 in control group; 11 of 16 in low-dose group) and renal adenomas (1 of 37 in control group; 5 of 16 in low-dose group) than controls. No renal tumors were observed in females in any group.

Blakley (1984) exposed female Swiss mice to MeHg chloride (approximately 0, 0.03, 0.07, or 0.27 mg/kg per day) in drinking water for 15 weeks. After 3 weeks of exposure, mice were given urethane in a single intraperitonal dose of 1.5 mg/kg. No more than one tumor per mouse was seen in the absence of urethane. With urethane, a statistically significant trend was seen for an increase in the size (0.7, 0.73, 0.76, and 0.76 millimeters (mm) at 0, 0.03, 0.07, and 0.27 mg/kg per day, respectively) and number of tumors per mouse (21.5, 19.4, 19.4, and 33.1 at 0, 0.03, 0.07, and 0.27 mg/kg per day, respectively). These findings suggest that MeHg may act as a tumor promoter.

In a follow-up study to Mitsumori et al. (1981), Hirano et al. (1986) fed MeHg chloride to ICR mice (60 per sex) at lower doses (males, 0, 0.03, 0.15, or 0.73 mg/kg per day; females, 0, 0.02, 0.11, or 0.6 mg/kg per day) for 104 weeks. Kidney and reproductive-system effects indicated that the MTD was exceeded at the highest dose. An increased incidence of renal epithelial tumors (adenomas and adenocarcinomas) occurred in males. In males in the high-dose group, 10 of the 13 tumors were adenocarcinomas; the incidence of renal epithelial adenomas was not increased. No renal tumors were seen in females.

The incidence of renal tumors was also increased in male $B6C3F_1$ mice following chronic exposure to MeHg chloride. Mitsumori et al. (1990) fed $B6C3F_1$ mice (60 per sex) MeHg chloride (males, 0, 0.03, 0.14, or 0.69 mg/kg per day; females, 0, 0.03, 0.13, or 0.60 mg/kg per day). Following 104 weeks of exposure, adverse effects were seen in the central nervous system, kidney, and testis. The MTD was achieved in males in the mid-dose group and in females in the high-dose group. The MTD was exceeded in males in the high-dose group. The incidence of renal epithelial carcinomas and renal adenomas was significantly increased in males in the high-dose group.

Although chronic exposure to MeHg increased the incidence of renal tumors in male mice in some studies, that effect was observed only at doses that were toxic to the kidneys and is thought to be secondary to cell damage and repair. Exposure to MeHg did not increase tumor rates in female mice or in rats of either sex.

GENOTOXICITY

Human Studies

Evidence that human exposure to Hg causes genetic damage is inconclusive. Several investigators have reported higher rates of chromosomal aberrations among workers who were exposed to elemental or inorganic forms of Hg (Popescu et al. 1979; Verschaeve et al. 1976; Barregard et al. 1991). However, questions have been raised regarding the influence of possible confounders, such as age or simultaneous exposure to other toxicants on these findings. In a recent occupational study, Queiroz et al. (1999) reported a significant increase in the percentage of micronuclei in Hg-exposed workers when compared with unexposed controls.

Skerfving et al. (1970, 1974) reported a positive correlation between blood Hg concentrations and chromosomal aberrations in the lymphocytes of 23 people who consumed Hg-contaminated fish. However, their findings have been questioned because of experimental problems, such as failure to identify smokers. In addition, significant effects were found only from lymphocyte cultures that were set up several days after collection, and the incidence of aneuploidy in the control and exposed groups was lower than expected. Wulf et al. (1986) reported an increased incidence of sister chromatid exchange in humans who ate Hg-contaminated seal meat. However, information on smoking status and exposure to other heavy metals was not provided for those individuals, making interpretation of the study difficult. More recently, Franchi et al. (1994) reported a correlation between the incidence of micronuclei in peripheral lymphocytes and blood Hg concentrations in a population of fishermen who had eaten Hg-contaminated seafood.

Animal Studies

A single dose of Hg chloride (HgCl) to male Swiss mice (2.2, 4.4, or 8.9 mg/kg) induced a dose-related increase in the frequency of chromosomal aberrations and the percentage of aberrant cells in bone marrow (Ghosh et al. 1991). Chronic exposure of cats to MeHg at doses of 0.0084, 0.02, or 0.046 mg/kg per day for 39 months produced a significant increase in the number of nuclear abnormalities in bone-marrow cells and inhibited DNA repair (Miller et al. 1979). The response, however, was not dose related.

In Vitro Studies

MeHg has been shown to cause DNA damage in cultured *Bacillus subtilis* (Kanematsu et al. 1980); chromosomal aberrations and aneuploidy in human lymphocytes (Betti et al. 1992); and DNA damage in cultured human nerve and lung cells, Chinese hamster V-79 cells, and rat glioblastoma cells (Fiskesjo 1979; Costa et al. 1991). Inorganic Hg concentrations greater than 10 μM have been shown to inhibit mammalian DNA polymerase activity in whole-cell extracts and in purified enzyme preparations (Williams et al. 1987; Robison et al. 1984). Sekowski et al. (1997) demonstrated the ability of mercuric ion to impair the fidelity of synthesome-mediated DNA replication at HgCl concentrations as low as 1 μM.

IMMUNOTOXICITY

The immune system appears to be sensitive to Hg. Although there are no data on the effect of MeHg on immune function in humans, occupational studies indicate that Hg compounds can affect the immune system. Animal studies have demonstrated MeHg effects on immune-cell ratios, cellular responses, and the developing immune system. Autoimmune effects have also been associated with exposure to elemental Hg.

Human Studies

The effect of MeHg on the human immune system has not been studied. However, occupational exposure to elemental Hg has been found to alter certain immune parameters. Queiroz and Dantas (1997a, b) evaluated B- and T-lymphocyte populations among 33 workers in a Brazilian Hg production facility. At the time of the study, all the workers had urinary Hg concentrations below 50 μg/g of creatinine. Analysis of T-cell populations found a reverse $CD4^+$-to-$CD8^+$ ratio that was characterized by a reduction in the number of $CD4^+$ lymphocytes. That alteration was significantly correlated with urinary Hg concentrations. B-lymphocyte counts were also significantly reduced in this cohort; however, that effect was not correlated with urinary Hg concentrations. Analysis of serum antibody levels found increased immunoglobulin E levels but did not detect anti-DNA or anti-nucleolar antibodies. The researchers reported a moderate negative correlation between length of exposure to Hg and IgE levels (Dantas and Queiroz 1997).

Moszczynski et al. (1995) studied lymphocyte subpopulations (T cells, T-helper cells, T-suppressor cells, and natural killer cells) in the peripheral blood of 81 men occupationally exposed to metallic Hg vapors and 36 unexposed men. The average Hg concentration in the workplace air was 0.0028 mg/m^3. Urinary Hg concentrations ranged from 0 to 240 μg/L, and concentrations in the blood varied from 0 to 30 μg/L. Stimulation of T-lymphocytes — manifested by an increased number of T cells, T-helper cells, and T-suppressor cells — was observed.

Animal Studies

Effects on the Adult Immune System

Work in animals has demonstrated that Hg can effect immune function (see Table 5-3). Ilbäck (1991) found that oral exposure to MeHg altered the ratio of lymphocyte subpopulations, enhanced lymphoproliferation in response to B- and T-cell mitogens, and depressed natural-killer-cell activity in mice. Exposure of female Balb/c mice to MeHg (3.9 ppm) in the diet (equivalent to 0.5 mg/kg per day) for 12 weeks significantly decreased thymus weight (22%) and cell number (50%). Lymphoproliferation in response to T- and B-cell mitogens was increased, and natural-killer-cell activity was decreased in exposed mice. Red- blood-cell counts were slightly higher in exposed mice than in unexposed mice, and white-blood-cell counts were unaffected.

Thompson et al. (1998) evaluated the effects of low-dose MeHg exposure in mice. Mice were exposed to MeHg at 0, 3, or 10 ppm in the drinking water for 4 weeks. MeHg altered the proportion of splenocyte and thymocyte subpopulations and caused dose-dependent decreases in splenocyte glutathione concentrations and mitogen-stimulated calcium flux.

Rats were exposed to MeHg (chloride or sulfide; concentrations of 5 or 500 μg/L) in drinking water for 8 or 16 weeks (Ortega et al. 1997). An 8-week exposure to both concentrations of MeHg sulfide enhanced the lymphocyte response to conconavalin A. However, only the 5-μg/L concentration of MeHg chloride had that effect. At 16 weeks, lymphocyte proliferation decreased in the rats exposed to MeHg chloride but increased in those exposed to MeHg sulfide. Those data indicate that the effects of MeHg on T-cell proliferation are dependent upon the dose, duration, and chemical form of the MeHg exposure.

Prolonged exposure to MeHg increased the susceptibility of mice to

TABLE 5-3 Summary of Immunological Studies in Animals

Species	NOAEL	LOAEL	Effect	Reference
Rat	None	3.9 ppm in diet of dams	Reduced NK cell activity in pups	Ilbäck et al. 1991
Rat	None	5 ppb in water	Altered mitogen response	Ortega et al. 1997
Rat	None	5 ppb in water of dams	Increased thymic weight in pups	Wild et al. 1997
Mouse	None	1 ppm in diet	Increased mortality when infected with encephalitis virus	Koller 1975
Mouse	None	3.9 ppm in diet (0.5 mg/kg/d)	Reduced NK cell activity; decreased thymus weight.	Ilbäck 1991
Mouse	None	3.69 ppm in diet of dams	Reduced resistance to Coxsackie B3	Ilbäck et al. 1996
Mouse	None	0.5 ppm in diet of dams	Altered immune effects in pups	Thuvander et al. 1996
Mouse	None	3 ppm in water	Altered B-cell and T-cell subtypes; decreased GSH concentrations in splenocytes	Thompson et al. 1998
Mouse	None	0.3 mg/kg/d	Antinucleolar antibody production	Hultman and Hansson-Georgiadis 1999

Abbreviations: NOAEL, no-observed-adverse-effect level; LOAEL, lowest-observed-adverse-effect level; NK, natural killer; GSH, glutathione.

viral infections. Koller (1975) fed mice subtoxic doses of MeHg chloride(1 or 10 mg/kg) for 84 days and saw significantly higher mortality after inoculation with encephalomyocarditis virus in exposed mice than in unexposed mice. In the same report, MeHg exposure did not alter the course of neoplasia in mice inoculated with Rauscher leukemia virus. MeHg (3.69 mg/g of diet) also did not alter the lethality of myocarditic coxsackie virus B3 in Balb/c mice but did increase heart tissue damage and viral persistence (Ilbäck et al. 1996).

Effects on the Developing Immune System

Prenatal and perinatal exposure to MeHg has long-term effects on the developing immune system. Ilbäck et al. (1991) reported alterations in white-blood-cell counts, natural-killer-cell activity, and the response of thymocytes and splenocytes to T-cell mitogen in Sprague-Dawley rats following prenatal and postnatal exposure of rat pups to MeHg. Wild et al. (1997) exposed rats, in utero and during the nursing period to MeHg (maternal drinking-water concentrations of MeHg chloride at 5 or 500 μg/L, or MeHg sulfide at 5 μg/L). At 6 weeks of age, total body and splenic weights were significantly increased in both MeHg-chloride-exposed groups. Rats exposed to MeHg sulfide had a significant increase in thymic weight at 6 weeks of age. Splenocyte response to pokeweed mitogen was enhanced at 6 and 12 weeks in both MeHg-chloride-exposed groups but was unaffected by MeHg sulfide. Natural-killer-cell activity was not affected in any exposure group at 6 weeks of age but was decreased by 57% in both groups exposed to MeHg chloride at age 12 weeks.

Similar effects have been demonstrated in mice. Female Balb/c mice were fed diets containing MeHg (0, 0.5 or 5 mg/kg) for 10 weeks before mating, throughout gestation, and up to day 15 of lactation (Thuvander et al. 1996). Blood Hg concentrations in the offspring were increased on day 22 (0.5-mg/kg group) and on days 22 and 50 (5-mg/kg group). The number of splenocytes and thymocytes increased, and the antibody response to a viral antigen was stimulated in the offspring of the 0.5-mg/kg group. The response of splenocytes to B-cell mitogen increased in offspring of the 5-mg/kg group. Lymphocyte subpopulations in the thymus were altered at both doses.

In Vitro Studies

The effects of MeHg on lymphocyte function have been studied in cell-culture systems in an attempt to elucidate the mechanisms involved in its ability to modulate immune function. Exposure of cultured lymphocytes to MeHg has been shown to inhibit mitogen-induced DNA synthesis, cell proliferation, and antibody synthesis. Electron microscopic analysis of MeHg-exposed lymphocytes revealed nuclear changes characterized by hyperchromaticity and fragmentation. MeHg exposure also induced a rapid and sustained increase in intracellular calcium levels (Nakatsuru et al. 1985; Shenker et al. 1993). Shenker et al. (1999) investigated the mechanism by which MeHg chloride induces human T-cell apoptosis. They reported that the earliest detectable event following MeHg exposure was at the level of the mitochondria. Exposure of T-cells to MeHg chloride caused a decrease in the overall size of mitochondria and changes in the structure of the cristae. Cellular thiol reserves were depleted and mitochondrial cytochrome c was translocated to the cytosol.

Autoimmune Response

Human Studies

There is some evidence that human exposure to metallic Hg can induce an autoimmune response. Renal biopsies of two Hg-exposed workers who had developed proteinuria revealed deposits of IgG and complement C3 in the glomeruli (Tubbs et al. 1982). Examination of 10 patients who complained of illnesses after they received dental amalgams found that 3 of them had antiglomerular basement membrane antibodies, and 2 had elevated antinucleolar antibodies (Anneroth et al. 1992). In addition to those reports, Cardenas et al. (1993) reported high anti-DNA antibody titers in 8 of 44 workers from a chloralkali plant. No studies were located that evaluated autoimmunity in humans following exposure to organic forms of Hg.

Animal Studies

Hg is one of the few chemicals which is able to induce loss of tolerance to self-antigens in animals. This effect is human leukocyte antigen (HLA) dependent and has been demonstrated in genetically susceptible strains of rats and mice. Brown-Norway rats injected with Hg chloride ($HgCl_2$) produce antilaminin antibodies, which attack the kidneys, causing an autoimmune glomerulonephritis (Druet et al. 1994). The autoimmune response observed following Hg exposure has been linked to a T-cell dependent polyclonal B-cell activation (Hua et al. 1993). Hu et al. (1999) found that Hg exposure induced an autoimmune response in C57BL/6(H-2b) wild-type and interlukin-4 (IL-4)-deficient mice. Antibodies of all classes were induced by Hg treatment, except that in the IL-4-deficient mice, no immunoglobulin E (IgE) and very little IgG1 were produced.

REPRODUCTIVE EFFECTS

Human Studies

In occupational exposure studies, paternal exposure to metallic Hg does not appear to cause infertility or malformations (Alcser et al. 1989; Lauwerys et al. 1985). However, a study of pregnancy outcomes among the wives of 152 Hg-exposed men revealed an increased incidence of spontaneous abortions (Cordier et al. 1991). Preconception paternal urinary Hg concentrations above 50 μg/L were associated with a doubling of the spontaneous abortion risk.

The effect of elemental Hg on fertility and reproductive success has also been examined among occupationally exposed women. The results of various studies are conflicting but are suggestive of an effect on fertility. Elghany et al. (1997) compared the pregnancy outcomes of 46 Hg-exposed workers to those of 19 women who worked in nonproduction areas of the same factory. Among cases and controls during the study period (1948-1977), 104 pregnancies were recorded. Women exposed to inorganic Hg had a higher rate of congenital anomalies. Concentrations were up to 0.6 mg/m^3. No significant differences in stillbirth or miscarriage rates were noted between the two groups of women. Rowland et al. (1994) found that the probability of conception among female dental hygienists who prepared more than 30 amalgams per week and had at least five poor hygiene practices when handling Hg was only 63% of that among unexposed controls. Women with lower exposures, however, were more fertile than unexposed controls. A large study conducted in Norway compared reproductive success rates among 558 female dental surgeons with those of 450 high-school teachers (Dahl et al. 1999). They concluded that exposure to Hg, benzene, and chloroform was not associated with decreased fertility except for a possible Hg effect on the last pregnancy of multiparous dental surgeons.

No studies were identified that specifically evaluated human reproductive success following exposure to MeHg. However, in a study that described the clinical symptoms and outcomes of more than 6,000 Iraqi citizens who were severely poisoned by bread that had been prepared with MeHg-treated wheat, Bakir et al. (1973) commented on the low number of pregnant women in the cohort. Their report states, in part, that "The admissions frequency of affected pregnant females was remarkably low. One would expect to find approximately 150 pregnant females with diagnosable poisoning in the 6350 cases admitted to hospitals, yet only 31 such females were reported." Although no explanation was offered for the small number of pregnancies among the exposed population, the report provides evidence of a possible effect of MeHg on human fertility.

Animal Studies

The reproductive effects of MeHg exposure in animals are summarized in Table 5-4. Abortion and decreased litter size are the most commonly reported reproductive effects of MeHg in animal studies. Pre- and post-implantation losses have been experimentally induced in rats, mice, guinea pigs, and monkeys exposed to MeHg.

In rats, an oral dose of MeHg at 7.5 mg/kg on gestational days 7-14 resulted in increased fetal deaths and an increased incidence of malformations. A dose of 5 mg/kg was also associated with an increased incidence of malformations as well as reduced fetal weight (Fuyuta et al. 1978).

In Fischer 344 rats, oral doses of MeHg chloride at 10, 20, or 30 mg/kg administered on day 7 of gestation decreased fetal survival by 19.1%, 41.4%, and 91.1%, respectively (Lee and Han 1995). Compared with control animals, implantation sites in the three groups were decreased by 5.9%, 13.7% and 22.5%, respectively. The median lethal dose for fetuses was 16.5 mg/kg.

Oral doses of MeHg hydroxide at 3, 5, or 10 mg/kg on day 8 of gestation in mice caused a significant dose-related decrease in litter size. No effects were seen at 2 mg/kg (Hughes and Annau 1976).

TABLE 5-4 Summary of Reproductive Studies in Animals

Species	NOAEL (mg/kg/d)	LOAEL (mg/kg/d)	Effect	Reference
Monkey	None	0.05	Abnormal sperm	Mohamed et al. 1987
Monkey	0.05	0.07	Low conception rate	Burbacher et al. 1988
Rat	2.5	5 (males)	Reduced litter size	Khera 1973a
Rat	None	10 on GD 7	Decreased fetal survival	Lee and Han 1995
Mouse	2	3 on GD 8	Decreased fetal survival	Hughes and Annau 1976
Mouse	None	5 on GD 6-13	Fetal malformations	Fuyuta et al. 1978
Mouse	None	10 on GD 10	Embryo resorption	Fuyuta et al. 1979
Mouse	None	0.73	Low sperm counts Tubular atrophy of testes	Hirano et al. 1986 Mitsumori et al. 1990
Guinea pig	None	11.5 on GD 21, 28, 35, or 42	Fetal abortions	Inouye and Kajiwara 1988

Abbreviations: NOAEL, no-observed-adverse-effect level; LOAEL, lowest-observed-adverse-effect level; GD, gestation day.

Fuyuta et al. (1978) reported that an oral dose of MeHg chloride at 7.5 mg/kg on gestational days 6-13 in mice was embryocidal, and doses of 5 or 6 mg/kg reduced fetal weights and increased the incidence of malformations (cleft palate and fused thoracic vertebrae).

Fuyata et al. (1979) also dosed mice with a single oral dose of MeHg at 10, 15, 20, or 25 mg/kg on gestational day 10. An increase in resorbed embryos occurred at 25 mg/kg. At the doses of 15, 20, and 25 mg/kg, fetuses weighed less than controls and had an increase in malformations.

A single dose of MeHg chloride at 11.5 mg/kg administered to pregnant guinea pigs on day 21, 28, 35, or 42 of gestation caused half of the litters to be aborted (Inouye and Kajiwara 1988).

Reproductive problems, including decreased conception rates, early abortions, and stillbirths were seen following exposure of female *Macaca fascicularis* monkeys to MeHg hydroxide at 50, 70, or 90 μg/kg per day for 4 months (Burbacher et al. 1988). Although no effects were observed on the menstrual cycle, the number of conceptions decreased with increasing dose (93% for controls, 81% for group at 50 μg/kg per day, 71% for group at 70 μg/kg per day, and 57% for group at 90 μg/kg per day). A significant reduction in the percentage of viable offspring was observed for the groups at 70 and 90 μg/kg per day (83% for controls, 69% for group at 50 μg/kg per day and 29% for groups at 70 or 90 μg/kg per day). The effects on reproduction were observed at a maternal blood concentration greater than 1.5 ppm. Maternal toxicity was also observed in the doses of 70 and 90 μg/kg per day following prolonged MeHg exposure (½ year to over 1 year), typically at maternal blood concentrations greater than 2 ppm. Maternal toxicity was not seen in monkeys exposed at 50 μg/kg per day.

Effects on reproduction have also been seen following paternal exposure to MeHg. Exposure of male rats to high doses of MeHg chloride (5 to 7 daily doses of 1, 2.5, or 5 mg/kg) before mating with unexposed females produced a dose-related increase in post-implantation losses and reduced litter size (Khera 1973a). Exposure of male mice to those doses had no effect on reproductive success (Khera 1973a). Mohamed et al. (1987) examined the testicular functions of male *Macaca fascicularis* following oral exposure to MeHg hydroxide at 50 or 70 μg/kg per day for 20 weeks. Although there was no significant decrease in sperm counts, MeHg exposure was associated with a decrease in the percentage of motile sperm, a reduction in sperm speed, and an increase in the number of abnormal sperm (primarily bent or kinked tails). No effects were observed on serum testosterone concentrations, and no histological abnormalities were detected in testicular biopsies. Sperm motility returned to normal soon after the cessation of MeHg exposure, and sperm morphology remained abnormal. Chronic exposure to MeHg chloride at 0.73 mg/kg per day decreased spermatogenesis and produced tubular atrophy of the testis in mice (Hirano et al. 1986; Mitsumori et al. 1990). That dose caused renal damage, indicating that it exceeded the MTD.

RENAL TOXICITY

Human Studies

The kidney is sensitive to metallic Hg following inhalation exposure, possibly due to accumulation of Hg. High exposures have resulted in mild transient proteinuria, gross proteinuria, hematuria, and oliguria. Kidney biopsies from workers with proteinuria revealed proximal tubular and glomerular changes (Kazantzis et al. 1962). Several investigations have found renal changes in workers chronically exposed to Hg vapor (Danziger and Possick 1973; Buchet et al. 1980; Barregard et al. 1988; Cardenas et al. 1993).

However, renal toxicity has rarely been reported following human exposure to organic forms of Hg (see Table 5-5). All cases in which renal damage was confirmed following exposure to organic Hg involved severe poisonings in which neurological symptoms were also present. An autopsy of a man who died following an acute exposure to alkyl Hg vapor revealed necrosis of the tubule epithelium, swollen granular protoplasm, and nonstainable nuclei in the kidneys (Höök et al. 1954). Jalili and Abbasi (1961) described the clinical course of several victims of the Iraqi poisoning incident who displayed symptoms of renal damage, including polyuria, polydypsia, and albuminuria. Similar symptoms were observed in two children who had consumed ethyl-Hg-contaminated pork over a period of several weeks (Cinca et al. 1979). Laboratory analyses conducted shortly after their illnesses began indicated elevated blood urea, urinary protein, and urinary sediment. Both children died of cardiac arrest, and their autopsies revealed severe nephritis and myocarditis.

The only evidence of a renal effect following ingestion of Hg-contaminated fish comes from a death-certificate review conducted by Tamashiro et al. (1986). They evaluated causes of death among residents of a small area of Minamata City that had the highest prevalence of MD using age-specific rates for the entire city as a standard. Between 1970 and 1981, the number of deaths attributed to nephritic diseases was higher than expected among women who resided in that region (SMR, 276.5) but was within the expected range (0.80) among men who resided in this region.

Animal Studies

Although it is well known that the kidney is the target organ for inorganic Hg (Samuels et al. 1982), several reports from animal studies have also described MeHg- induced renal toxicity (see Table 5-6). A

TABLE 5-5 Summary of Renal Studies in Humans Exposed to Various Organic Mercurials

Exposure source	Effects	Reference
Occupational exposure to alkyl Hg vapors	Necrosis of renal tubules	Höök et al. 1954
Occupational exposure to Hg vapors	Albuminuria, tubular changes	Kazantzis et al. 1962
Occupational exposure to Hg vapors	Proteinuria	Danziger and Possick 1973
Occupational exposure to Hg vapors (urinary > 50 μg/g creatinine)	Albuminuria	Buchet et al. 1980
Occupational exposure to Hg vapors	Increased N-acetyl-B-glucosaminidase	Barregard et al. 1988
Occupational exposure to Hg vapors	Tamm-Horsfall protein, tubular antigen	Cardenas et al. 1993
Ingestion of mercuric chloride (30 mg/kg)	Oliguria, proteinuria	Afonso and deAlvariz 1960
Ingestion of mercuric chloride	Fatal acute renal failure	Murphy et al. 1979
Dermal application of mercuric ammonium chloride	Impaired renal function	Barr et al. 1972
Dermal application of mercuric ammonium chloride	Impaired renal function	Dyall-Smith and Scurry 1990
Ingestion of ethyl-Hg-contaminated pork	Elevated blood urea, urinary protein, urinary sediment	Cinca et al. 1979
Ingestion of MeHg-treated wheat	Polyuria, albuminuria	Jalili and Abbasi 1961
Ingestion of MeHg-contaminated fish	Increase in deaths due to nephritic diseases among women	Tamashiro et al. 1986

report by Fowler (1972) described the presence of large numbers of spherical masses containing bundles of smooth endoplasmic reticulum in the pars recta of the kidney proximal tubules in rats following a 12-week exposure to MeHg at 2 ppm (0.08 mg/kg per day). Those effects were observed in female rats only. The authors indicated that the sex-specific effects were most likely due to sex differences known to exist in the activity of kidney enzymes associated with MeHg metabolism. In

TABLE 5-6 Summary of Renal Studies in Animals

Species	Duration	NOAEL (mg/kg/d)	LOAEL (mg/kg/d)	Effect	Reference
Rat	3-12 wk	None	0.84	Fibrosis, inflammation, large foci in renal cortex	Magos and Butler 1972
Rat	12 wk		0.08	Cytoplasmic masses in proximal tubules	Fowler 1972
Rat	2 yr	0.02	0.1	Increased renal weights Decreased renal enzymes	Verschuuren et al. 1976
Rat	0-21 days of age	None	1	Altered renal function and renal hypertrophy	Slotkin et al. 1985
Rat	2 yr		0.4	Nephrosis	Solecki et al. 1991
Mouse	26 wk	0.15	0.6	Degeneration of proximal tubules	Hirano et al. 1986
Mouse		0.03 (males) 0.13 (females)	0.14 (males) 0.6 (females)	Chronic nephropathy, interstitial fibrosis	Mitsumori et al. 1990
Mouse	Once	8 (males) 24 (females)	16 (males) 32 (females)	Decreased phenolsulfonphthalein excretion, increased serum creatinine, swollen tubuler epithelium	Yasutake et at. 1991

Abbreviations: NOAEL, no-observed-adverse-effect level; LOAEL, lowest-observed-adverse-effect level.

a similar study, Magos and Butler (1972) reported fibrosis in the renal cortex of female rats following 12 weeks of MeHg exposure at the lowest dose studied (0.84 mg/kg per day). Increased kidney weight and decreased proximal convoluted tubule enzymes were seen in rats given MeHg chloride in the diet (0.1 mg/kg per day) for 2 years. No histopathological changes were observed (Verschuuren et al. 1976). Subsequent studies of rats and mice reported nephrosis following long-term exposure to MeHg (Mitsumori et al. 1983, 1984, 1990; Solecki et al. 1991). Nephrosis was also observed in rats exposed to phenylmercuric acid in drinking water for 2 years (Solecki et al. 1991).

Degeneration of the proximal tubules was observed in mice given MeHg chloride in the diet (0.11 mg/kg per day) for 2 years (Hirano et al. 1986). Epithelial degeneration and regeneration of the proximal tubules and interstitial fibrosis were noted in both male and female mice following almost 2 years of exposure to MeHg in the diet (estimated dose associated with effects was approximately 0.2 mg/kg per day) (Mitsumori et al. 1990). Yasutake et al. (1991) showed in mice that a single oral dose of MeHg (16 mg/kg) impaired renal function, causing increased plasma creatinine concentrations and swelling of tubular epithelium with exfoliation of cells into the tubular lumen. No effects were observed after a single gavage dose of Hg at 8 mg/kg.

A study by Slotkin et al. (1985) examined the renal effects of MeHg exposure during the neonatal period. Rats exposed to daily doses of 1 or 2.5 mg/kg per day from birth to 21 days of age (weaning) exhibited renal hypertrophy and altered renal function (elevated fractional excretions of water, glucose, sodium, chloride, osmotic particles), which peaked at approximately 20 days of age. The authors indicated that the results reflected effects on tubular function and that tests conducted in conjunction with physiological challenge might reveal even greater impairment.

CARDIOVASCULAR EFFECTS

Numerous studies have examined fish consumption and cardiovascular disease risk, and there are strong indications of protective effects of fish. These effects could be due to a number of components in fish, such as omega-3 fatty acids and selenium and might also indicate a different style of eating (diets lower in red meats).

Although inclusion of fish in the diet is generally beneficial, some fish contain agents such as MeHg and PCBs that have been associated with adverse cardiovascular effects. Therefore, future studies should control for co-exposure to these common contaminants in their analyses of the beneficial effects of fish intake.

Hg accumulates in the heart, and exposures to organic and inorganic forms of this metal have been associated with blood-pressure alterations and abnormal cardiac function. Numerous reports of human poisonings have described marked hypertension and abnormal heart rate among victims. Autopsies of two boys who died of cardiac arrest after they were fed ethylmercury-contaminated pork over a period of several weeks revealed myocarditis. Two recent epidemiological studies have found associations between dietary exposure to very low levels of MeHg and cardiovascular effects. One of those studies found evidence of an effect of prenatal MeHg exposure on heart function at age 7. Additional studies are needed to better characterize the effect of MeHg exposure on blood pressure and cardiovascular function at various stages of life.

Human Studies

The cardiovascular effects of Hg exposure in humans are summarized in Table 5-7. Warkany and Hubbard (1953) reported several cases in which children developed tachycardia and elevated blood pressure after they were treated with mercurous chloride-containing medications for worms or teething discomfort. Increases in blood pressure and heart rate have also been reported following inhalation of high concentrations of metallic Hg (Hallee 1969; Soni et al. 1992; Bluhm et al. 1992). In one of the cases, the increase in heart rate was described as a sinus tachycardia (Soni et al. 1992). Marked hypertension (160/120 mm Hg) and tachycardia (120 beats per minute) were also described in an 11-year old girl who was hospitalized with a diagnosis of acute Hg intoxication (Wössmann et al. 1999). Vroom and Greer (1972) reported a high incidence (five of nine workers) of hypertension among workers in a thermometer plant.

Exposure to organic Hg has also been associated with cardiovascular changes. Three clinical case reports and two epidemiological investigations have reported similar effects. The first evidence of cardiovascular abnormalities following exposure to organic Hg was provided by Jalili and Abbasi's (1961) description of patients who were hospitalized during the Iraqi grain poisoning epidemic. Abnormalities seen in severely poisoned patients included irregular pulse and electrocardiograms showing ventricular ectopic beats, prolongation of the Q-T interval, depression of the S-T segment and T inversion. Electrocardiograms of four family members who consumed ethylmercury-contaminated pork revealed similar findings, including abnormal heart rhythms with S-T segment depression and T-wave inversion (Cinca et al 1979). Deaths of two children in this family were attributed to cardiac arrest, and their autopsies revealed myocarditis. A child who was diagnosed with acrodynia following exposure to vapors from a paint that contained

Table 5-7 Summary of Cardiovascular Studies in Humans

Exposure source	Effects	Reference
Mercurous Chloride medications	Tachycardia and increased blood pressure in children	Warkany and Hubbard 1953
Occupational exposure to alkyl Hg vapors	Increased blood pressure	Höök et al. 1954
Alkyl Hg-contaminated wheat	Irregular heart rate	Jalili and Abbasi 1961
Ethylmercury-contaminated meat	Irregular heart rate, cardiac arrest, myocarditis	Cinca et al. 1979
Phenylmercuric acetate vapors	Hypertension and rapid heart rate	Aronow et al. 1990
Metallic Hg vapors	Increased blood pressure and heart rate	Hallee 1969, Bluhm et al. 1992, Soni et al. 1992, Vroom and Greer 1972
Dental amalgams	Increased blood pressure	Siblerud 1990
Frequent fish consumption	Higher cardiovascular death rates	Salonen et al. 1995
Hg intoxication (source unspecified)	Marked hypertension in child	Wössmann et al. 1999
Unspecified	High Hg concentrations in myocardium of IDCM patients	Frustaci et al. 1999
Prenatal exposure	Increased blood pressure and decreased heart rate variability at age 7	Sørensen et al. 1999

phenylmercuric acetate exhibited a rapid heart beat and hypertension (Aronow et al. 1990).

Two recent epidemiological investigations have found associations between exposure to low levels of MeHg and adverse cardiovascular effects. A recent study by Sørensen et al. (1999) showed an association between prenatal exposure to MeHg and cardiovascular function at age 7. The study of 1,000 children from the Faroe Islands found that diastolic and systolic blood pressures increased by 13.9 and 14.6 mm Hg, respectively, as cord-blood Hg concentrations rose from 1 to 10 µg/L. In boys, heart-rate variability, a marker of cardiac autonomic control, decreased by 47% as cord-blood Hg concentrations increased from 1 to 10 μg/L.

Salonen et al. (1995) compared dietary intake of fish and Hg, and compared Hg concentrations in hair and urine with the prevalence of acute myocardial infarction (AMI) and death from coronary heart disease or cardiovascular disease in a cohort of 1,833 Finnish men. All study participants were free of clinical heart disease, stroke, claudication, and cancer at the beginning of the study. Daily fish intake ranged from 0 to 619.2 g (mean = 46.5 g per day) and hair Hg concentrations ranged from 0 to 15.67 ppm (mean = 1.92 ppm). Dietary Hg intake ranged from 1.1 to 95.3 μg per day (mean = 7.6 μg per day). Over a 7-year observation period, men in the highest tertile (at or more than 2 ppm) of hair Hg content had a 2.0-fold higher risk of AMI than men in the two lowest tertiles. The relative risk was similar for coronary deaths and cardiovascular deaths, although the difference for coronary deaths did not reach statistical significance due to small numbers. Men who consumed at least 30 g of fish a day had a 2.1-fold higher risk of AMI. For each additional 10 g of fish consumed, there was an increment of 5% in the 5-year risk of AMI.

Trace elements were measured in myocardial and muscle-tissue samples from 13 patients diagnosed with idiopathic dilated cardiomyopathy (IDCM). The subjects had no history of Hg exposure. Findings were compared with Hg concentrations measured in myocardial and muscle biopsies from age-matched patients with valvular (12 patients) or ischemic heart disease (13 patients), papillary and skeletal-muscle biopsies from 10 patients with mitral stenosis, and left-ventricle endomyocardial biopsies from 4 normal subjects. Hg concentrations in myocardial samples collected from patients with IDCM were 22,000 times higher than those in control samples. Antimony, gold, chromium, and cobalt concentrations were also higher in IDCM patients, but the greatest differences were for Hg (178,400 ng/g versus 8 ng/g) and antimony (19,260 ng/g versus 1.5 ng/g). The investigators concluded that the increased concentration of trace elements found in patients with IDCM might adversely affect mitochondrial activity and myocardial metabolism and worsen cellular function (Frustaci et al. 1999). Matsuo et al. (1989) analyzed Hg concentrations in human autopsy tissues collected from 46 cadavers. The subjects (32 males and 14 females aged

4 months to 82 years) were residents of metropolitan Tokyo and had no known exposure to Hg. The average total Hg content in heart tissue was 43 ng/g, with 80% of this being in the form of MeHg.

Animal Studies

Effects of MeHg on the heart and circulatory system have been observed in several animal models (see Table 5-8). A report by Shaw et al. (1979) described cerebrovascular lesions in four nonhuman primates following long-term exposure to near-toxic to toxic doses of MeHg hydroxide (90 to 120 µg/kg per day). Lesions were similar to those observed in humans with hypertension; intimal thickening, smooth-muscle cell proliferation, and adventitial fibrosis were reported.

Mitsumori et al. (1983, 1984) fed Sprague-Dawley rats diets containing MeHg chloride (males, 0, 0.011, 0.05, or 0.28 mg/kg per day; females, 0.014, 0.064, or 0.34 mg/kg per day) for up to 130 weeks. Polyarteritis nodosa and calcification of the arterial wall were seen at the highest

TABLE 5-8 Summary of Cardiovascular Studies in Animals

Species	NOAEL (mg/kg/d)	LOAEL (mg/kg/d)	Effects	Reference
Monkeys	None	0.09	Cerebrovasular changes, hypertension, intimal thickening	Shaw et al. 1979
Rat	0.05 (males) 0.06 (females)	0.28 (males) 0.34 (females)	Polyarteritis nodosa, calcification of arterial wall	Mitsumori et al. 1983, 1984
Spontaneous hypertensive rat	None	2 (26 d)	Increased blood pressure in females	Tamashiro et al. 1986
Rat	None	0.4	Hypertension	Wakita 1987
Rat	None	12 (2 d)	18% decrease in heart rate	Arito and Takahashi 1991

Abbreviations: NOAEL, no-observed-adverse-effect level; LOAEL, lowest-observed-adverse-effect level.

dose. Histological examination revealed evidence of hemosiderosis and extramedullary hematopoiesis of the spleen.

Tamashiro et al. (1986) reported an increase in blood pressure in spontaneous hypertensive rats (SHR) exposed to MeHg at 2 mg/kg per day for 26 consecutive days. That effect was sex specific, being observed only in females. Considerable variation was observed in blood pressure for both the MeHg-exposed and the control rats. Differences were observed at only two time points, week 3 and week 5 of the study.

In Wistar rats, hypertension was induced after a 30-day exposure to MeHg chloride at 0.4 or 1.2 mg/kg per day (Wakita 1987). The onset of hypertension occurred 42 days after the exposure period ended, and the effect persisted for more than 1 year. In rats, a decrease in heart rate (18%) was observed following 2 daily doses of MeHg at 12 mg/kg per day (Arito and Takahashi 1991).

HEMATOLOGICAL EFFECTS

Hematological changes have not been reported following human exposure to Hg. Studies conducted in animals suggest that Hg exposure might pose a risk of anemia and clotting disorders. Those animal studies are summarized in Table 5-9.

Munro et al. (1980) exposed rats to Hg at 0.25 mg/kg per day for up to 26 months. Exposed males had decreased hematocrit and hemoglobin values, as well as overt signs of neurotoxicity and increased mortality

TABLE 5-9 Summary of Hematological Studies in Animals

Species	NOAEL (mg/kg/d)	LOAEL (mg/kg/d)	Effect	Reference
Rat	None	0.25 for 26 mon.	Decreased hematocrit and hemoglobin values	Munro et al. 1980
Rat	None	8.0	Decreased clotting time	Kostka et al. 1989
Rat	None	4.2	Decreased hematocrit and hemoglobin values	Solecki et al. 1991

Abbreviations: NOAEL, no-observed-adverse-effect level; LOAEL, lowest-observed-adverse-effect level.

compared with unexposed controls. Hematological changes were not observed in exposed female rats.

Kostka et al. (1989) examined the coagulability of blood in rats exposed to either a single dose of MeHg chloride at 17.9 mg/kg per day or 5 consecutive days of dosing at 8 mg/kg per day. Blood coagulation was measured 1, 3, and 7 days after administration of the single dose or 24 hr after the 5 consecutive days of dosing. A reduction in clotting time and an increase in the fibrinogen concentrations in plasma were observed in both MeHg dose groups. Reduced clotting time was observed in the single-dose group 1 day after exposure.

Decreased hemoglobin, hematocrit, and red-blood-cell counts were seen in rats exposed to phenylmercuric acetate in drinking water (4.2 mg/kg per day) for 2 years (Solecki et al. 1991). The anemia might have been secondary to blood loss associated with ulcerative lesions seen at that dose in the large intestine. Polycythemia developed in rats exposed in utero to a combinations of MeHg chloride, ethylurea, and sodium nitrate. The polycythemia occurred as early as 1 month of age in as many as 24% of the offspring. Many features of this condition were similar to the features of polycythemia vera in man (elevated hematocrit, white- and red-blood-cell counts, splenomegaly, and hyperplasia of bone marrow) (Koller et al. 1977). Because that study involved concurrent exposure to MeHg, ethylurea, and sodium nitrite, the observed effects cannot be attributed to MeHg.

DEVELOPING CENTRAL-NERVOUS-SYSTEM TOXICITY

Human Studies

The central-nervous-system (CNS) effects of MeHg in humans have been extensively studied following accidental poisoning incidents and low-dose exposures. In this section, the Minamata and Iraqi Hg poisoning episodes are reviewed, documenting the severe neurological dysfunctions and developmental abnormalities that occur in children exposed in utero to high doses of MeHg. That review is followed by a review of the effects of low-dose prenatal MeHg exposure on neurological status, age at achievement of developmental milestones, infant and preschool development, childhood development, sensory and neurophysiological functions, and other end points in children; and neurological, neurophysiological and sensory functions in adults.

High-Dose Poisonings

Poisoning Episode in Japan

The mass poisoning of residents living near Minamata Bay in Japan in the 1950s first raised awareness of the severe neurological sequelae associated with MeHg poisoning, particularly when it occurs prenatally. The primary route of exposure in that episode was the consumption of fish contaminated with MeHg that bioaccumulated as it ascended the aquatic food chain. According to Harada (1995), all children identified as suffering from the most severe form of congenital Minamata disease (CMD) expressed mental retardation, primitive reflexes, cerebellar ataxia, disturbances in physical growth, dysarthria, and limb deformities. Most of the affected children also expressed hyperkinesis (95%), hypersalivation (95%), seizures (82%), strabismus (77%), and pyramidal signs (75%). The incidence of cerebral palsy among children with CMD was also increased (9% of 188 births in three villages versus a national incidence of 0.2% to 2.3%). Some signs and symptoms decreased over time (e.g., paroxysmal events, hypersalivation, primitive reflexes, and ataxia), although others (e.g., reduced intelligence and dysarthria) did not (Harada 1995). Most of the patients with the severe form of CMD were unable to function successfully in society.

It is difficult to reconstruct the MeHg doses in the CMD patients. Measurements of Hg in hair and blood were not made until 1959, several years after the poisoning episode was identified. The Hg concentrations in maternal-hair samples taken 5 to 8 years after giving birth to infants with CMD ranged from 1.8 to 191 ppm (Harada 1995). Analyses of the Hg concentrations in 151 archived umbilical-cord tissue samples dating from 1950 to 1969 confirmed that exposures increased during this period (Harada et al. 1999). Concentrations were highest in patients with CMD, intermediate in patients with acquired MD, and lowest in asymptomatic individuals. On the basis of these data, Akagi et al. (1998) estimated that the mean maternal-hair Hg concentration in CMD patients was approximately 41 ppm (range 3.8 to 133 pm). The uncertainty associated with that estimate, however, is likely to be substantial. Identification of cases was undoubtedly incomplete, particularly among individuals who suffered milder forms of CMD. For example, even excluding cases of known CMD, the prevalence of mental retardation among children born between 1955 and 1958 in the contaminated area was 29%, far higher than that expected as a background prevalence. That finding suggests that many children with less severe forms of CMD were undiagnosed. Thus, the data cannot provide precise estimates of the minimum dose of MeHg required to produce CMD.

Several observations associated with MD suggest that neurological deficits might emerge decades after exposure to MeHg has ended and that the severity of deficits might increase as a patient ages. It is difficult, however, to definitively rule out continued Hg exposure in adulthood as having a role in progressive neurological disorders. Harada (1995) distinguished three groups of patients with atypical, incomplete, or slight symptoms: (1) gradually progressive type, (2) delayed- onset type, and (3) escalator-progressive type. Evidence consistent with delay in the expressions of MeHg neurotoxicity was reported in a long-term follow-up study of 90% of diagnosed MD patients at least 40 years of age (1,144 patients). Kinjo et al. (1993) found not only that the prevalence of deficits in "activities of daily living" (i.e., eating, bathing, and dressing) was greater among cases than among age- and sex-matched controls but also that the difference between the prevalence rates of the two groups increased significantly with age. Increased deficits with age and delayed effects were also seen in animal studies (Spyker et al. 1972; Rice 1996, 1998; see section on Animal Studies).

Poisoning Episode in Iraq

A second episode of mass MeHg poisoning occurred in Iraq in the early 1970s when seed grain treated with a MeHg-containing fungicide was ground into flour and consumed. Those MeHg exposures were most likely more acute and involved higher exposures than those experienced by the residents of Minamata Bay. Early studies of the most severely affected children exposed to MeHg during fetal development were concordant with the Minamata findings. Those children manifested severe sensory impairments (blindness and deafness), general paralysis, hyperactive reflexes, cerebral palsy, and impaired mental development (Amin-Zaki et al. 1974). Several follow-up studies of the exposed population were conducted. Marsh et al. (1987) identified 81 children who had been in utero at the time of the episode and collected information from two sources on children's neurodevelopmental outcomes: neurological examination of each child and a maternal interview regarding the age at which the child achieved standard developmental milestones, such as walking and talking. Maximum maternal-hair Hg concentrations during the time when the study child was in utero served as the index of fetal exposure and ranged from 1 to 674 ppm. Developmental retardation was defined as a child's failure to walk a few steps unaided by 18 months of age or to talk (two or three meaningful words) by 24 months of age. A point system was devised for scoring the neurological examination; a score greater than 3 indicated a definite abnormality. There was a dose-response relationship between the prevalence of those indicators of poor outcomes and maternal-hair Hg concentrations. The most frequent neurological findings were increased limb tone and deep tendon reflexes with persisting extensor plantar responses. Ataxia, hypotonia, and athetoid movements were also reported. Boys appeared to be more severely affected than girls. Seven of the 28 children with the highest exposures had seizures (versus none of the 53 children with the lowest exposures). For those seven children, maternal-hair Hg concentrations ranged from 78 to 674 ppm. Many children of mothers with hair concentrations exceeding 100 ppm had normal neurological scores and achieved milestones at the expected times. Moreover, many of the women who had very high hair Hg concentrations and whose infants did poorly experienced only mild and transient signs or symptoms of MeHg toxicity.

Additional analyses of that data set were conducted in an attempt to identify more precisely the shape of the dose-response relationship and, in particular, the threshold for adverse neurodevelopmental effects, if indeed such a threshold exists. Cox et al. (1989) obtained more accurate estimates of peak exposure during pregnancy by applying an X-ray fluorescent method to single strands of maternal hair. Using a variety of statistical models (logit, hockey-stick, and nonparametric kernel-smoothing methods), they estimated a population threshold of approximately 10 ppm for the outcomes investigated (see Figures 5-1, 5-2, and 5-3). However, the uncertainty associated with that estimate is heavily

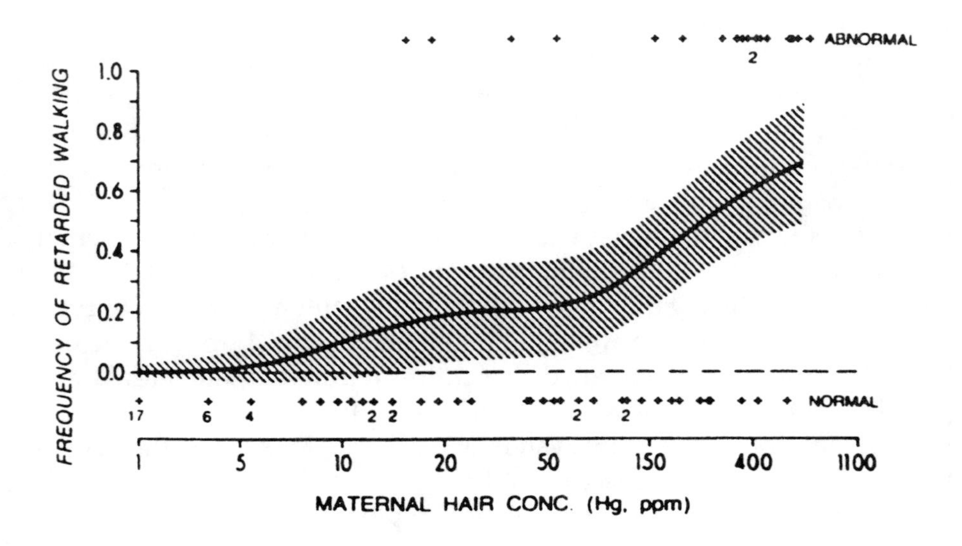

FIGURE 5-1 Nonparametric kernel-smoothing analysis of the relationship between maternal-hair concentration of Hg and retarded walking in the offspring. Maternal-hair concentrations were estimated using XRF single-strand analysis. The exposure value is the maximum level during gestation based on the growth rate of the hair and the birth date of the child. Results from multiple strands were averaged for the final exposure value. The shaded area denotes nonsimultaneous 95% confidence limits for individual points on the smoothed curve (for details, see text). Maternal-hair concentrations for normal and abnormal infants are plotted below and above the graph, respectively. Source: Cox et al. 1989. Reprinted with permission from *Environmental Research*; copyright 1989, Academic Press.

dependent on the assumed background prevalence of the poor outcomes. (No data were available on the true background prevalence of the poor outcomes among Iraqi children.) For example, for motor retardation, the upper bound of the 95% CI increases from 14 to 190 ppm when the estimate of background prevalence is changed from 0% to 4%. For neurological abnormality, the upper bound of the 95% CI for the threshold estimate is 287 ppm (assuming a 9% background prevalence). In re-analyses of those data, Crump et al. (1995) and Cox et al. (1995) showed that the estimate of population threshold is highly model de-

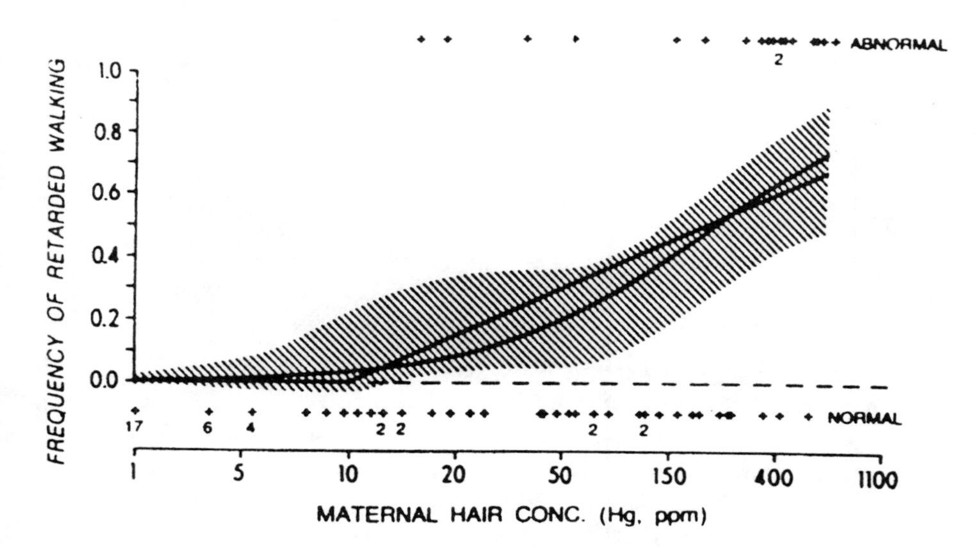

FIGURE 5-2 Plots of the logit and hockey-stick dose-response analysis of the relationship between retarded walking and maternal-hair concentrations during gestation. The two dose-response curves are shown by solid lines. The shaded area represents the 95% confidence limits from kernel smoothing. Source: Cox et al. 1989. Reprinted with permission from *Environmental Research*; copyright 1989, Academic Press.

pendent, sensitive to the definition of abnormality, and, in the case of delayed walking, heavily influenced by only four cases of delayed walking among children of women with hair Hg concentrations below 150 ppm. The statistical variability of the threshold estimates appears likely to be considerably greater than that provided by Cox et al. (1989). Crump et al. (1995) concluded that the Iraqi data do not provide convincing evidence of any adverse neurodevelopmental effects of MeHg below maternal-hair concentrations of 80 ppm.

In evaluating the Iraqi data, it is important to note that the interviews were conducted when the children were a mean age of 30 months. However, some children must have been considerably older, as the ages at which children in the sample were reported to have walked or talked were as high as 72 months. In addition, birth dates are generally not

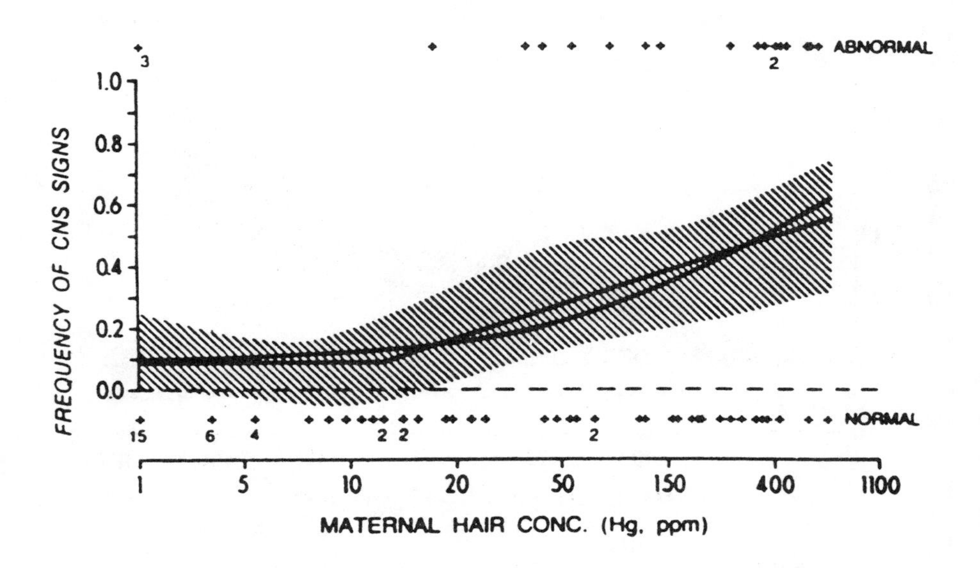

FIGURE 5-3 Plots of the logit and hockey-stick dose-response analysis of the relationship between CNS signs and maternal-hair concentrations during gestation. The two dose-response curves are shown by solid lines. The shaded area represents the 95% confidence limits from kernel smoothing. Source: Cox et al. 1989. Reprinted with permission from *Environmental Research*; copyright 1989, Academic Press.

important among Iraqi nomads. Therefore, maternal recollection of ages at which children achieved milestones had to be referenced to external events, such as the poisoning. The extent of the imprecision in those data is suggested by the strong digit preferences in the mothers responses. For instance, for 70 of the 78 children, the estimated age at walking was an even number of months. Furthermore, 75% of the estimates were multiples of 6 months. For age of talking, 70 of the 73 responses were an even number of months. (It should be noted, however, that the neurological scores were assigned to the children on the basis of a clinical examination and, therefore, were not subject to recall bias.) Finally, the extent of selection bias in this cohort cannot be characterized, because the size of the base population from which it was drawn and the referral mechanism that brought mothers and children to medical attention are both unknown. For instance, women who knew that they had consumed large amounts of contaminated grain and had concerns about their children's welfare might have come forward, and women who consumed equally large amounts of contaminated grain but whose children were developing well might not have come forward. That issue is critical, because the calculation of a threshold, a reference dose, or a benchmark dose requires a denominator (i.e., the size of the exposed population) as well as the background prevalence of the adverse outcomes to estimate the added risk associated with the exposure of interest. It appears that the background prevalence of developmental abnormality was extremely high among the Iraqi children who participated in the follow-up studies. The prevalence of delayed walking among children whose mothers had hair Hg concentrations below 10 ppm (and can be viewed essentially as a control group for the purpose of estimating background prevalence) was 36% (11 of 31). In contrast, among the population of U.S. children on whom the Bayley Scales of Infant Development (first edition) were standardized, the prevalence of delayed walking by that criterion was approximately 5%. Similarly, the prevalence of delayed talking (two or three words by 24 months) among the Iraqi children was 22% (6 of 27), and 95% of 24-month-old U.S. children in the standardization sample of the MacArthur Communicative Development Inventory were producing approximately 50 words (Fenson et al. 1993).

Chronic Low-Dose-Exposure Epidemiological Studies

A number of epidemiological studies have been carried out on populations exposed chronically to low doses of MeHg. Table 5-10 summarizes some key methodological aspects of those studies. In this section, those studies are discussed in terms of the end points assessed. End points discussed are status on neurological examination, age at achievement of developmental milestones, infant and preschool development, childhood development, sensory, and neurophysiological functions, and other end points in children.

TABLE 5-10 Summary of Developmental Neurotoxicity Studies in Humans

Study Site	Size of Cohort Identified and Enrolled	Exposure Biomarker	MeHg or Total Hg Concentration	Age at Assessment	Number of Children Assessed	End Points	Reference
Amazon	351	Child hair	Mean, 11.0 ppm; 80% > 10 ppm	7-12 years	354 (Table 1)	Fingertapping, WISC-III: digit span forward, Santa Ana dexterity test, Stanford-Binet: copying, bead memory	Grandjean et al. 1999
Ecuador	75 (36 children, 39 adults)	Blood	Mean, 17.5 mg/L (3.0 in 34 controls)	Children: 3-15 years Adults: 16-57 years	19-40	Pure tone conduction threshold BAEP	Counter et al. 1998
Faroe Islands	182	Maternal hair	Geometric mean, 4.1 ppm; range, 2.5-7.4 ppm	2 wk, adjusted for gestational age	182	Neurological exam	Steuerwald et al. 2000
		Umbilical cord blood	Geometric mean, 20.4 μg/L; range, 11.8-40.0 μg/L				
		Umbilical cord serum	Geometric mean, 2.5 μg/L; range, 1.7-3.7 μg/L				
	1,023	Maternal hair	Geometric mean, 4.3 ppm; interquartile range, 2.6-7.7 ppm	Maternal interview "during the first year"	583	Developmental milestones	Grandjean et al. 1995

French Guiana	Approx. 400	Maternal hair	Median, 6.6 ppm range, 2.6-17.8 35%>10 ppm	9 month-12 years	248 (neuro exam)	neurological exam fingertapping, Stanford-Binet: block design, copying, bead memory	Cordier and Garel 1999
					206 (psychological exam)	MSCA: numerical memory, leg coordination	
Madeira	149	Maternal hair	Geometric Mean, 9.6 ppm Range, 1.1-54.4 52%> 10ppm	6-7 years	146-149	BAEP, VP NES: fingertapping, hand eye coordination, continuous performance test WISC-R: digit span, block design Stanfford-Binet: bead memory	Murata et al. 1999a
Mancora, Peru	369	Maternal hair	Geometric mean, 7.1 ppm; geometric SD, 2.1; range, 0.9-28.5	?	194 (131 with both exposure and outcome data)	Neurological exam Developmental milestones	Marsh et al. 1995a
Northern Quebec	247	Maternal hair	Mean, 6 ppm; 6%, >20 ppm	12-30 mon	234	Neurological exam	McKeown-Eyssen et al. 1983
New Zealand	10,930 mothers screened, 935 "high" fish consumers identified, 73 "high" Hg mothers identified	Maternal hair	"High" Hg defined as >6 ppm; mean, 8.3 ppm in "high" Hg group; range, 6-86 ppm; only 16 values >10 ppm	4 yr	74; 38 "high" Hg, 36 "low" Hg, including 30 matched pairs	DDST, vision, functional neurological exam	Kjellstrom et al. 1986
				6 yr	237; 57 complete sets of 1 "high" Hg child, 3 matched controls, and 4 incomplete sets	WISC-R, TOLD, MSCA, CDS, BWRT, KMDAT, PPVT, EBRS	Kjellstrom et al. 1989

TABLE 5-10 *(Continued)*

Study Site	Size of Cohort Identified and Enrolled	Exposure Biomarker	MeHg or Total Hg Concentration	Age at Assessment	Number of Children Assessed	End Points	Reference
Seychelles Islands (pilot)	804	Maternal hair	Median, 6.6 ppm; range, 0.6-36.4 ppm, interquartile range: 6.1	5-109 wk	789	Neurological exam, DDST-R	Myers et al. 1995a
				66 mon	217	MSCA, PLS, WJTA:LWI, WJTA:AP	Myers et al. 1995c
Seychelles Islands (main)	779	Maternal hair	Median, 5.9 ppm; interquartile range, 6.0 ppm; all values, <30 ppm	6.5 mon	712-737	Neurological exam, DDST-R, FTII, visual attention	Myers et al. 1995b
				19 mon	738	Developmental milestones	Myers et al. 1997; Axtell et al. 1998
						BSID	Davidson et al. 1995b
				29 mon	736	BSID	Davidson et al. 1995b
				66 mon	711	MSCA, PLS, B-G, WJTA:LWI, WJTA:AP, CBCL	Davidson et al. 1998

Abbreviations: DDST, Denver Developmental Screening Test (DDST-R is revised version); MSCA, McCarthy Scales of Children's Abilities; PLS, Preschool Language Scale; WJTA, Woodcock Johnson Test of Achievement (AP, applied problems; LWI, letter-word identification); FTII, Fagan Test of Infant Intelligence; BSID, Bayley Scales of Infant Development; B-G, Bender-Gestalt Test; CBCL, Child Behavior Checklist; WISC-R, Wechsler Intelligence Scale for Children-Revised; TOLD, Test of Language Development; CDS, Clay Diagnostic Survey; BWRT, Burt Word Recognition Test; KMDAT, Key Math Diagnostic Arithmetic Test; PPVT, Peabody Picture Vocabulary Test; EBRS, Everts Behaviour Rating Scale; NES, Neurobehavioral Evaluation System; CVLT-C, California Verbal Learning Test-Children; BNT, Boston Naming Test; POMS, Profile of Mood States; VEP, Visual Evoked Potentials; BAEkP, Brainstem Auditory Evoked Responses.

Status on Neurological Examination

McKeown-Eyssen et al. (1983) studied 234 12- to 30-month-old Cree children (95% of eligible children) for whom prenatal MeHg exposure was estimated on the basis of maternal-hair samples. The subjects lived in four communities in northern Quebec. For 28% of the mothers, hair samples were collected during pregnancy; for the balance of the cohort, prenatal exposure was estimated on the basis of hair segments assumed to date from the time the study child was in utero. The measure of exposure used was the maximum concentration of Hg in the segment of hair corresponding most closely to the period from 1 month before conception to 1 month after delivery. The mean maternal-hair Hg concentration was approximately 6 ppm, with 6% of samples exceeding 20 ppm. One of four pediatric neurologists blinded to individual Hg-exposure status, measured height, weight, and head circumference, identified dysmorphologies, and conducted a neurological examination (assessing coordination, cranial nerves, muscle tone, and reflexes). The neurologist made a summary clinical judgment as to the presence of a neurological abnormality. No child was judged to have any abnormal physical findings. Overall, 3.5% (4) of the boys and 4.1% (5) of the girls were considered to have a neurological abnormality. The most frequent abnormality involved tendon reflexes, seen in 11.4% (13) of the boys and 12.2% (14) of the girls. The only neurological findings significantly associated with prenatal MeHg exposure, either before or after adjustment for confounding, were abnormalities of muscle tone or reflexes in boys. Two boys had increased tone in the legs only, five had isolated decreased reflexes, six had generalized decreased reflexes, and two had generalized increased reflexes ($p = 0.05$). The risk of an abnormality of tone or reflexes increased 7 times with each 10-ppm increase in prenatal MeHg exposure (95% CI 1.0-51.0). With log transformation of prenatal MeHg exposure, however, the p value associated with the risk of an abnormality due to MeHg exposure increased to 0.14. When exposure was categorized, the prevalence of tone or reflex abnormality did not increase in a clear dose-response manner across categories (i.e., 15.8%, 5.6%, 26.3%, 0%, 7.1%, and 38.5%). In girls, the only association identified was in the unexpected direction between prenatal MeHg exposure and incoordination (60% decrease in probability of incoordination for each 10-ppm increment; odds ratio (OR), 0.3; 95% CI, 0.1-0.9; $p = 0.02$).

The authors noted five caveats about the one significant adverse association identified: (1) the abnormalities of muscle tone and reflexes in boys were isolated, mild, and of doubtful clinical importance; (2) children exposed to very high MeHg doses manifested as severe generalized neurological disease, including increases in tone and reflexes, rather than the mild, isolated muscle tone and reflex abnormalities (mostly decreased) seen in Cree children; (3) the absence of a coherent dose-response relationship; (4) the absence of consistency across sex; and (5) the possibility that the finding reflects chance, lack of normality in the distribution of the exposure index, or residual confounding.

Infants' status on neurological examination was also evaluated as an end point in a study of 194 children in Mancora, Peru. Although the study was conducted in the early 1980s, it was not published until 1995 (Marsh et al. 1995a). Fish consumption was the primary route of MeHg exposure, and maternal hair was used as the index of exposure (geometric mean, 7.05 ppm; range, 0.9 to 28.5 ppm). Geometric-mean peak hair MeHg concentration was similar (8.34 ppm; range, 1.2 to 30.0 ppm), suggesting that the women were in steady state due to stability in their fish-consumption patterns. Maternal-hair samples and data on child neurological status were available for 131 children. Several elements of the study design are not described, including the size of the eligible population from which the 131 children were sampled, the specific elements of the neurological assessment conducted, and the ages at which the children were examined. However, frequencies are reported for the following end points: tone decreased (two children), tone increased (none), limb weakness (one child), reflexes decreased (one child), reflexes increased (four children), Babinski's sign (an indication of a pyramidal-tract abnormality) (one child), primitive reflexes (none), and ataxia (none). No end point was significantly associated with either mean or peak maternal-hair Hg concentration.

A cross-sectional pilot study was carried out for the Seychelles Child Development Study (SCDS) (Myers et al. 1995a). For 2 years before the start of the study, all women attending an antenatal clinic were asked to provide one or more hair samples during and after pregnancy. A total of 804 infants were subsequently enrolled in the study, and tested during three visits over 2 months in 1987-1988. No data are provided on the size of the population from which that sample was drawn. Fifteen infants were excluded due to maternal illnesses during pregnancy (e.g., diabetes or eclampsia) or to newborn characteristics thought to place a child at developmental risk (e.g., low birth weight or maternal alcohol ingestion during pregnancy) (Marsh et al. 1995b). A total of 789 infants and children were evaluated between the ages of 5 and 109 weeks by one blinded pediatric neurologist. Mean maternal-hair Hg concentration in the cohort was 6.1 ppm (range, 0.6 to 36.4 ppm). The end points assessed were mental status, attention, social interactions, vocalizations, behavior, coordination, postures and movements, cranial nerves (II-XII), muscle strength and tone, primitive and deep tendon reflexes, plantar responses, and age-appropriate abilities such as rolling, sitting, pulling to stand, walking, and running. The statistical analyses focused on three end points selected due to their apparent sensitivity to prenatal MeHg exposure in the Iraq and Cree studies: overall neurological examination, increased muscle tone, and deep tendon reflexes in the extremities. The overall examination was considered to be abnormal if any findings judged to be pathological were present or if the examiner judged the child's speech or functional abilities to be below age level. Pathological findings included abnormalities of cranial nerves (pupils, extraocular muscles, facial or tongue movement, swallowing, or hearing), alteration in muscle tone or deep tendon reflexes (increase or decrease), incoordination, and involuntary movements. Findings that were not considered to be either normal or pathological were categorized as questionable. Because of the low frequency of abnormal examinations (2.8%), the questionable (11.3%) and abnormal categories were combined. No association was evident between maternal-hair Hg concentration and questionable and abnormal results. The frequency of those results ranged from 16.5% in the group with Hg at 0-3 ppm to 11.7% in the group with Hg at more than 12 ppm. The frequencies of abnormalities of limb tone or deep tendon reflexes were about 8%, and the frequency of both end points did not vary with maternal-hair Hg concentrations in a dose-dependent manner.

The main cohort of the SCDS consisted of 779 mother-infant pairs, representing approximately 50% of all live births during the recruitment period. The final sample size was 740. In addition to 18 infants being excluded for the criteria used in the pilot study, 15 were excluded because of insufficient maternal-hair samples, and 6 were excluded for being a twin. When the infants were 6.5 months old, one blinded pediatric neurologist administered essentially the same neurological examination that had been used in the pilot phase (Myers et al. 1995b). The overall examination was considered abnormal if changes in muscle tone, deep tendon reflexes, or other neurological features were pathological or if functional abilities were not considered appropriate for the age. An examination could also be coded as questionable. A total of 3.4% (25) of the children had overall neurological scores considered abnormal or questionable, a frequency too low to permit statistical analysis of the overall neurological examination. The frequency of abnormalities was 2% for both limb tone and abnormal deep tendon reflexes. Questionable limb tone was identified in approximately 20% of the children, and questionable deep tendon reflexes, in approximately 15%. Although such findings were not considered pathological, they were combined with abnormal findings for statistical analyses. The frequency of abnormal and questionable findings for limb tone or deep tendon reflexes was not significantly associated with maternal-hair Hg concentrations.

Steuerwald et al. (2000) recruited a cohort of 182 singleton, full-term infants born in the Faroe Islands and evaluated the associations between neurological function at 2 weeks of age and various dietary contaminants and nutrients. The cohort represents 64% of all births in the catchment area. The primary outcome variable was the neurological optimality score (NOS), which reflects an infant's functional abilities, reflexes, responsiveness, and stability of state. In addition, two subscores were generated (muscle tone and reflexes). A variety of thyroid-function indices considered to be outcomes were also assessed. The exposure biomarkers measured were Hg concentration in maternal hair, cord whole blood, and cord serum. Measurements were also taken of 18 pesticides (or metabolites) and 28 polychlorinated biphenyl (PCB) congeners in maternal serum (lipid adjusted) and breast milk, selenium in cord whole blood, and fatty acids (arachidonic, eicosapentanoic, docosahexaenoic, and total omega) in cord serum. There was a significant inverse relationship between NOS scores and cord-whole-blood Hg concentrations. The mean concentration was 20.4 μg/L (range, 1.9-102 μg/L) (see Figure 5-4). Although the unadjusted correlation between cord-whole-blood Hg concentration and NOS score was modest (−0.16), a 10-fold increase in cord-whole-blood Hg was associated with the equivalent of a 3-week reduction in gestational age based on NOS score. Adjustments for total PCBs and fatty acid concentrations did not appreciably affect the results. Selenium did not appear to function as an effect

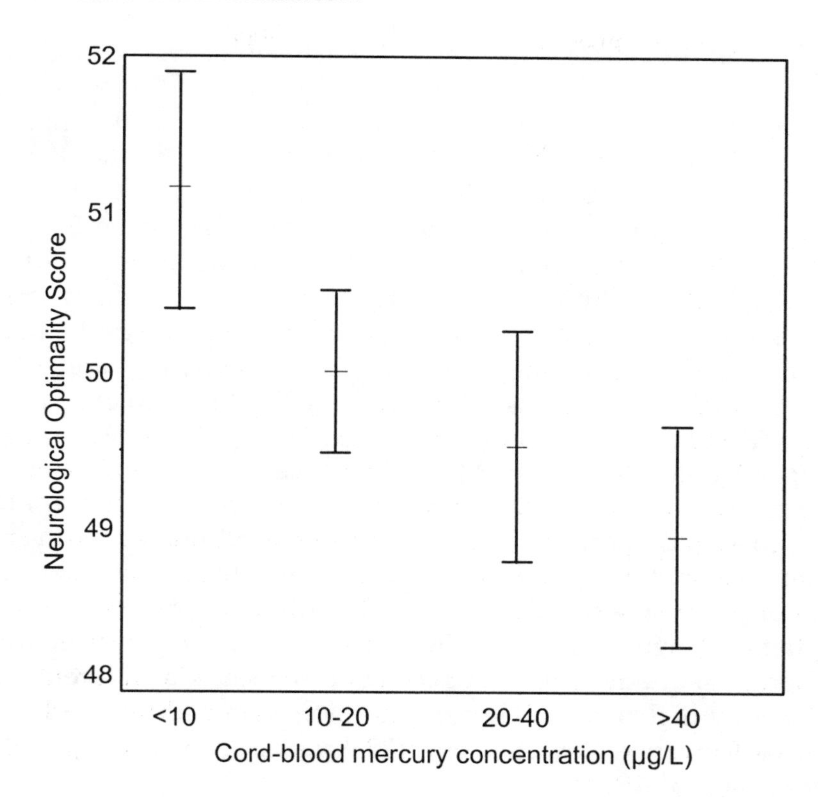

FIGURE 5-4 Neurological optimality score (mean ± standard error of the mean) in relationship to cord-blood Hg concentrations in approximate quartile groups. Source: Stewerwald et al. 2000. Reprinted with permission from the *Journal of Pediatrics*; copyright 2000, Mosby, a Harcourt Health Sciences Company.

modifier. Muscle-tone and reflexes subscores were not significantly associated with any exposure biomarker. Maternal hair-Hg concentrations (mean, 4.08 ppm; range, 0.36-16.3 ppm; 10.4%, more than 10 ppm) were not significantly associated with NOS scores.

A functional neurological examination, as part of a general physical examination, was administered at age 7 years to another cohort of children from the Faroe Islands. The cohort consisted of consecutive deliveries at three hospitals during a 21-month period between 1986 and

1987. Of 1,386 infants born, cord-blood and maternal-hair samples were obtained from 1,022 singleton births (75%) and 917 children were examined (66%) (Grandjean et al. 1992). The mean cord-blood Hg concentration was 22.9 μg/L; the mean maternal-hair Hg concentration was 4.3 ppm. In particular, the examination of the cohort at 7 years of age focused on motor coordination and perceptual-motor performance (Dahl et al. 1996). The coordination tests included diadochokinesia (fast pronation and supination), reciprocal coordination (alternately closing and opening the fists), and finger opposition (the pulpa of the thumb touching the pulpa of the other fingers of the same hand). The perceptual-motor tests included catching a ball with a diameter of 15 cm thrown from a distance of 4 m, finger agnosia, and double finger agnosia. Results were scored as automatic or questionable and poor. Hg concentration was not significantly associated with the number of tests on which a child's performance was considered automatic. On the tests of reciprocal motor coordination, simultaneous finger movement, and finger opposition, fewer than 60% of the children achieved a score of automatic. Finger opposition, however, was the only test in which children with questionable and poor performance (425 children) had a significantly higher mean Hg concentration than children with automatic performance (465 children) (23.9 versus 21.8 μg/L, $p = 0.04$) (Grandjean et al. 1997).

Cordier and Garel (1999) recently reported on the association between MeHg exposure and neurological status in 9-month-old to 6-year-old children living in gold-mining regions of French Guiana. The concentrations of Hg in samples of hair collected from children's mothers at the time of the study were used as a surrogate for exposure during pregnancy. The median concentration was 6.6 ppm (range, 2.6-17.8 ppm; 35%, greater than 10 ppm). Among children 2 years of age and older, the prevalence of increased reflexes was significantly higher with increased Hg concentrations in maternal hair, the association being stronger in boys than in girls. When 10 children who had been found to have increased reflexes were re-examined 9 months later by a different examiner, only three were considered to have increased reflexes. Therefore, the investigators advised caution in interpreting those data.

Overall, the evidence that children's neurological status is associated with low-dose prenatal Hg exposure consists of four findings: (1) an increased prevalence of tone or reflex abnormalities (most often decreased) in boys with increased maternal-hair Hg concentrations, although that effect is not dose dependent (McKeown-Eyssen et al. 1983); (2) an inverse association between newborns' NOS and umbilical-cord Hg concentration in the Faroe Islands (Steuerwald et al. 2000); (3) a modest but statistically significant increase (2.1 μg/L) in the mean cord-blood Hg samples of 7 year olds who performed suboptimally on a finger opposition test compared with children whose performance was normal (Grandjean et al. 1997); and (4) an association, especially in boys, between increased reflexes and higher maternal-hair Hg concentrations in a cohort of 9-month-old to 6-year-old children in French Guiana (Cordier and Garel 1999). One limitation in the use of neurological status as an end point is its categorical nature; a child either expresses a particular abnormality or does not. In the SCDS main cohort, the prevalence of abnormal neurological findings was quite low (i.e., 3.4% for abnormal or questionable findings), limiting the statistical power of hypothesis testing. Although the high-dose exposure episodes that occurred in Minamata and Iraq produced classic signs of neurological dysfunction in children exposed in utero, the low doses of MeHg to which the cohorts in the epidemiological studies were exposed prenatally appeared to be associated with subtle neurological effects that are of uncertain clinical significance (e.g., tone or reflex abnormalities). Research on other environmental toxicants such as lead and PCBs has shown, however, that it is important to distinguish individual risk from population risk. A decrement in mean function that is too small to be clinically significant for the individual child might be quite important when it is considered from the standpoint of the impact on the population distribution of the affected function (Weiss 1998).

Age at Achievement of Developmental Milestones

The association between the achievement of developmental milestones and prenatal MeHg exposure was evaluated in the main cohort of the SCDS by Myers et al. (1997) and Axtell et al. (1998). The ages at which a child was able to walk without support and to say words other than "mama" or "dada" were determined by an interview with a child's primary caregiver (person with whom the child spent 5 or more nights per week) conducted at the 19-month evaluation. Those data were available for 738 of the 779 children enrolled. Prenatal MeHg exposure was estimated as the total Hg in the single longest hair segment dating from the time the study child was in utero (mean, 5.8 ppm; range, 0.5 to 26.7 ppm; 22%, greater than or equal to 10 ppm). Several statistical approaches were carried out, including a standard multiple regression of a log transformation of the age at milestone achievement, hockey-stick models estimating the threshold maternal-hair Hg concentration associated with delay in milestone achievement, and logistic regression analyses of delayed walking, a binary variable in which an abnormal response was defined as greater than 14 months. The mean age at which a child was considered to talk was not significantly associated with maternal-hair Hg in any of the models tested. In regressions stratified by child sex, a positive association was found between age at walking and maternal-hair Hg in boys ($p = 0.043$) but not in girls. A term for the interaction between Hg and sex was not statistically significant in the analyses of the complete cohort, however. The magnitude of the delay in the age at which boys walked was viewed by the authors as clinically insignificant; a 10-ppm increase in maternal-hair Hg was associated with an approximate 2-week delay in walking (see Figure 5-5). The association in boys was not significant when four statistical outliers were excluded from the analysis. Hockey-stick models provided no evidence of a threshold, as the fitted curves were essentially flat. A child's risk of delayed walking was not associated with maternal-hair Hg concentration. Axtell et al. (1998) re-analyzed those milestone data, applying semiparametric generalized additive models, which use smoothing techniques to identify nonlinearities. Those models are less restrictive than the approaches used by Myers et al. (1997), whose approaches make strong assumptions about the true functional form of a relationship.

The major finding of the analyses of Axtell et al. (1998) was that the association between age at walking and maternal-hair Hg in boys was nonlinear, walking appearing at a later age as concentrations increased from 0 to 7 ppm but at a slightly earlier age as Hg concentration increased beyond 7 ppm. The size of the effect associated with the increase from 0 to 7 ppm was very small, corresponding to a delay of less than 1 day in the achievement of walking. Because a coherent dose-response relationship did not hold above 7 ppm, the authors expressed

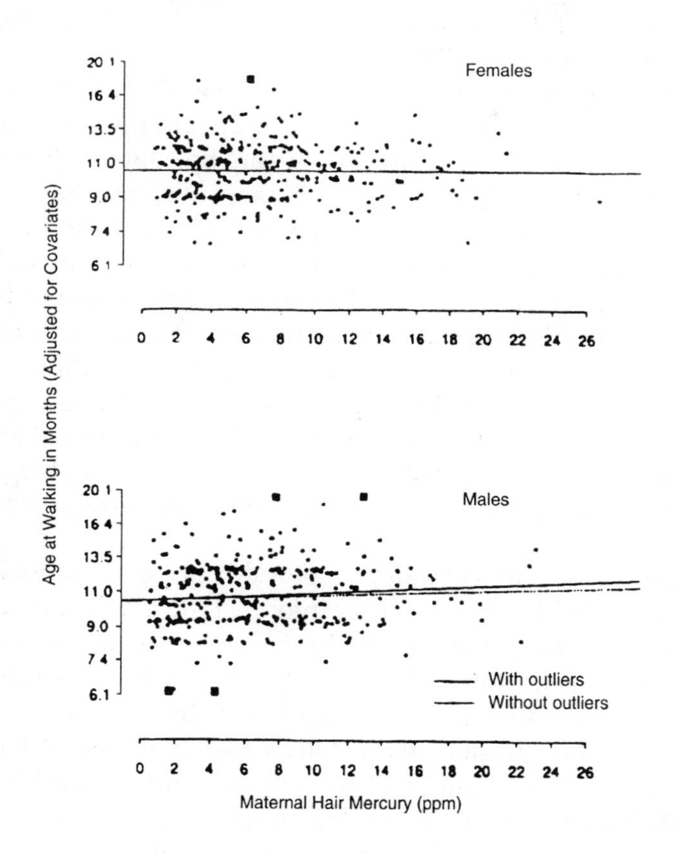

FIGURE 5-5 Plot of partial residuals for log of age at walking versus maternal-hair Hg concentrations for the reduced model with Hg by gender interaction. The partial residual is the natural log of the subject's age at walking adjusted for all variables in the model except Hg. It is computed by adding the Hg effect (estimated coefficient on Hg times the Hg value for that subject) to the raw residual (observed value minus predicted value) obtained from the reduced model. The partial residuals have been rescaled by adding the mean value of log of age at walking to each partial residual. The axis for age at walking (in months) is on a log scale. Outliers are identified on the plot with a different symbol (solid squares). The slope of the solid lines is the regression coefficient for Hg in the originial regression analysis with statistical outliers. The dotted line is the slope with those outliers removed. Source: Myers et al. 1997. Reprinted with permission from *NeuroToxicology*; copyright 1997, Intox Press.

doubt that the association found below 7 ppm reflected a causal effect of Hg exposure on age at walking.

Data on developmental milestones were collected in the Peruvian study conducted by Marsh et al. (1995a). The ages of the children when mothers were queried about the milestones are not stated, although the study was conducted prospectively and data were apparently collected in an ongoing manner over the course of a woman's visits to a postnatal clinic. Regression analyses, including analyses stratified by child sex, did not reveal any significant associations between maternal-hair Hg concentrations and the ages at which children sat, stood, walked, or talked. The geometric mean maternal-hair Hg concentration was 7.05 (S.D. = 2.06). The rates of developmental retardation, especially in speech (13 of 131), were substantial, although the criteria used to define that outcome were not provided. Children's birth weight, height, and head circumference were also unrelated to maternal-hair Hg concentrations.

Ages at milestone achievement of motor development were investigated in a 21-month birth cohort (1022 infants, 1986-1987) of children in the Faroe Islands (Grandjean et al. 1995). Milestone data were obtained from maternal interviews and the observations of district health nurses who visited the homes on several occasions during the children's first year of life. Hg concentrations were determined in maternal-hair samples at delivery, infants' umbilical cord blood, and in children's hair samples obtained at about 12 months of age. Complete data were available for 583 children (57% of the complete cohort). Three motor-development milestones commonly achieved between 5 and 12 months of age were selected for analysis: "sits without support," "creeps," and "gets up into standing position with support." The age at achievement was not significantly associated with either index of prenatal Hg exposure (cord-blood or maternal-hair concentrations) for any of the three milestones. For all three milestones, however, a significant inverse association was found between age at achievement and children's hair Hg concentration at 12 months. Children's hair Hg concentration was interpreted as an index of children's postnatal exposure to MeHg. Nursing was associated with both higher hair Hg concentrations in children at 12 months of age and with more rapid achievement of milestones. Therefore, the authors concluded that the inverse association reflected residual confounding by duration of breast feeding. That finding suggests that the beneficial effects of nursing on early motor development are sufficient to compensate for any slight adverse impact that low-dose prenatal MeHg exposure might have on the end points.

In conclusion, recent epidemiological studies provide scant evidence that prenatal MeHg exposures, at least those resulting in maternal-hair Hg concentrations below 30 ppm, are associated with the ages at which children achieve developmental milestones. Although the mean age at walking in the SCDS cohort was later among children whose mothers had high hair Hg concentrations, that association was limited to boys, and the risk of late walking did not appear to be dose related. The association was apparent only at concentrations below 7 ppm, and increases in maternal-hair Hg concentrations above 7 ppm were not associated with further delay in walking age of boys. In the Faroe Islands cohort, a negative association was found between children's hair Hg concentration at 12 months and age at achievement of three motor-development milestones. That finding might be due to higher Hg exposure among breast-fed children and might actually reflect beneficial nutritional effects from breast milk. Those recent data are consistent with re-analyses of the Iraqi data (Cox et al. 1995; Crump et al. 1995), suggesting that the population thresholds for delayed achievement of milestones that were originally calculated might be too low. The thresholds appear to be highly dependent on the assumptions made about background prevalence of delay, the definition of late achievement used, and the influence exerted by a small number of data points.

Infant and Preschool Development

In several epidemiological studies, the association between low-dose prenatal MeHg exposure and early child development has been assessed using several widely used standardized tests.

In the Cree study reported by McKeown-Eyssen et al. (1983), the Denver Developmental Screening Test (DDST) was administered to the 12- to 30-month-old children in the cohort. Scores were reported as the percentage of items passed on each subscale (gross-motor, fine-motor, language, and personal and social subscales) and on the entire test.

Although quantitative estimates are not provided for the associations between test scores and maternal-hair Hg concentrations (mean, 6 ppm; 6% greater than 20 ppm), the authors reported that they did not find any significant associations in a direction compatible with an adverse effect of MeHg either before or after adjustment for confounding variables.

Kjellström et al. (1986) studied a cohort of New Zealand children for whom prenatal MeHg exposure was estimated on the basis of maternal-hair samples as well as dietary questionnaires collected during the period when the study child was in utero. Although exposure information was collected on nearly 11,000 women, the authors focused on 935 women who reported eating fish more than three times per week during pregnancy. Seventy-three women had hair Hg concentrations greater than 6 ppm. The 74 children of those women were designated as the "high-Hg group." Efforts were made to match each child in the high-Hg group with a reference child on the basis of maternal ethnicity, hospital of birth, maternal age, and child age. In the follow-up evaluations completed when children were 4 years old, 38 exposed and 36 reference children were tested, including 30 complete matched pairs. On the DDST, the primary outcome used at this age, 52% of the children in the high-Hg group had an abnormal or questionable result compared with 17% of the children in the control group ($p < 0.05$). That result corresponds to an odds ratio of 5.3. Results were similar when pairs that were poorly matched on ethnicity were excluded. It was not possible to identify the specific developmental domains in which performance was most strongly associated with maternal-hair Hg concentrations.

In the SCDS pilot cross-sectional study, a revised version of the DDST (the DDST-R) was administered blindly by one examiner to 789 children between the ages of 1 and 25 months (Myers et al. 1995a). No association was found between maternal-hair Hg concentration during pregnancy (mean 6.6 ppm) and DDST-R results when normal and questionable examinations were combined in the conventional manner, although the prevalence of abnormal findings was so low (three children, less than 1%) that statistical analysis was not meaningful. When abnormal and questionable results were grouped (in 65 children, 8%), as was done in the New Zealand study (Kjellström et al. 1986), however, high maternal-hair Hg concentrations were significantly associated with poor outcomes ($p = 0.04$, one-tailed test). That result was largely attributable to the higher frequency of abnormal and questionable results (approximately 13%) among children in the highest hair-Hg category (greater than 12 ppm), in contrast to the frequency of approximately 7% among children in each of the other four Hg groups (0-3, 3-6, 6-9, and 9-12 ppm).

In the main SCDS study, the DDST-R was administered by one blinded examiner to a cohort of 740 children at age 6.5 months (mean maternal-hair Hg concentration during pregnancy, 5.9 ppm; interquartile range, 6.0 ppm) (Myers et al. 1995b). The frequencies of examinations considered to be abnormal (three children, 0.4%) or questionable (11 children, 1.5%) were very low, precluding meaningful statistical analysis of the DDST-R data. The Fagan Test of Infant Intelligence (FTII), an assessment of visual-recognition memory or novelty preference, was also administered at 6.5 months to 723 children. The mean percent novelty preference in the entire cohort was 60%, similar to that observed in many other cohorts, and varied by less than 1% across categories of maternal-hair Hg concentration (less than 3 ppm to greater than 12 ppm). Visual attention (the time required to reach visual-fixation criterion on familiarization trials) also was unrelated to maternal-hair Hg concentrations.

The Bayley Scales of Infant Development (BSID) was administered by blinded examiners to children in the SCDS cohort at ages 19 and 29 months (738 at 19 months and 736 at 29 months) (Davidson et al. 1995b). In addition, at 29 months, six items of the Infant Behavior Record (IBR), a rating scale, were completed by the examiner, assessing activity level, attention span, responsiveness to examiner, response to caregiver, cooperation, and general emotional tone. The BSID yield two primary scores: the mental development index (MDI) and psychomotor development index (PDI). At both ages, MDI scores (97.5 and 100.4 at 19 and 29 months, respectively) were similar to the expected mean for U.S. children of 100 ± 16. At both ages, however, the SCDS children performed markedly better on PDI (with scores averaging 126.7 and 121.1 at 19 and 29 months, respectively) than the expected mean for U.S. children. In fact, at 19 months, approximately 200 children in the SCDS cohort achieved the highest possible PDI score of 150 (Davidson et al. 1995a), a finding that most likely reflects the ethnic composition of the cohort. Because of this skew, PDI scores at both ages were expressed as a binary variable, splitting the distribution at the median score. The MDI scores at 19 or 29 months were not significantly associated with maternal-hair Hg concentration during pregnancy (see Figure 5-6). Similar results were obtained in a secondary analysis that included only children with the lowest (less than or equal to 3 ppm) or highest (greater than 12 ppm) maternal-hair Hg concentrations. Assessments of perceptual skills at 19 months (Kohen-Raz method), dichotomized due to skewing, were not associated with Hg exposure. Scores on that test at 29 months could not be evaluated because of a pronounced ceiling effect. Risk of a PDI score below the median was not significantly associated with maternal-hair Hg concentration in the full logistic regression model but was associated ($p = 0.05$) with this exposure index in a reduced model in which adjustment was made for a smaller number of covariates selected on an a priori basis. The secondary analysis of the PDI scores of children with the lowest and highest Hg concentrations was not conducted, because the full logistic regression model was not statistically significant.

In the analyses of the six IBR items, maternal-hair Hg concentration was significantly associated only with examiner ratings of activity level during the test session and only in males. The score decreased 1 point (on a 9-point scale) for each 10 ppm. Additional analysis of the data of the main SCDS study cohort failed to identify significant effect modification by factors such as caregiver intelligence, H.O.M.E. score, family income, and gender (Davidson et al. 1999).

Among the four studies that used the DDST (or DDST-R) as a measure of infant development, only in the New Zealand study and the SCDS pilot phase did children's scores appear to be associated with prenatal Hg concentrations, at least when the questionable and abnormal findings were combined. One factor that might partially account for the differences between the findings of those studies is the age at which the examinations were conducted (4 years in the New Zealand study, 1 to 25 months in the SCDS pilot phase and 6.5 months in the SCDS main phase). Another factor is the different rates of abnormal or questionable examinations (50% of the New Zealand group with prenatal maternal-hair concentrations greater than 6 ppm and 17% of controls; 8% of the Seychelles complete cohort in the SCDS pilot phase; and 1.9% of the cohort in the SCDS main phase). That difference is large enough to raise the possibility that the test items were either administered differently in

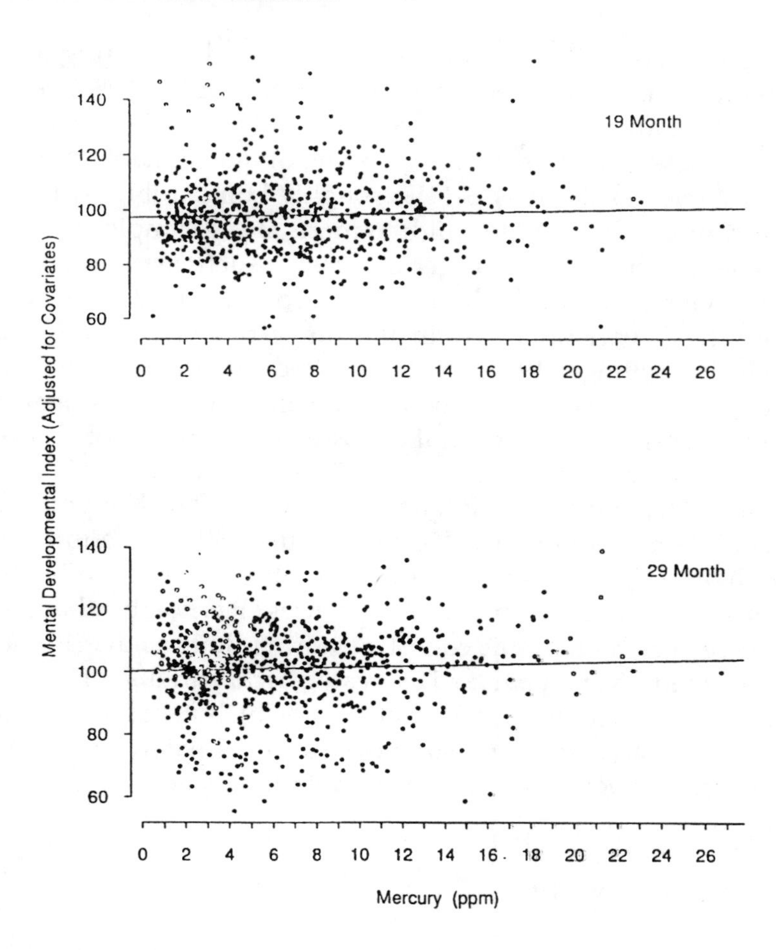

FIGURE 5-6 The 19-month and 29-month mental-developmental-index (MDI) partial residuals from the Bayley Scales of Infant Development. Each data point represents the overall cohort MDI mean plus the partial residual. The partial residual is defined as the subject's MDI score adjusted for all variables in the reduced model except Hg (computed by adding the Hg effect to the residual from the reduced model). The MDI has a U.S. mean of 100 (standard deviation, 16). Scores are plotted as a function of maternal-hair Hg in parts per million. The slope for the 19-month MDI, shown in the upper graph, was 0.125. The slope for the 29-month MDI, shown in the lower graph, was 0.149. Neither effect was significant. Source: Davidson et al. 1995b. Reprinted with permission from *NeuroToxicology*; copyright 1995, Intox Press.

the two studies or that different criteria were used in judging whether an individual passed or failed the tests. Kjellström et al. (1986) reported, however, that among 3- and 4-year-old children in South Auckland routinely assessed with the DDST, the rate of questionable, abnormal, or not testable results was 8-14%, roughly comparable to the rates observed among the low-Hg children in the study sample.

In general, the use of screening tests, such as the DDST, in neurobehavioral toxicology studies is not recommended because of their insensitivity to variations within the range of normal performance (Dietrich and Bellinger 1994). More detailed instruments, such as the BSID, currently considered to be the best in infant assessment, have proved to be sensitive to prenatal exposures to a variety of neurotoxicants, including lead (Bellinger et al. 1987; Dietrich et al. 1987; Wasserman et al. 1992) and PCB's (Rogan and Gladen 1991; Koopman-Esseboom et al. 1996). Among the Hg studies, the BSID was administered only in the SCDS, and no significant associations were found between children's scores and their prenatal exposures. It is notable that the PDI scores were very high in this cohort, requiring that the distribution be split at the median and analyzed as a categorical variable. The median value is not provided by Davidson et al. (1995b), but, based on Figure 3 in Davidson et al. (1995a), appeared to be approximately 130, or 2 standard deviations above the expected population mean.

Childhood Development

Children in the New Zealand cohort were followed up at 6 years of age. In that phase of the study, three controls were matched to each high-Hg child on the basis of ethnic group, sex, maternal age, maternal smoking, area of maternal residence, and the duration of maternal residence in New Zealand. One of the controls for each subject had a hair Hg concentration of 3 to 6 ppm, and the other two controls had hair Hg concentrations of 0 to 3 ppm. For one of the two low-Hg controls, maternal fish consumption was high (more than three times per week), and for the other, it was low. Fifty-seven fully matched groups of four children each and four incomplete sets (resulting in a cohort of 237 children) participated in a follow-up evaluation of neurodevelopmental status at 6 years of age (Kjellström et al. 1989). In the high-Hg group, the mean maternal-hair Hg concentration was 8.3 ppm (range 6-86 ppm, with all but 16 between 6 and 10 ppm). Extensive information was collected on possible confounding factors, such as social class, medical history, and nutrition. A battery of 26 psychological and scholastic tests was administered, assessing the domains of general intelligence, language development, fine- and gross-motor coordination, academic attainment, and social adjustment. Multiple regression analyses of five primary end points were carried out: the Test of Language Development—spoken language quotient (TOLD_SL), the Wechsler Intelligence Scale for Children-Revised (WISC-R) performance IQ, the WISC-R full-scale IQ, the McCarthy Scales of Children's Abilities perceptual-performance scale (MC_PP), and the McCarthy Scales motor scale. Analyses were adjusted for potential confounders, including maternal ethnic group, maternal age, maternal smoking and alcohol use during pregnancy, length of maternal residence in New Zealand, social class, primary language, siblings, sex, birth weight, fetal maturity, Apgar score, and duration of breast feeding. In addition, robust regression methods were applied, involving the assignment of a weight (0 to 1) to an observation depending on the degree to which it was an outlier. In the robust regressions, maternal-hair Hg concentration was associated with poorer scores (p values ranging from 0.0034 to 0.074) on full-scale IQ, language development (spoken language quotient), visual-spatial skills (perceptual-performance scale) and gross-motor skills (motor scale). The unweighted regression analyses yielded findings that were similar in direction, although generally less statistically significant. The poorer mean scores of the children in the high-Hg group appeared to be largely attributable to the children whose mothers had hair Hg concentrations above 10 ppm (for whom the mean average hair Hg concentration during pregnancy was 13 to 15 ppm and the mean of the peak monthly hair segments was about 25 ppm). Maternal-hair Hg concentrations accounted for relatively small amounts of variance in the outcome measures and generally accounted for less than covariates, such as social class and ethnic group. In additional analyses of that data set, Crump et al. (1998) found that when maternal-hair Hg was expressed as a continuous rather than a binary variable, none of the 5 primary end points studied by Kjellström et al (1989) were associated with Hg at $p < 0.10$. The results were heavily influenced, however, by the data of a child with a maternal-hair Hg concentration of 86 ppm (more than 4 times the next highest concentration), despite the fact that the child's test scores were not outliers by the usual technical criteria. When the data for this child were excluded, scores on the TOLD_SL and MC_PP were inversely associated with maternal-hair Hg concentration at $p < 0.05$. These associations were diminished somewhat in statistical significance, although not in the magnitude of the coefficient, when parental education and child's age at testing were included in the regressions. When these regressions were repeated on all 26 scholastic and psychological tests, 6 were associated with maternal hair-Hg (excluding the child with a level of 86 ppm) at $p < 0.10$: Clay Reading Test—concepts, Clay Reading Test—letter test, McCarthy Scales—general cognitive index, McCarthy Scales—perceptual-performance scale, Test of Language Development—grammar completion, and Test of Language Development—grammar understanding).

Several features of the New Zealand study are noteworthy, including the efforts made to collect data on potential confounding variables and the broad battery of standardized outcome measures administered by trained examiners. In contrast to the acute high-dose exposures experienced by the Iraqi population, the MeHg exposures of the New Zealand cohort were chronic, low dose, and most likely fairly constant over time, reflecting well-established food consumption patterns. In addition, the maternal-hair Hg concentrations were measured prospectively. As part of the SCDS pilot phase, children from the pilot cohort of 789 who turned 66 months old within a 1-year time window underwent developmental assessments (Myers et al. 1995c). Of the 247 eligible children, 217 (87.9%) were administered a test battery consisting of the McCarthy Scales of Children's Abilities, the Preschool Language Scale, and two subtests of the Woodcock-Johnson Tests of Achievement: letter-word identification and applied problems. All 73 children with maternal-hair Hg concentrations greater than or equal to 9 ppm or less than or equal to 4 ppm were assessed. The median maternal-hair Hg concentration in that subsample of the pilot cohort was 7.1 ppm (1.0 to 36.4). The frequency of missing values was substantial for some end points (e.g., 34% for the summary score of the general cognitive index (GCI) yielded by the McCarthy scales). Increased maternal-hair Hg concentrations were associated with significantly lower GCI scores ($p = 0.024$). Scores declined approximately 5 points between the lowest (3 ppm or less) and highest (greater than 12 ppm) exposure categories. A similar association was found on the perceptual-performance scale of the McCarthy scales ($p = 0.013$). Children's scores on the auditory comprehension scale of the Preschool Language Scale were also inversely associated with maternal-hair Hg concentrations ($p = 0.0019$). Scores declined approximately 2.5 points across the range of Hg concentrations. Additional analyses identified several outlier or influential data points, whose exclusion from the analyses reduced the estimates of the Hg effect substantially, sometimes to nonsignificance. In the pilot phase of the SCDS, information was not collected on several key variables that frequently confound the association between neurotoxicant exposures and child development. Those variables are socioeconomic status, caregiver intelligence, and quality of the home environment.

In the main SCDS, 711 children (91.2%) from the original cohort of 779 were evaluated at 66 months of age (5.5 years) ± 6 months using a battery of standardized neurodevelopmental tests (Davidson et al. 1998). The major domains assessed (and the tests used) were general cognitive ability (McCarthy Scales of Children's Abilities), expressive and receptive language (Preschool Language Scale), reading achievement (letter-word recognition subtest of the Woodcock-Johnson Tests of Achievement), arithmetic (applied problems subtest of the Woodcock-Johnson Tests of Achievement), visual-spatial ability (Bender Gestalt Test), and social and adaptive behavior (Child Behavior Checklist). Total Hg in a segment of maternal hair taken during pregnancy was the measurement of prenatal MeHg exposure (mean, 6.8 ppm; range, 0.5-26.7 ppm). Total Hg in a 1-cm segment of hair obtained from a child at 66 months served as the measurement of postnatal MeHg exposure (mean, 6.5 ppm; range, 0.9-25.8 ppm). The pattern of scores of the six primary end points did not suggest an adverse effect of either prenatal or postnatal Hg exposure. The associations that were found were consistent with enhanced performance among children with increased exposure to MeHg (see Figures 5-7 and 5-8). For the total score on the Preschool Language Scale, increased prenatal and postnatal Hg concentrations were significantly associated with better scores (both $p = 0.02$). For the applied problems score, increased postnatal Hg concentrations were associated with better scores ($p = 0.05$). Among boys, increased postnatal Hg concentrations were associated with fewer errors on the Bender Gestalt Test ($p = 0.009$) (see Figure 5-8).

The R^2 (square of the multiple correlation coefficient) value (0.10) of

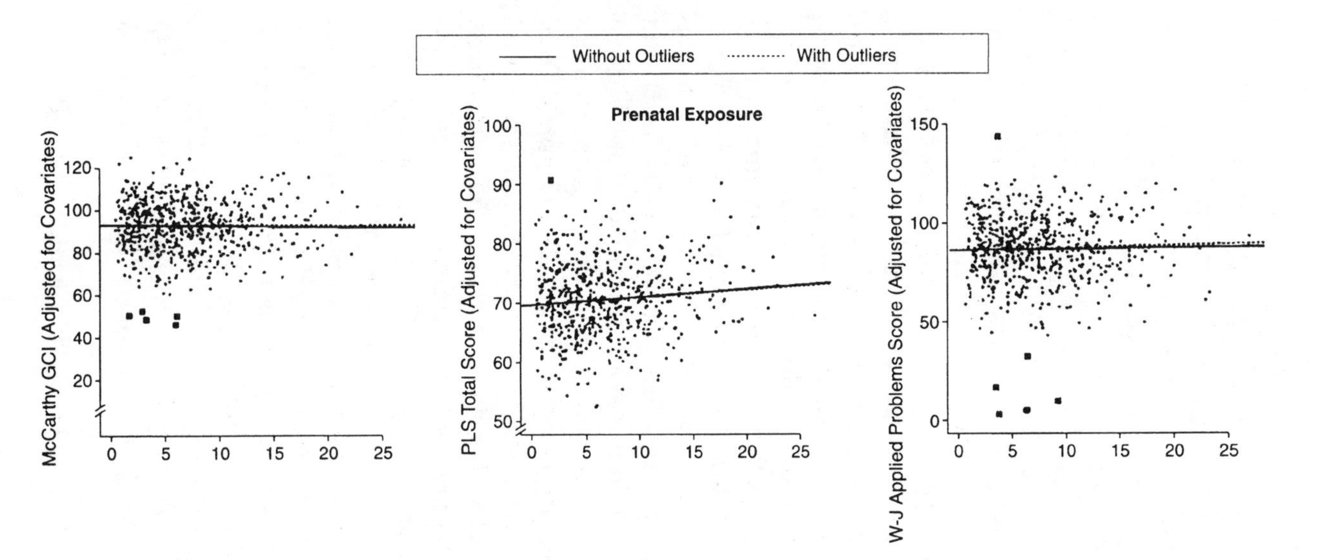

FIGURE 5-7 Partial residuals for prenatal exposure. The measures are the McCarthy Scales of Children's Abilities general cognitive index (GCI), the Preschool Language Scale (PLS) total score, and the Woodcock-Johnson (W-J) applied problems subtest. Each test score was added to the resulting partial residual. The slope of the line in the plot is the regression coefficient for the multiple regression model. Slopes are shown for the model with and without outliers. Black squares indicate outliers. Source: Davidson et al. 1998. Reprinted with permission from the *Journal of the American Medical Association*; copyright 1998, American Medical Association.

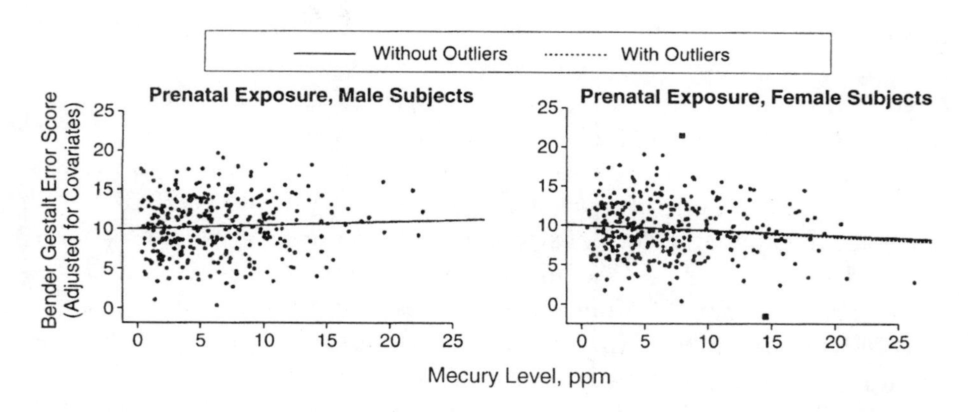

FIGURE 5-8 Partial residuals for prenatal exposure. The measures are Bender Gestalt error scores for male and female subjects. Each test score was adjusted for all reduced model predictors except the exposure value used in the plot. For graphical representations, the overall mean test score was added to the resulting partial residual. The slope of the line in the plot is the regression coefficient for the multiple regression model. Slopes are shown for the model with and without outliers. Black squares indicate outliers. Source: Davidson et al. 1998. Reprinted with permission from the *Journal of the American Medical Association*; copyright 1998, American Medical Association.

the reduced regression model for the GCI score in the main SCDS study was identical to that in the pilot study. That also appeared to be true for scores on the Preschool Language Scale (R^2 of 0.12 for the auditory comprehension scale in the pilot study and 0.14 for total score in the main study). That finding is puzzling because the pilot-study models, as noted previously, did not include several key covariates, including socioeconomic status, caregiver intelligence, and the quality of the home environment, and because the regression coefficients for socioeconomic status and caregiver intelligence were statistically significant for total scores of the GCI and the Preschool Language Scale in the main study cohort. Those differences suggest that maternal-hair Hg concentration is very highly confounded with those key covariates in the Seychelles population, or they suggest that the associations between child neurodevelopment and the covariates differ substantially in the pilot and main study cohorts, or both.

In the Faroe Islands cohort, comprehensive evaluations were conducted at approximately 7 years of age on 917 (90.3%) of the surviving members of a 1986-1987 birth cohort of 1,022 singleton births (Grandjean et al. 1997). The neuropsychological battery included three computer-administered tests from the Neurobehavioral Evaluation System (NES) (finger tapping, hand-eye coordination, and continuous performance test), the Tactual Performance Test, three subtests of the WISC-R (digit span, similarities, and block design), the Bender Gestalt Test, the California Verbal Learning Test—Children, the Boston Naming Test, and the Nonverbal Analogue Profile of Mood States. Parents were administered selected items from the Child Behavior Checklist. The primary measure of MeHg exposure was the concentration of Hg in umbilical cord blood (geometric mean, 22.9 μg/L; interquartile range, 13.4-41.3 μg/L; 894 measurements). Measurements were made of the concentration of Hg in maternal hair at parturition (geometric mean, 4.3 ppm; interquartile range, 2.6-7.7 ppm; N = 914), child hair at 12 months of age (geometric mean, 1.1 ppm; interquartile range, 0.7-1.9 ppm, N = 527), and child hair at 7 years (geometric mean, 3.0 ppm; interquartile range, 1.7-6.1 ppm, N = 903).

Not all children were able to complete all tests, and, in some cases, failure was associated with significantly increased Hg concentrations (e.g., finger opposition test and mood test). In multiple regression analyses, increased cord-blood Hg concentration was significantly associated with worse scores on finger tapping (preferred hand, $p = 0.05$), continuous performance test in the first year of data collection (false negatives, $p = 0.02$; mean reaction time, $p = 0.001$), WISC-R digit span ($p = 0.05$), Boston Naming Test (no cues, $p = 0.0003$; with cues, $p = 0.0001$), and the California Verbal Learning Test—Children (short-term reproduction, $p = 0.02$; long-term reproduction, $p = 0.05$). On the basis of the regression coefficients for Hg and age, the investigators estimated that a 10-fold increase in cord Hg concentration was associated with delays of 4 to 7 months in those neuropsychological domains (thus a doubling of Hg with delays of 1.5-2 months).

For two end points (WISC-R block design, Bender Gestalt Test errors), associations indicating adverse Hg effects ($p < 0.05$) were found when an alternative approach to adjustment for confounders (Peters-Belson method) was applied. Results were similar when the 15% of the cohort with maternal-hair Hg concentrations greater than 10 ppm were excluded from the analyses. A term for the interaction between Hg and sex was not statistically significant, indicating that the Hg effects were similar among boys and girls. In general, children's test scores were more strongly associated with cord-blood Hg concentration than with maternal-hair Hg concentration or with Hg concentrations in samples of children's hair collected at 1 and 7 years of age. Five tests were selected, on the basis of high psychometric validity, to represent key domains of cognitive function: motor, attention, visual-spatial ability, language, and memory. For the tests selected to represent attention, language, and memory, the percentages of children with adjusted scores in the lowest quartile increased significantly as cord-blood Hg concentration increased (see Figure 5-9).

In an additional set of analyses (Grandjean et al. 1998), the investigators compared the neuropsychological scores of two groups of children: a case group of 112 children with maternal-hair concentrations of 10 to 20 ppm (median, 12.5 ppm) at parturition, and a control group of 272 children with maternal-hair Hg concentrations less than 3 ppm (median, 1.8 ppm) matched to cases on age, sex, year of examination, and caregiver intelligence. Median cord-blood Hg concentrations also differed substantially (59.0 μg/L in the case group versus 11.9 μg/L in the control group). On 6 of the 18 end points, the case group scored significantly lower than the control group (one-tailed p value of 0.05). Those end points were finger tapping (both hands), hand-eye coordination (average of all trials), WISC-R block design, Boston Naming Test (no cues, cues), and California Verbal Learning Test—Children (long-term reproduction). The results of those analyses differ in certain respects from those of the main analyses. First, the set of end points on which the cases and controls differed is similar but not identical to the set of end points that were significantly associated with cord Hg concentration found in the main analyses. Moreover, in contrast to the main analyses, a term for the interaction between Hg and sex was statistically significant for several scores, including the Bender Gestalt Test error score, short-term reproduction on the California Verbal Learning Test—Children, all three finger-tapping conditions, continuous-performance-test reaction time, and average hand-eye coordination score. For all scores, adverse Hg effects were noted for boys but not girls.

Grandjean and colleagues assembled an additional study cohort of 351 children 7 to 12 years old from four riverine communities in Amazonian Brazil (Rio Tapajos) with increased exposures to MeHg due to the

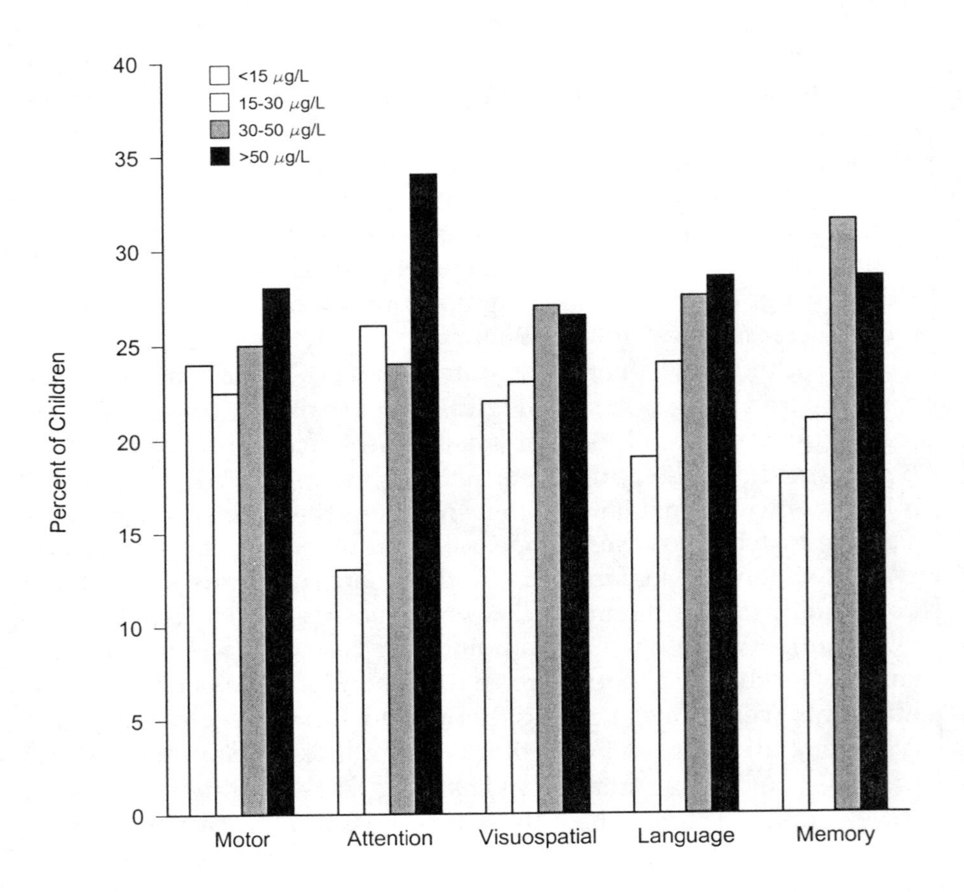

FIGURE 5-9 Prenatal Hg exposure concentrations (in quartile groups) of Faroe Island children with scores in the lowest quartile after adjustment for confounders. For each of five major cognitive functions, one neuropsychological test with a high psychometric validity was selected. Motor: neurobehavioral evaluation system 2 (NES2) finger tapping with preferred hand (p value for trend, 0.23). Attention: Reaction time on the NES2 continued performance test ($p = 0.003$). Visual-spatial: Bender visual motor gestalt test error score ($p = 0.16$). Language: Boston Naming Test score after cues ($p = 0.02$). Memory: California Verbal Learning Test—Children long-delay recall ($p = 0.004$). Source: Adapted from Grandjean et al. 1997.

consumption of fish contaminated by upstream gold-mining activities (Grandjean et al. 1999). Among children, the mean hair Hg concentration ranged from a geometric mean (range) of 11.9 (35.1) ppm for the lowest exposed of the three communities on the Rio Tapajos to 25.4 (82.9) ppm for the highest; 80% of the children in these villages exceeded 10 ppm. Most of the children reportedly ate fish for two meals per day. In the hair samples available for 63% of the children's mothers, the mean Hg concentration was 11.6 ppm. The battery of neurobehavioral tests administered to the children focused on motor function, attention, visual-spatial function, and short-term memory (finger tapping, Santa Ana dexterity test, WISC-III digit span forward, and Stanford-Binet copying (including recall), and bead memory subtests. In three villages, the tests were administered in Portuguese, although in the fourth (Village D), administration required the services of a Mundurucu interpreter. (The finger tapping, and Santa Ana dexterity tests could not be administered to children in that village.) Combining all four villages, children's hair Hg concentrations were significantly associated with their scores on finger tapping (both preferred and other hand; both $p < 0.001$), Santa Ana dexterity test (preferred hand, $p = 0.005$; other hand, $p = 0.05$), WISC-III digit span ($p = 0.001$), Stanford-Binet copying ($p < 0.001$) and recall ($p < 0.001$), and Stanford-Binet bead memory ($p < 0.001$). Adjustment for community generally reduced the magnitude of the associations, sometimes dramatically (e.g., from $p < 0.001$ to $p = 0.99$ for finger tapping preferred hand). Hair Hg concentrations and village of residence were so highly confounded, however, that adjustment for village might be inappropriate.

In the French Guiana cohort assembled by Cordier and Garel (1999) 206 children 5 to 12 years old were administered a battery of neuropsychological tests that included finger tapping, three subtests from the Stanford-Binet (block designs, copying designs, and bead memory), and two subtests from the McCarthy scales (numerical memory and leg coordination). Median maternal-hair concentration was 6.6 ppm (range, 2.6-17.8 ppm). With adjustment for potential confounders, increased Hg concentrations were associated with copying-design score especially in boys. The findings are complicated, however; when only the children living in the region that has higher exposures were considered and the analyses were stratified by sex, increased Hg concentrations were associated with poorer leg coordination in boys and poorer block-design scores in girls.

Sensory, Neurophysiological, and Other End Points in Children

In the Faroe Islands cohort, the 7-year evaluation included, in addition to the neuropsychological tests, assessments of visual acuity, near contrast sensitivity, otoscopy and tympanometry, and neurophysiological tests (pattern-reversal-visual-evoked potentials at 30′ and 15′, brainstem auditory-evoked potentials at 20 and 40 clicks per second (Hz), and postural sway) (Grandjean et al. 1997). Visual acuity, contrast sensitivity, auditory thresholds, and visual-evoked potentials were not significantly associated with prenatal MeHg exposures. For brainstem auditory-evoked potential, peaks I, III, and V were slightly delayed at increased cord-blood Hg concentrations at both 20 Hz and 40 Hz (p values, 0.01 to 0.10), although interpeak latencies were not associated with Hg at either frequency. In additional analyses reported separately (Murata et al. 1999b), in which data collected during the second year of this phase of the study were excluded due to concerns about the electromyograph used, higher maternal hair and cord-blood Hg concentrations were associated with lower peak III latencies, as well as with longer peak I-III latencies. Of the four conditions under which postural sway was assessed, only when subjects stood on the platform without foam under it with their eyes closed did the results approach significance ($p = 0.09$). Visual acuity and contrast sensitivity were not related to Hg exposure.

In a cross-sectional study of 149 6- to 7-year-old children living in a fishing village on Madiera, many of the same neurophysiological tests were administered (Murata et al. 1999a). Because patterns of fish consumption were considered to be stable, current maternal-hair Hg concentration was used as a measurement of a child's prenatal Hg exposure (mean, 9.6 ppm; range, 1.1-54.4 ppm). With respect to brainstem auditory evoked potential, maternal-hair Hg was significantly associated with I-III and I-V interpeak latencies at both 20 and 40 Hz, as well as with total latencies for peaks III and V at both frequencies. Those results are similar to the findings in the Faroe Islands cohort, at least among the children who were tested in the first year (see above). With respect to visual-evoked potentials on a pattern-reversal task, maternal-hair Hg concentration was significantly associated with one of the three latencies measured (N145 at 15′), as well as with the N75-N145 and P100-N145 latencies (15′ only). As noted above, VEP latencies were unrelated to Hg concentrations in the Faroe Islands cohort.

The relationship between blood Hg concentrations and auditory function in children and adults was investigated by Counter et al. (1998). The study sample consisted of 75 individuals (36 children and 39 adults) from a gold-mining region in Ecuador (study area) and 34 individuals (15 children and 19 adults) from a control area. Blood Hg concentrations were significantly higher in individuals from the gold-mining area than in individuals from the control region (17.5 μg/L versus 3.0 μg/L). Neuro-otological examinations were carried out on all individuals. Audiological evaluations, consisting of determinations of pure tone air-conduction thresholds in each ear at 0.25, 0.5, 1, 2, 3, 4, 6, and 8 kHz, were carried out on 40 individuals in the study area. Brainstem auditory-evoked-potential studies were carried out on 19 subjects in the study area. The absolute latencies of waves I, II, III, IV, and V and the interpeak latencies of I-III, III-V, and I-V were measured for left and right sides. Blood Hg concentration was significantly associated with hearing threshold at 3 kHz in the right ear only and for children only. A borderline association was found between blood Hg concentration and I-III interpeak transmission time on the left side. The authors concluded that although the end points assessed were generally unaffected at the blood Hg concentrations represented in the cohort of adults and children in the study area, the associations found were consistent with an effect of Hg at the level of the auditory nerve and the cochlear nuclear complex.

Animal Studies

Developmental Effects in Animals

The results of nearly 30 years of experimental studies using various animal models have helped characterize the neurotoxic effects of MeHg following in utero or early postnatal exposures (see Table 5-11). Several excellent reviews on the topic have been published over the years (WHO 1976, Chang 1977, Inskip and Piotrowski 1985, IPCS 1990, Burbacher et al. 1990, Gilbert and Grant-Webster 1995, Clarkson 1997) including a recent "Toxicological Profile for Hg" published by the Agency for Toxic Substances and Disease Registry (ATSDR 1999). In general, experimental studies have reported a continuum of neurodevelopmental effects similar to those reported in studies of humans exposed to MeHg (see

TABLE 5-11 Neurobehavioral Effects of Developmental MeHg Exposure in Animals

Species	Exposure Time	NOAEL (mg/kg/d)	LOAEL (mg/kg/d)	Effect	Reference
Monkey (*M. fascicularis*)	Birth to 7 yr old	No NOAEL	0.05	Decreased visual-contrast sensitivity (spatial) thresholds at 3-4 yr old	Rice and Gilbert 1982
Monkey (*M. fascicularis*)	In utero	No NOAEL	0.05-0.07	Retarded Object Permanence Development	Burbacher et al. 1986
Monkey (*M. fascicularis*)	In utero	No NOAEL	0.05-0.07	Impaired visual-recognition memory in offspring (abstract)	Gunderson et al. 1986
Monkey (*M. fascicularis*)	In utero	No NOAEL	0.05-0.07	Impaired visual-recognition memory in offspring (social)	Gunderson et al. 1988
Monkey (*M. fascicularis*)	Birth to 7 yr old	No NOAEL	0.05	Increased clumsiness in exercise cage at 13 yr old	Rice 1989
Monkey (*M. fascicularis*)	In utero	No NOAEL	0.05	Reduced social play Increased nonsocial behaviors	Burbacher et al. 1990
Monkey (*M. fascicularis*)	In utero plus 4 yr postnatally	No NOAEL	0.01-0.05	Decreased visual-contrast sensitivity (spatial) thresholds at 5 yr old	Rice and Gilbert 1990
Monkey (*M. fascicularis*)	Birth to 7 yr old	No NOAEL	0.05	Increased pure-tone thresholds at 14 yr old	Rice and Gilbert 1992
Monkey (*Saimiri iureus*)	Gestation wk 11-14.5 to birth (22 wk)	No NOAEL	Not stated (dosed for stable blood concentration)	Retarded schedule-control behavior during transitions	Newland et al. 1994

Monkey (*M. fascicularis*)	Birth to 7 yr old	No NOAEL	0.05	Increased vibration-sensitivity thresholds at 18 yr old	Rice and Gilbert 1995
Monkey (*M. fascicularis*)	In utero plus 4 yr postnatally	No NOAEL	0.01-0.05	Increased pure-tone thresholds at 19 yr old	Rice 1998
Monkey (*M. fascicularis*)	In utero	No NOAEL	0.05-0.09	Decreased visual-contrast sensitivity (spatial)	Burbacher et al. 1999
Rat	Gestation d 1 to postnatal d 42 (68 d)	No NOAEL	0.10	Abnormal swimming; delayed righting reflex; impaired learning in maze	Olson and Bousch 1975
Rat	Gestation d 1 to birth; birth to postnatal d 21; postnatal d 21 to d 30	No NOAEL	2.5	Impaired learning in water T-maze for all 3 groups	Zenick 1974
Rat	Gestation d 10 (in utero only)	6	8	Decreased activity during the postweaning period	Su and Okita 1976
	Gestation d 10-12 (in utero only)	No NOAEL	4	Decreased activity during the postweaning period	
Rat	Gestation d 1 to birth (in utero plus lactational)	No NOAEL	2.5	Abnormal visual-evoked potentials	Zenick 1976
Rat	Gestation d 7 (in utero plus lactational)	No NOAEL	5	Abnormal visual-evoked potentials	Dyer et al. 1978

Table 5-11 *(Continued)*

Species	Exposure Time	NOAEL (mg/kg/d)	LOAEL (mg/kg/d)	Effect	Reference
Rat (Wistar)	Gestation d 6-9 (in utero plus lactational)	No NOAEL	0.05	Reduced behavioral performance on DRH operant test	Musch et al. 1978
Rat (Wistar)	Gestation d 6-9 (in utero only)	0.005	0.01	Reduced behavioral performance on DRH operant test	Bornhausen et al. 1980
Rat (Sprague-Dawley)	Gestation d 4 (in utero plus lactational)	No NOAEL	10	Impaired avoidance learning	Schalock et al. 1981
Rat	Gestation d 8 or 15 (in utero plus lactational)	No NOAEL	5	Increased activity during the preweaning period; impaired learning in shuttle box	Eccles and Annau 1982a,b
Rat	Gestation d 6-9 (in utero plus lactational)	No NOAEL	2	Increased auditory startle response; increased activity in figure-8 maze	Buelke-Sam et al. 1985
Rat (Sprague-Dawley)	Gestation d 6-9 (in utero plus lactational)	2	6	Delayed surface righting; delayed swimming ontogeny; decreased postweaning figure-8 activity; increased time and errors in Biel maze	Vorhees 1985
Rat (Sprague-Dawley)	Gestation d 6-9 (in utero only) (cross-fostered)	1.25	2.5	Delayed surface righting; delayed negative geotaxis; delayed swimming ontogeny; decreased activity in open field	Geyer et al. 1985
Rat	2 wk before mating through weaning		0.08	Impaired tactile-kinesthetic function in offspring	Elsner 1991

Rat (Sprague-Dawley)	Gestation d 6-9 (in utero plus lactational)	2 for MeHg No NOAEL (with Hg^0)	2	Potentiated effects of Hg^0: increased activity; increased swimming speed; increased errors in radial maze	Fredriksson et al. 1996
Rat	4 wk before mating through postnatal d 16	6.4	No effects	Normal behavioral performance on DRH operant test	Rasmussen and Newland 1999
Mouse (129/SvS1)	Gestation d 7 or 9 (in utero plus lactational)	No NOAEL	8	Abnormal swimming behavior	Spyker et al. 1972
Mice	Gestation d 8 (in utero only)	No NOAEL	3	Impaired avoidance learning	Hughes and Annau 1976

Abbreviations: NOAEL, no-observed-adverse-effect level; LOAEL, lowest-observed-adverse-effect level; Hg^0, elemental mercury; DRH

Table 5-10). Those effects are largely dependent on the dose, timing, and duration of the MeHg exposure.

Fetal Minamata Disease

Experimental studies using nonhuman primates, cats, and rodent models exposed to high doses of MeHg have reported some or all of the cluster of neuropathological effects consistent with fetal Minimata disease (MD) that were first described from human autopsy cases following the catastrophic exposures in Minamata, Japan, and Iraq (Matsumoto et al 1965; Takeuchi 1968; Choi et al 1978). Those effects include microcephaly, degeneration and atrophy of cortical structures, loss of cells in the cerebrum and cerebellum, a reduction of myelin, ventricular dilation, gliosis, disorganized cell layers, and ectopic cells. In addition, seizures, spasticity, blindness, and severe learning deficits have been reported. In nonhuman primates, maternal doses above 100 μg/kg per day were associated with MD in offspring autopsied during infancy (Mottet et al. 1987; Burbacher et al. 1990). Similar effects were observed at doses as low as 1 to 1.5 mg/kg per day in mice and rats (Khera and Tabacova 1973; O'Kusky 1983), 2 mg/kg per day in golden hamsters (Reuhl et al. 1981a,b), 12 mg/kg per day in guinea pigs (Inouye and Kajiwara 1988), and 0.25 mg/kg per day in cats (Khera 1973b). Differences in the lowest-observed-adverse-effect levels (LOAELs) for signs consistent with MD do not necessarily represent true species differences in susceptibility to MeHg, because the choices of doses and exposure periods used across the studies are not comparable.

Neurobehavioral Effects

The focus of many of the developmental studies using animal models has been to define the effects of MeHg at exposures that are not associated with gross signs of toxicity. Studies using nonhuman primate and rodent models have reported numerous subclinical effects at doses below those associated with overt maternal or offspring toxicity (see Table 5-11). In a long-term study examining the effects of maternal oral doses of MeHg hydroxide at 50, 70, or 90 μg/kg per day before and during pregnancy in nonhuman primates (*Macaca fascicularis*), maternal toxicity (blindness and motor incoordination) was observed at the highest dose tested (90 μg/kg per day), and reproductive effects (nonconceptions, abortions, and stillbirths) were observed at 90 and 70 μg/kg per day (Burbacher et al. 1988). The maternal dose of MeHg hydroxide at 50 μg/kg per day was associated with developmental effects in offspring, but not with maternal toxicity or reproductive effects. Impairments in perceptual-cognitive functioning (Fagan Test and Object Permanence Test) and the development of species-specific social behavior were observed in offspring during infancy (Gunderson et al. 1986, 1988; Burbacher et al. 1986, 1990). A significant reduction in the weight gain of exposed males beginning at 2.5 years of age was also observed (Burbacher et al. 1993). That sex-specific effect appeared to be related to the adolescent growth spurt, because adult weight was not affected by MeHg. Studies conducted when the monkeys were adults indicated significant effects due to MeHg exposure on spatial vision (visual contrast-sensitivity functions). Although there were overall group differences in spatial vision, there were large individual differences in the response of the MeHg-exposed monkeys (Burbacher et al. 1999). Tests of adult learning and memory did not indicate significant effects due to MeHg exposure (Gilbert et al. 1993, 1996) and pure-tone auditory thresholds appeared normal when the monkeys were tested at approximately 12 to 15 years of age (Burbacher et al. 1999).

In another series of studies, *Macaca fascicularis* were orally exposed to MeHg at 50 μg/kg per day from birth to 7 years of age (Rice 1989) or at 0, 10, 25, or 50 μg/kg per day in utero plus 4 years postnatally (Rice and Gilbert 1990). No effects of MeHg were observed in the tests on infant or juvenile learning and memory for the in utero plus postnatally exposed animals (Rice 1992). However, impaired spatial vision was observed in monkeys from both dose groups when they were tested on a contrast-sensitivity task between 3 and 5 years of age (Rice and GIlbert 1982; Rice and Gilbert 1990). At 13 years of age, overt toxicity (clumsiness) was observed in some of the monkeys exposed postnatally to MeHg for 7 years (Rice 1989). Tests of those monkeys at 14 years of age indicated impaired high-frequency hearing in four of the five MeHg-exposed monkeys, and tests at age 18 indicated impaired somatosensory function (vibration sensitivity) in the same four monkeys (Rice and Gilbert 1992, 1995). Two of four monkeys exposed to MeHg in utero plus 4 years postnatally also demonstrated impairments on the vibration-sensitivity test when tested at 15 years of age. Auditory testing of the monkeys exposed to MeHg in utero plus 4 years postnatally at 11 and 19 years of age indicated elevated pure-tone thresholds throughout the full range of frequencies tested (0.125 to 31.5 kHz) at 19 years of age (Rice 1998). Although both controls and MeHg-exposed monkeys showed higher thresholds at 19 years of age compared with 11 years, MeHg-exposed monkeys showed a greater deterioration in auditory function with increasing age. Across studies, MeHg effects were observed in individual monkeys at maternal doses of 10 μg/kg per day to 50 μg/kg per day or a dose of 50 μg/kg per day postnatally. However, the numbers of monkeys in the studies were small (one at 10 μg/kg per day), allowing only individual comparisons.

Newland et al. (1994) examined the effects of in utero exposure to MeHg in squirrel monkeys. Maternal exposure to MeHg varied to provide steady-state blood Hg concentrations between 0.7 and 0.9 ppm. At 5 to 6 years of age, offspring were trained to lever press under concurrent schedules of reinforcement. The results of the study indicated that MeHg-exposed monkeys were not able to change their response rates consistent with changes in reinforcement contingencies. Those effects were most prominent during transitions in reinforcement schedules.

Many of the studies using rodent models have also focused on examining the effects of MeHg exposure on neurobehavioral development. One of the largest studies to examine the effects of developmental exposure to MeHg in rats was the "Collaborative Behavioral Teratology Study" (CBTS), which was performed to compare the results of a standard behavioral test battery across several laboratories (Buelke-Sam et al. 1985). Maternal rats were exposed to MeHg at 0, 2, or 6 mg/kg per day via gavage on gestation days 6-9. Offspring were exposed in utero and during lactation (no cross-fostering). Behavioral assessments of offspring indicated an increase in auditory startle-response habitation, mostly at the high dose. Maze activity increased with increasing MeHg exposure, and performance on a visual discrimination test was affected at the high dose. Two parallel studies, Vorhees (1985) and Geyer et al. (1985), reported similar findings, as well as delayed surface righting and swimming ontogeny. Retarded maze performance at the highest dose tested (Vorhees 1985) and retarded negative geotaxis and pivoting at a dose of 2.5 mg/kg per day on gestation days 6-15 were also reported (Geyer et al. 1985). The effects observed by Geyer et al. (1985) were related to gestational exposure alone, because the MeHg offspring were cross-fostered to nonexposed dams at birth. Consistent with the results of the nonhuman primate studies, two studies have reported effects of early MeHg exposure on visual functions in rats (Zenick 1976; Dyer et al. 1978). Abnormal visual-evoked potentials were reported following in utero and lactational exposure to MeHg following a single maternal dose of 5 mg/kg per day on gestation day 7 (Dyer et al. 1978) or continuous maternal exposure at 2.5 mg/kg per day (Zenick 1976). Other studies of rats and mice have reported MeHg effects on motor performance and measures of activity and learning. MeHg effects on the swimming behavior of mice was reported following in utero and lactational MeHg exposure (maternal dose of 8 mg/kg on day 7 or 9 of gestation) (Spyker et al. 1972) or following exposure from gestation day 1 to postnatal day 42 (0.1 mg/kg per day) (Olson and Bousch 1975). Reports indicated increased activity in rats during the preweaning period following in utero and lactational MeHg exposure (5 or 8 mg/kg per day on gestation day 8 or 15) (Eccles and Annau 1982a,b), and decreased activity was reported in mice tested postweaning following in utero MeHg exposure alone (8 or 12 mg/kg per day on gestation day 10, or 4 mg/kg per day on gestation days 10-12 with cross-fostering at birth) (Su and Okita 1976). Learning deficits have been reported in both rats and mice following in utero or early postnatal MeHg exposure. Rats exposed prenatally to MeHg displayed impaired learning on a shuttle-box avoidance test (Eccles and Annau 1982a,b), and a single dose of 10 mg/kg per day on gestation day 4 with no cross-fostering of dams (in utero plus lactational exposure) was associated with decreased escape, avoidance, and appetitive learning (Schalock et al. 1981). Olson and Bousch (1975) reported impaired learning in rats on a maze task following exposure to Hg at 0.1 mg/kg per day via a fish diet from gestation day 1 to postnatal day 42. In mice, a maternal dose of 3 mg/kg on gestation day 8 (Hughes and Annau 1976) retarded both active and passive avoidance learning. The results suggested that the effects were due to exposure in utero, because the effects were observed in exposed offspring cross-fostered to control females at birth (no lactational exposure). Zenick (1974) compared the learning performance of rats following prenatal, lactational, or 9 days postweaning exposure to MeHg (2.5

mg/kg per day) in a water T maze. Deficits in learning were observed in the prenatal and postweaning exposure groups but not in the lactational exposure group. Thus far, the Differential Reinforcement of High Rates (DRL) test has proved to be the most sensitive of the behavioral tests used with rodents for detecting effects of in utero MeHg exposure (Müsch et al. 1978; Bornhausen et al. 1980). Rats exposed in utero to maternal doses of MeHg at 0.01 to 0.05 mg/kg per day and cross-fostered at birth displayed abnormal response patterns on the DRL task when tested at 4 months of age. Using a similar DRL paradigm, Rasmussen and Newland (1999), however, were not able to replicate that finding. A procedure designed to measure tactile-kinesthetic function in rodents has also been shown to be sensitive to MeHg exposure. Elsner (1991) reported a decrease with that procedure in the performance of rats following in utero and lactational MeHg exposure at a very low maternal dose (0.08 mg/kg per day, 2 weeks before mating and throughout gestation).

In 1996, Fredricksson et al. reported interactive behavioral effects following exposure of rats to MeHg and metallic Hg vapor. Exposure to Hg vapor at 1.8 mg/m^3 for 1.5 hours per day on gestation days 14-19 was related to hyperactivity and decreased spatial learning. While exposure to MeHg at 2 mg/kg per day on gestation days 6-9 was not related to adverse behavioral effects, co-exposure to MeHg and Hg vapor potentiated the activity and spatial learning effects observed with Hg vapor exposure alone. The reported Hg vapor effects were consistent with previous reports (Danielsson et al. 1993, Fredriksson et al. 1992, 1993). This is the first report, however, of an interactive effect of in utero exposure toHg vapor and MeHg. The results indicate that total exposure to the different forms of Hg during pregnancy is critical in evaluating the effects on the fetus.

Finally, the results of a few of the studies using animal models have provided some preliminary data on the potential effects of early-developmental exposure to MeHg on the functional status of aging animals. An early report by Spyker (1975) summarized the effects of MeHg observed over the lifetime of mice exposed in utero and during lactation. Offspring were normal at birth but exhibited effects of exposure on exploratory activity and swimming ability at 1 month of age and neuromuscular and immune effects after 1 year of age. Those findings provide the first evidence of delayed neurotoxicity in MeHg-exposed animals. The effects of in utero plus postnatal MeHg exposure described by Rice (1989) also support the notion of delayed neurotoxicity, which might be related to increased functional impairment with aging. Monkeys exposed to MeHg at 50 µg/kg per day from birth to 7 years of age were observed in an exercise cage throughout their life. Obvious motor incoordination was observed only after the monkeys reached 14 years of age (Rice 1989). Subsequent testing of those monkeys indicated higher thresholds for vibration sensitivity, indicating effects on somatosensory functioning (Rice and Gilbert 1995). More recently, monkeys exposed to MeHg at 10-50 µg/kg per day in utero plus 4 years postnatally showed a greater deterioration in auditory function with increasing age when tested at 11 and 19 years of age (Rice 1998). Whether those effects are related to cumulative damage from early MeHg exposure and aging or to a continuous process from long-term retention of inorganic Hg in the brain following MeHg exposure is not known. The results clearly indicate, however, that the health risks associated with early MeHg exposure could last a lifetime (Harada 1995; Kinjo et al 1993).

ADULT CENTRAL-NERVOUS-SYSTEM TOXICITY

Adult Human Neurological, Neurophysiological, and Sensory Function

Several neurological signs and symptoms are among the cardinal features of chronic high-dose exposures to MeHg in adults. As no pathognomonic test is available to confirm the diagnosis of Minamata disease, cases were identified on the basis of a characteristic combination of symptoms (Harada 1997; Uchino et al. 1995; Tsubaki and Takahashi 1986). These included peripheral neuropathy (e.g., sensory impairment of the extremities of the glove-stocking type and perioral dysesthesia), dysarthria, tremor, cerebellar ataxia, gait disturbance, ophthalmological impairment (e.g., visual-field constriction and disturbed ocular movements), audiological impairment (e.g., hearing loss), disturbance of equilibrium (e.g., vertigo, dizziness and fainting), and subjective symptoms such as headache, muscle and joint pain, forgetfulness, and fatigue. In patients with classic Minamata disease, many of those signs and symptoms were still evident after 20 years. Later studies of patients with Minamata disease reported increased pain thresholds in the body (truncal hypesthesia) and distal extremities (Yoshida et al. 1992).

To evaluate the WHO (IPCS 1990) estimate that 5% of adults with a blood Hg concentration of 200 ppb would manifest paresthesia, Kosatsky and Foran (1996) reviewed 13 studies of neurological status in long-term fish consumers. Although they identified pervasive weaknesses in study design (e.g., crude measures of exposure and outcome, possible selection bias, and absence of blinding), the authors concluded that the studies suggested neurological effects in as few as 11% (95 % confidence interval, 4-22) and as many as 31% (95% confidence interval, 19-45) of adults with a blood Hg concentration of 200 ppb or more. Thus, they argued that these data do not support the WHO (IPCS 1990) conclusion that a blood Hg concentration of 200 ppb (corresponding to a hair Hg concentration of 50 ppm) represents a LOAEL for adult paresthesia and identified a need for additional research to define the lower portion of the dose-response curve (20-200 ppb).

Important data on the impact of chronic low-dose MeHg exposures on adult neurological and sensory function are being generated in ongoing studies of fish-eating populations living in the Amazon Basin, where gold is extracted from soil or river sediments and Hg is released. Lebel et al. (1996) studied 29 young adults (ages, 15-35 years; 14 females and 15 males) randomly selected from participants in a previous survey. The geometric-mean hair Hg concentration was 14.0 ppm (range, 5.6 to 38.4 ppm). Subjects underwent a battery of quantitative behavioral, sensory, and motor tests, including tests of visual functions (near and far acuity, chromatic discrimination, near contrast sensitivity, and peripheral visual fields) and motor functions (maximum grip strength and manual dexterity). Individuals with increased hair Hg concentrations had reduced chromatic discrimination. Three individuals with hair Hg concentrations above 24 ppm demonstrated reduced contrast sensitivity, and individuals with concentrations above 20 ppm tended to demonstrate reductions in peripheral visual fields. An increase from 10 to 20 ppm was associated with about a 10 degree difference. Highly exposed women tended to have lower scores than low-exposed women on both manual dexterity and grip strength. Such a tendency was not seen in men, indicating that association between hair Hg concentration and motor function was sex-specific.

In a subsequent study, Lebel et al. (1998) assembled another sample of 91 individuals (ages 15-81 years), representing approximately 38% of the adult population of the study village. Four measures of exposure were derived based on the Hg concentration in a hair sample (length not specified): mean total hair Hg averaged over all 1-cm segments of the sample (up to 24 segments), total Hg in the first centimeter, maximum total Hg in any segment, and MeHg in the first centimeter. Individuals for whom at least 1 cm of hair contained MeHg at more than 50 ppm were excluded. The mean hair MeHg concentration was approximately 13 ppm. The assessments included the same tests of motor (maximum grip strength and manual dexterity) and visual functions (acuity, chromatic discrimination, and near contrast sensitivity) that were used in the previous study. In addition, a clinical neurological examination was administered to a random sample of the cohort (59 subjects). That examination included the Branches Alternate Movement Task (BAMT), which requires imitation of a prescribed sequence of hand movements. Abnormal performance on the BAMT was significantly associated with all measures of Hg exposure, and abnormal visual fields were associated with mean hair Hg and peak Hg concentrations. Hyper-reflexia (patellar and bicepital) was not associated with any Hg measurement. Increased hair Hg concentrations, most notably peak Hg, were associated with poor scores on the intermediate and higher frequencies of near visual-contrast sensitivity (in the absence of near visual-acuity loss), with poor scores on the manual dexterity test, and with increased muscular fatigue. In women, but not in men, grip strength varied with peak Hg concentration. For many end points, the associations between hair Hg concentration and performance were stronger in younger subjects (less than 35 years) than in older ones. The authors stress that the dose-related decrements in visual and motor functions were associated with hair Hg concentrations below 50 ppm, a range in which clinical signs of Hg intoxication are not apparent. The Hg exposure of the cohort is presumed to have resulted from fish-consumption patterns that are stable and thus relevant to estimating the risk associated with chronic, low-dose MeHg exposure. In fact, the possibility cannot be excluded that the neurobehavioral deficits of the adult subjects were due to increased prenatal, rather than ongoing, MeHg exposure.

Beuter and Edwards (1998) investigated the prevalence and severity of three types of subtle motor deficits in a cohort of 36 adult Cree (mean age 56 years), comparing them with patients with Parkinson's disease (PD) (21 subjects), cerebellar deficit (6 subjects), essential tremor (3 subjects), or controls (30 subjects). The mean of the annual maximum hair Hg concentration over a period of 25 years varied from 2.2 to 31.1 ppm. Ten of 14 static tremor end points (with visual feedback) and 5 of 8 kinetic tremor end points (during voluntary finger movements) assessed were significantly related to group (i.e. Cree versus PD versus control). Nested analyses were carried out in which the six Cree with the highest hair Hg concentrations (mean of annual maximum hair concentrations greater than 24 ppm; range, 24.34-31.10 ppm) were matched to six Cree with low hair Hg concentrations (mean of annual maximum hair concentrations 6.02-11.89 ppm) and six controls to get a better idea of whether the group differences were likely to be due to Hg or some confounding factor associated with group membership. Despite the reduced number, significant group differences were still found on several end points. Overall, the performance characteristics that best discriminated groups were drift (static tremor), event index (static tremor), mean tracking error (kinetic tremor), and the center of mass harmonicity (static tremor).

The same groups of subjects were administered a test of rapid, precise promixo-distal movements (i.e. eye-hand coordination) (Beuter et al. 1999a). A eurythmkinesiometer recorded subjects' efforts to strike targets with a stylus, yielding measures of precision, imprecision, tremor, Fitts' constant (an index of the trade-off between speed and accuracy), and irregularity. The Cree subjects' performance was more than 1 standard deviation worse than the controls' performance on tremor, Fitts' constant, and irregularity. In the same type of nested analyses carried out in the study of tremor described above, the order of group scores, from best to worst, was control better than low Hg better than high Hg for all three end points, the group differences on Fitts' constant and irregularity being significant.

Rapid alternating movements (diadochokinesia) were assessed in those groups of subjects by asking them to rotate two foam spheres under three conditions: (1) both hands, natural cadence; (2) right and left hands separately, fast cadence (i.e., as fast as possible); and (3) both hands, fast cadence (Beuter et al. 1999b). Seven dimensions of performance were measured: duration, range, maximum slope, similarity in shape, smoothness, sharpness, and coherence. Significant group difference were found on most end points, and the results of nested analyses provided additional evidence that group differences in Hg concentrations were probably contributory.

Neurotoxic Effects in Adult Animals

Experimental studies of the effects of MeHg exposure on adult animals have reported neurological effects similar to those reported for adult humans. Studies using monkeys, rodents, and cats have reported effects consistent with adult MD (Harada 1995). Some of those studies are summarized in Table 5-12. Neurotoxic signs reported reflect the regional specificity of the neuropathological effects observed in adult subjects. Signs of ataxia, constriction of the visual field, and sensory disturbances are commonly associated with pathological lesions in the calcarine cortices, dorsal root ganglia, and cerebellum (Chang 1980).

A study of macaque and squirrel monkeys has reported ataxia, tremor, and constriction of visual fields in animals with blood Hg concentrations between 1 and 2 ppm (Evans et al. 1977). The latency for the onset of symptoms in that study was 135 to 140 days.

Constriction of the visual field was reported in macaques following variable dosing schedules that produced blood MeHg concentrations from 1.5 to 3 ppm. The onset of visual-field disturbances preceded overt signs of toxicity (Merigan et al. 1983).

Ataxia, tremor, and apparent blindness was reported in adult female macaques exposed orally to doses of MeHg hydroxide at 70 µg/kg per day and above (Burbacher et al. 1988). The durations to onset of symptoms ranged from 177 days to 392 days.

In rodents, several studies have reported severe neurological effects, such as ataxia, paralysis, spasms, and hindlimb crossing in adult rats and mice, from exposure to MeHg (see Table 5-12). In general, the onset of symptoms is dependent on the dose and duration of exposure. In rats, overt signs of neurotoxicity were reported at doses ranging from 0.8 mg/kg per day for 6 weeks (Chang and Hartmann 1972) to 10 mg/kg

TABLE 5-12 Neurological Effects of MeHg Exposure in Adult Animals

Species	Exposure Time	NOAEL (mg/kg/d)	LOAEL (mg/kg/d)	Effect	Reference
Monkeys (macaque and squirrel)	135-1,000 d (variable dosing)	NS	NS	Ataxia, tremor, constriction of visual field	Evans et al. 1977
Monkey (macaque)	20-73 wk (variable dosing)	NS	NS	Tremor, constriction of visual field	Merigan et al. 1983
Monkey (*Macaca fascicularis*)	177-392 d	0.05	0.09	Ataxia, tremor, blindness	Burbacher et al. 1988
Rat	29 d		2.4	Ataxia, paralysis	Hunter et al. 1940
Rat	1-6 wk		0.8	Ataxia, degeneration of cerebellum and dorsal root ganglia	Chang and Hartmann 1972
Rat	3-12 wk	0.84	1.68	Ataxia, edema and necrosis of cerebellum	Magos and Butler 1972
Rat (Wistar)	2 d		10	Impaired performance in tilting plane test	Fehling et al. 1975
Rat (Wistar)	9, 13, or 21 d	2	4	Hindlimb crossing	Inouye and Murakami 1975
Rat (Wistar)	Gestation d 7-14	4	6	Spasms, gait disturbances	Fuyuta et al. 1978
Rat (Sprague-Dawley)	2 d	1.32	4	Altered sleep cycles	Arito and Takahashi 1991
Rat	15 d (dosed every 3 d)		10	Hindlimb crossing, flailing	Leyshon and Morgan 1991

Mouse	60 d	0.25	1.0	Hindleg weakness, microgliocytosis and cerebellar degeneration	Berthoud et al. 1976
Mouse	24 wk		1.9	Paralysis	MacDonald and Harbison 1977
Mouse (C57B1/6)	17 d 6 d		4 8	Increased auditory brainstem response thresholds	Wassick and Yonovitz 1985
Mouse ($B6C3F_1$)	2 yr		0.6	Paralysis, neuropathy	Mitsumori et al. 1990
Rabbit (New Zealand)	7 d		30	Ataxia, decreased muscle tone, reduced splay reflex	Jacobs et al. 1977
Cat			0.05	Ataxia	Khera et al. 1974
Cat	60 wk		0.046	Impaired hopping	Charbonneau et al. 1976

Abbreviations: NOAEL, no-observed-adverse-effect level; LOAEL, lowest-observed-adverse-effect level; NS, not stated.

per day given every 3 days for 15 days (Leyshon and Morgan 1991). Studies in mice reported that doses from 0.6 mg/kg per day for approximately 2 years (Mitsumori et al. 1990) to 1.9 mg/kg per day for 24 weeks (MacDonald and Harbison 1977) caused paralysis. A study by Jacobs et al. (1977) using New Zealand rabbits reported ataxia and decreased muscle tone following a dose of 30 mg/kg per day for 7 days. Two studies using cats reported ataxia and impaired hopping after long-term exposure at approximately 0.05 mg/kg per day (Khera et al. 1974; Charbonneau et al. 1976).

A few studies using rodents have reported less severe symptoms, such as altered sleep cycles, hindlimb weakness, or increased brainstem-auditory-response thresholds following exposure to MeHg. Altered sleep cycles in rats were reported by Arito and Takahashi (1991) following 2 days of exposure at 4 mg/kg per day. Hindlimb weakness in mice was reported by Berthoud et al. (1976) following exposure at 1 mg/kg per day for 60 days. Wassick and Yonovitz (1985) reported increased brainstem-auditory thresholds in mice following 17 days of exposure at 4 mg/kg per day or 6 days of exposure at 8 mg/kg per day.

In summary, reports from animal models of adult MD have provided supportive evidence for the neurological signs reported in humans. These studies have also provided detailed descriptions of the associated neuropathological effects from high-dose MeHg exposures (Chang 1979, 1990). Studies using adult animal models of chronic low-dose MeHg effects have been sparse, most likely because of the focus on neurodevelopmental effects following in utero or early postnatal MeHg exposure.

CONCLUSIONS

- MeHg is highly toxic. The data reviewed in this chapter indicate that the adverse effects of MeHg exposure can be expressed in multiple organ systems throughout the lifespan.
- Studies in humans on the carcinogenic effects of MeHg are inconclusive. Renal tumors have been seen in male mice but only at or above the MTD of MeHg.
- The effect of MeHg on the human immune system is poorly understood. However, studies in vitro and in animals suggest that

exposure to MeHg could increase human susceptibility to infectious diseases and autoimmune disorders by damaging the immune system.

- The reproductive effects of MeHg have not been fully evaluated in humans, but animal studies, including work in nonhuman primates, indicate that MeHg causes functional reproductive effects.
- Damage to the renal tubules and nephron has been observed following human exposure to inorganic and organic forms of Hg. However, symptoms of renal damage have been seen only at Hg exposures that also caused neurological effects. In animals, similar effects have been observed as well as altered renal function and renal hypertrophy have been observed following early postnatal exposure to MeHg.
- Although the data base is not as extensive for cardiovascular effects as it is for other end points (i.e., neurotoxic effects), the cardiovascular system appears to be a target for MeHg toxicity in both humans and animals. Evidence suggests that adverse health effects can occur at very low Hg exposures.
  - Exposure to elemental and organic forms of Hg alters blood-pressure regulation. That effect has been documented in children and adults who were exposed to toxic and subtoxic doses of Hg and have been induced experimentally in rats.
  - Prenatal exposure to MeHg has been shown to alter blood-pressure regulation and heart-rate variability in children. Those effects were observed at cord-blood Hg concentrations that have not been associated with other developmental effects (less than 10 μg/L).
  - Men who consumed at least 30 g of fish per day or had a hair Hg concentration of 2 ppm or more had a higher risk of suffering a fatal or nonfatal acute myocardial infarction. Mercury exposure was also correlated with an increased risk of dying from coronary heart disease or cardiovascular heart disease. A hair Hg concentration of 2 ppm has not been associated with other adverse health effects.
- The human data base on the neurodevelopmental effects of MeHg is extensive, and includes studies of populations following high-dose Hg poisonings and chronic low-dose Hg exposure. Some study results appear to be conflicting. Table 5-10 provides informa-

tion about the hair and blood Hg concentrations in the studies on which the following conclusions are based.

- Several studies have detected significant MeHg-associated increases in the frequency of abnormal and questionable findings on standardized neurological examinations, although the functional importance of the apparent effects is uncertain.
- Recent epidemiological studies provide little evidence that the ages at which children achieve major language and motor milestones are affected appreciably by low-dose prenatal MeHg exposure.
- Two out of four studies using the Denver Developmental Screening Test reported an association of low-dose MeHg exposure on early childhood development.
- Of the three major prospective long-term studies, the Faroes study reported associations between low-dose prenatal MeHg exposure and children's performance on standardized neurobehavioral tests, particularly in the domains of attention, fine-motor function, confrontational naming, visual-spatial abilities, and verbal memory, but the Seychelles study did not report such associations. The smaller New Zealand study also observed associations, as did a large pilot study conducted in the Seychelles.

• Recent studies in adults suggest that hair Hg concentrations below 50 ppm are significantly associated with disturbances of the visual system (chromatic discrimination, contrast sensitivity, and peripheral fields) and with neuromotor deficits (tremor, dexterity, grip strength, complex-movement sequences, hand-eye coordination, and rapid alternating movement). Those findings suggest that the current reference dose for adults based on 50 ppm in hair might not be sufficiently protective.
• Neurodevelopmental studies using animal models (nonhuman primates, rodents) exposed in utero and/or early postnatally to MeHg have reported a continuum of effects related to dose. Effects have been reported on sensory, sensorimotor, and cognitive development. Overall, sensory effects seem to be the most long-lasting.
• Experimental studies of adult animal models exposed to MeHg have also reported a continuum of effects associated with dose. The

effects are similar to those observed in cases of human MeHg poisoning.

- Neurodevelopmental effects are the most extensively studied sensitive end point for MeHg toxicity and are appropriate for use in establishing an RfD. New data are emerging, however, indicating that there might be important adverse effects on other end points (e.g., cardiovascular and immune systems) in the same exposure range. Those effects should be considered as the data become available.

RECOMMENDATIONS

- Epidemiological research is needed to evaluate the prevalence of chromosomal aberrations and cancer, especially leukemia and renal tumors, among populations that have chronic exposure to MeHg through ingestion of contaminated fish.
- The ability of MeHg to cause chromosomal damage and promote tumor growth should be considered in the establishment of exposure guidelines.
- Research is needed to determine the effects of MeHg exposure on the immune system, including the effects on the developing immune system, resistance to microbial pathogens, and autoimmunity. Mechanisms by which the immune system is involved in the target-organ toxicity of Hg should also be examined.
- Research is needed to assess the effects of MeHg on reproduction, including the effects on fertility indicators, such as sperm production, conception rates, and pregnancy outcomes.
- Research is needed to evaluate the impact of dietary exposure to MeHg on the prevalence of hypertension and cardiovascular disease in the United States. The risk of fatal and nonfatal heart disease must be considered in the development of a reference dose for this contaminant.
- Research is needed to determine the long-term implications of the neuropsychological and neurophysiological effects of low-level prenatal MeHg exposure detected in children, specifically whether

they are associated with an increased risk for later neurological diseases.

- Research using animal models is needed to better define the immediate and long-term effects of early chronic low-level MeHg exposure. Studies should focus on several important issues:
  - Critical periods for MeHg effects (in utero or postnatal).
  - Low-level dose-response relationships (ppb range).
  - MeHg demethylation in the brain following early MeHg exposure.
  - Synergistic effects of early MeHg and Hg vapor exposure.
  - Neurodegenerative disorders related to early MeHg exposure.
- Animal studies should be conducted to examine the neurodevelopmental effects of continuous versus peak MeHg exposures.

REFERENCES

Afonso, J.F., and R.R. de Alvarez. 1960. Effects of mercury on guman gestation. Am. J. Obstet. Gynec. 80(July):145-154.

Akagi, H., P. Grandjean, Y. Takizawa, and P. Weihe. 1998. Methylmercury dose estimation from umbilical cord concentrations in patients with Minamata disease. Environ. Res. 77(2):98-103.

Alcser, K.H., K.A. Brix, L.J. Fine, L.R. Kallenbach, and R.A. Wolfe. 1989. Occupational mercury exposure and male reproductive health. Am. J. Ind. Med. 15(5):517-29.

Amin-Zaki, L., S. Elhassani, M.A. Majeed, T.W. Clarkson, R.A. Doherty, and M. Greenwood. 1974. Intra-uterine methylmercury poisoning in Iraq. Pediatrics 54(5):587-95.

Anneroth, G., T. Ericson, I. Johansson, H. Mornstad, M. Ryberg, A. Skoglund, and B. Stegmayr. 1992. Comprehensive medical examination of a group of patients with alleged adverse effects from dental amalgams. Acta Odontol. Scand. 50(2):101-11.

Arito, H., and M. Takahashi. 1991. Effect of methylmercury on sleep patterns in the rat. Pp. 381-394 in Advances in Mercury Toxicology, T. Suzuki, N. Imura, and T.W. Clarkson, eds. New York: Plenum Press.

Aronow, R., C. Cubbage, R. Wiener, B. Johnson, J. Hesse, and J. Bedford. 1990. Mercury exposure from interior latex paint- Michigan. MMWR 39(8)125-136.

ATSDR (Agency for Toxic Substances and Disease Registry). 1999. Toxicological Profile for Mercury. (Update). U.S. Department of Health and Human Services, Agency for Toxic Substances and Disease Registry, Atlanta, GA.

Axtell, C.D., G.J. Myers, P.W. Davidson, A.L. Choi, E. Cernichiari, J. Sloane-Reeves, C. Cox, C. Shamlaye, and T.W. Clarkson. 1998. Semiparametric modeling of age at achieving developmental milestones after prenatal exposure to methylmercury in the Seychelles child development study. Environ. Health Perspect. 106(9):559-564.

Bakir, F., S.F. Damluji, L. Amin-Zaki, M. Murtadha, A. Khalidi, N.Y. al-Rawi, S. Tikriti, H.I. Dhahir, T.W. Clarkson, J.C. Smith, and R.A. Doherty. 1973. Methylmercury poisoning in Iraq. Science 181(96):230-241.

Barr, R.D., P.H. Rees, P.E. Cordy, A. Kungu, B.A. Woodger, and H.M. Cameron. 1972. Nephrotic syndrome in adult Africans in Nairobi. Br. Med. J. 2(806):131-4.

Barregard, L., B. Hultberg, A. Schutz, and G. Sallsten. 1988. Enzymuria in workers exposed to inorganic mercury. Int. Arch. Occup. Environ. Health 61(1-2):65-9.

Barregard, L., B. Hogstedt, A. Schutz, A. Karlsson, G. Sallsten, and G. Thiringer. 1991. Effects of occupational exposure to mercury vapor on lymphocyte micronuclei. Scand. J. Work Environ. Health 17(4):263-8.

Bellinger, D.C., A. Leviton, C. Waternaux, H. Needleman, and M. Rabinowitz. 1987. Longitudinal analyses of prenatal and postnatal lead exposure and early cognitive development. N. Engl. J. Med. 316(17):1037-1043.

Berthoud, H.R., R.H. Garman, and B. Weiss. 1976. Food intake, body weight, and brain histopathology in mice following chronic methylmercury treatment. Toxicol. Appl. Pharmacol. 36(1):19-30.

Betti, C., T. Davini, and R. Barale. 1992. Genotoxic activity of methylmercury chloride and dimethyl mercury in human lymphocytes. Mutat. Res. 281(4):255-260.

Beuter, A., and R. Edwards. 1998. Tremor in Cree subjects exposed to methylmercury: a preliminary study. Neurotoxicol. Teratol. 20(6):581-9.

Beuter, A., A. de Geoffroy, and R. Edwards. 1999a. Quantitative analysis of rapid pointing movements in Cree subjects exposed to mercury and in subjects with neurological deficits. Environ. Res. 80(1):50-63.

Beuter, A., A. de Geoffroy, and R. Edwards. 1999b. Analysis of rapid alternating movements in Cree subjects exposed to methylmercury and in subjects with neurological deficits. Environ. Res. 80(1):64-79.

Blakley, B.R. 1984. Enhancement of urethane-induced adenoma formation in Swiss mice exposed to methylmercury. Can. J. Comp. Med. 48(3):299-302.

Bluhm, R.E., R.G. Bobbitt, L.W. Welch, A.J. Wood, J.F. Bonfiglio, C. Sarzen, A.J. Heath, and R.A. Branch. 1992. Elemental mercury vapour toxicity, treatment, and prognosis after acute, intensive exposure in chloralkali plant workers. Part I: History, neuropsychological findings and chelator effects. Hum. Exp. Toxicol. 11(3):201-10.

Bornhausen, M., H.R. Müsch, and H. Greim. 1980. Operant behavior performance changes in rats after prenatal methylmercury exposure. Toxicol. Appl. Pharmacol. 56(3):305-10.

Buchet, J.P., H. Roels, A. Bernard, and R. Lauwerys. 1980. Assessment of renal function of workers exposed to inorganic lead, calcium or mercury vapor. J. Occup. Med. 22(11):741-50.

Buelke-Sam, J., C.A. Kimmel, J. Adams, C.J. Nelson, C.V. Vorhees, D.C. Wright, V. St Omer, B.A. Korol, R.E. Butcher, M.A. Geyer, J.F. Holson, C.L. Kutscher, and M.J. Wayner. 1985. Collaborative Behavioral Teratology Study: Results. Neurobehav. Toxicol. Teratol. 7(6):591-624.

Burbacher, T.M., K.S. Grant, and N.K. Mottet. 1986. Retarded object permanence development in methylmercury exposed Macaca fascicularis infants. Dev. Psychol. 22(6):771-776.

Burbacher, T.M., M.K. Mohamed, and N.K. Mottett. 1988. Methylmercury effects on reproduction and offspring size at birth. Reprod. Toxicol. 1(4):267-278.

Burbacher, T.M., P.M. Rodier, and B. Weiss. 1990. Methylmercury developmental neurotoxicity: a comparison of effects in humans and animals. Neurotoxicol. Teratol. 12(3):191-202.

Burbacher, T., P. Rodier, K. Grant-Webster, S. Gilbert, and N.K. Mottet. 1993. Pubertal growth retardation: a sex specific effect of in utero methylmercury exposure. Teratology 47(5):455.

Burbacher, T.M., K.S. Grant, S.G. Gilbert, and D.C. Rice. 1999. The effects of methylmercury exposure on visual and auditory functions in nonhuman primates. Toxicologist 48(1-S):362.

Cardenas, A., H. Roels, A.M. Bernard, R. Barbon, J.P. Buchet, R.R. Lauwerys, J. Rosello, G. Hotter, A. Mutti, I. Franchini, et al. 1993. Markers of early renal changes induced by industrial pollutants. I. Application to workers exposed to mercury vapour. Br. J. Ind. Med. 50(1):17-27.

Chang, L.W. 1977. Neurotoxic effects of mercury--a review. Environ. Res. 14(3):329-73.

Chang, L.W. 1979. Pathological effects of mercury poisoning. Pp. 519-580 in The Biogeochemistry of Mercury in the Environment, J.O. Nriagu, ed. New York: Elsevier.

Chang, L.W. 1980. Mercury. Pp. 508-526 in Experimental and Clinical Neurotoxicology, P.S. Spencer, and H.H. Schaumburg, eds. Baltimore, MD: Williams & Wilkins.

Chang, L.W. 1990. The neurotoxicology and pathology of organomercury, organolead, and organotin. J. Toxicol. Sci. 15(Suppl. 4):125-51.

Chang, L.W., and H.A. Hartmann. 1972. Ultrastructural studies of the nervous system after mercury intoxication. I. Pathological changes in the nerve cell bodies. Acta Neuropathol. (Berl) 20(2):122-38.

Charbonneau, S.M., I.C. Munro, E.A. Nera, F.A. Armstrong, R.F. Willes, F. Bryce, and R.F. Nelson. 1976. Chronic toxicity of methylmercury in the adult cat. Toxicology 5(3):337-349.

Choi, B.H., L.W. Lapham, L. Amin-Zaki, and T. Saleem. 1978. Abnormal neuronal migration, deranged cerebral cortical organization, and diffuse white matter astrocytosis of human fetal brain: a major effect of methylmercury poisoning in utero. J. Neuropathol. Exp. Neurol. 37(6):719-33.

Cinca, I., I. Dumetrescu, P. Onaca, A. Serbanescu, and B. Nestorescu. 1979. Accidental ethyl mercury poisoning with nervous system, skeletal muscle, and myocardium injury. J. Neurol. Neurosurg. Psychiatry 43(2):143-149.

Clarkson, T.W. 1997. The toxicology of mercury. Crit. Rev. Clin. Lab. Sci. 34(4):369-403.

Cordier, S., and M. Garel. 1999. Neurotoxic Risks in Children Related to Exposure to Methylmercury in French Guiana. INSERM U170 and U149 – Study financed by the Health Monitoring Institute (RNSP). National Institute of Health and Medical Research. April.

Cordier, S., F. Deplan, L. Mandereau, and D. Hemon. 1991. Paternal exposure to mercury and spontaneous abortions. Br. J. Ind. Med. 48(6):375-81.

Costa, M., N.T. Christie, O. Cantoni, J.T. Zelikoff, X.W. Wang, and T.G. Rossman. 1991. DNA damage by mercury compounds: An overview. Pp. 255-273 in Advances in Mercury Toxicology, T. Suzuki, N. Imura, and T.W. Clarkson, eds. New York: Plenum Press.

Counter, S.A., L.H. Buchanan, G. Laurell, and F. Ortega. 1998. Blood mercury and auditory neuro-sensory responses in children and adults in the Nambija gold mining area of Ecuador. Neurotoxicology 19(2):185-196.

Cox, C., T.W. Clarkson, D.O. Marsh, L. Amin-Zaki, S. Tikriti, and G.G. Myers. 1989. Dose-response analysis of infants prenatally exposed to methyl mercury: An application of a single compartment model to single-strand hair analysis. Environ. Res. 49(2):318-332.

Cox, C., D. Marsh, G. Myers, and T. Clarkson. 1995. Analysis of data on delayed development from the 1971-72 outbreak of methylmercury poisoning in Iraq: Assessment of influential points. Neurotoxicology 16(4):727-730.

Crump, K.S., T. Kjellström, A.M. Shipp, A. Silvers, and A. Stewart. 1998. Influence of prenatal mercury exposure upon scholastic and psychological test performance: benchmark analysis of a New Zealand cohort. Risk Anal. 18(6):701-713.

Crump, K., J. Viren, A. Silvers, H. Clewell 3rd, J. Gearhart, and A. Shipp. 1995. Reanalysis of dose-response data from the Iraqi methylmercury poisoning episode. Risk Anal. 15(4):523-532.

Dahl, J.E., J. Sundby, A. Hensten-Pettersen, and N. Jacobsen. 1999. Dental workplace exposure and effect on fertility. Scand. J. Work Environ. Health 25(3):285-90.

Dahl, R., R.F. White, P. Weihe, N. Sørensen, R. Letz, H.K. Hudnell, D.A. Otto, and P. Grandjean. 1996. Feasibility and validity of three computer-assisted neurobehavioral tests in 7-year-old children. Neurotoxicol. Teratol. 18(4):413-419.

Danielsson, B.R., A. Fredriksson, L. Dahlgren, A.T. Gardlund, L. Olsson, L. Dencker, and T. Archer. 1993. Behavioural effects of prenatal metallic mercury inhalation exposure in rats. Neurotoxicol. Teratol. 15(6):391-6.

Dantas, D.C., and M.L. Queiroz. 1997. Immunoglobulin E and autoantibodies in mercury-exposed workers. Immunopharmacol. Immunotoxicol. 19(3): 383-92.

Danziger, S.J., and P.A. Possick. 1973. Metallic mercury exposure in scientific glassware manufacturing plants. J. Occup. Med. 15(1):15-20.

Davidson, P.W., G.J. Myers, C. Cox, C. Shamlaye, O. Choisy, J. Sloane-Reeves, E. Cernchiari, D.O. Marsh, M. Berlin, M. Tanner, and T.W. Clarkson. 1995a. Neurodevelopmental test selection, administration, and performance in the main Seychelles child development study. Neurotoxicology 16(4):665-676.

Davidson, P.W., G.J. Myers, C. Cox, C.F. Shamlaye, D.O. Marsh, M.A. Tanner, M. Berlin, J. Sloane-Reeves, E. Cernichiari, O. Choisy, A. Choi, and T.W. Clarkson. 1995b. Longitudinal neurodevelopmental study of Seychellois children following in utero exposure to methylmercury from maternal fish ingestion: outcomes at 19 and 29 months. Neurotoxicology 16(4):677-688.

Davidson, P.W., G.J. Myers, C. Cox, C. Axtell, C. Shamlaye, J. Sloane-Reeves, E. Cernichiari, L. Needham, A. Choi, Y. Wang, M. Berlin, and T.W. Clarkson. 1998. Effects of prenatal and postnatal methylmercury exposure from fish consumption on neurodevelopment:outcomes at 66 monts of age in the Seychelles child development study. JAMA 280(8):701-707.

Davidson, P.W., G.J. Myer, C. Shamlaye, C. Cox, P. Gao, C. Axtell, D. Morris, J. Sloane-Reeves, E. Cernichiari, A. Choi, D. Palumbo, and T.W. Clarkson. 1999. Association between prenatal exposure to methylmercury and developmental outcomes in Seychellois children: Effect modification by social and environmental factors. Neurotoxicology 20(5):833-41.

Dietrich, K.N., and D. Bellinger. 1994. The assessment of neurobehavioral development in studies of the effects of prenatal exposure to toxicants. Pp. 57-85 in Prenatal Exposure to Toxicants: Developmental Consequences, H.L. Needleman, and D. Bellinger, eds. Baltimore, MD: Johns Hopkins University Press.

Dietrich, K.N., K.M. Krafft, R.L. Bornschein, P.B. Hammond, O. Berger, P.A. Succop, and M. Bier. 1987. Low-level fetal lead exposure effect on neurobehavioral development in early infancy. Pediatrics 80(5):721-730.

Druet, E., J.C. Guery, K. Ayed, B. Guilbert, S. Avrameas, and P. Druet. 1994. Characteristics of polyreactive and monospecific IgG anti-laminin autoantibodies in the rat mercury model. Immunology 83(3):489-494.

Dyall-Smith, D.J., and J.P. Scurry. 1990. Mercury pigmentation and high mercury levels from the use of a cosmetic cream. Med. J. Aust. 153(7):409-410; 414-415.

Dyer, R.S., C.U. Eccles, and Z. Annau. 1978. Evoked potential alterations following prenatal methyl mercury exposure. Pharmacol. Biochem. Behav. 8(2):137-41.

Eccles, C.U., and Z. Annau. 1982a. Prenatal methylmercury exposure: I. Alterations in neonatal activity. Neurobehav. Toxicol. Teratol. 4(3):371-376.

Eccles, C.U., and Z. Annau. 1982b. Prenatal methylmercury exposure: II. Alterations in learning and psychotropic drug sensitivity in adult offspring. Neurobehav. Toxicol. Teratol. 4(3):377-382.

Elghany, N.A., W. Stopford, W.B. Bunn, and L.E. Fleming. 1997. Occupational exposure to inorganic mercury vapour and reproductive outcomes. Occup. Med. (Lond) 47(6):333-6.

Elsner, J. 1991. Tactile-kinesthetic system of rats as an animal model for minimal brain dysfunction. Arch. Toxicol. 65(6):465-73.

Evans, H.L., R.H. Garman, and B. Weiss. 1977. Methylmercury: exposure duration and regional distribution as determinants of neurotoxicity in nonhuman primates. Toxicol. Appl. Pharmacol. 41(1):15-33.

Fehling, C., M. Abdulla, A. Brun, M. Dictor, A. Schutz, and S. Skerfving. 1975. Methylmercury poisoning in the rat: a combined neurological, chemical, and histopathological study. Toxicol. Appl. Pharmacol. 33(1):27-37.

Fenson, L., P.S. Dale, J.S. Reznick, D. Thal, E. Bates, J.P. Hartung, S. Pethick, and J.S. Reilly. 1993. MacArthur Communicative Development Inventory: User's Guide and Technical Manual. San Diego, CA: Singular Publishing Group.

Fiskesjo, G. 1979. Two organic mercury compounds tested for mutagenicity in mammalian cells by use of the cell line V 79-4. Hereditas 90:103-109.

Fowler, B.A. 1972. Ultrastructural evidence for nephropathy induced by long-term exposure tosmall amounts of methyl mercury. Science 175(23):780-781.

Franchi, E., G. Loprieno, M. Ballardin, L. Petrozzi, and L. Migliore. 1994. Cytogenetic monitoring of fishermen with environmental mercury exposure. Mutat. Res. 320(1-2):23-9.

Fredriksson, A., L. Dahlgren, B. Danielsson, P. Eriksson, L. Dencker, and T. Archer. 1992. Behavioural effects of neonatal metallic mercury exposure in rats. Toxicology 74(2-3):151-60.

Fredriksson, A., L. Dencker, T. Archer, and B.R. Danielsson. 1996. Prenatal coexposure to metallic mercury vapour and methylmercury produce interactive behavioural changes in adult rats. Neurotoxicol. Teratol. 18(2):129-34.

Fredriksson, A., A.T. Gardlund, K. Bergman, A. Oskarsson, B. Ohlin, B. Danielsson, and T. Archer. 1993. Effects of maternal dietary supplementation with selenite on the postnatal development of rat offspring exposed to methyl mercury in utero. Pharmacol. Toxicol. 72(6):377-82.

Frustaci, A., N. Magnavita, C. Chimenti, M. Caldarula, E. Sabbioni, R. Pietra, C. Cellini, G.F. Possati, and A. Maseri. 1999. Marked elevation of mycardial trace elements in idiopathic dilated cardiomyopathy compared with secondary cardiac dysfunction. J. Am. Coll. Cardiol. 33(6):1578-83.

Fuyuta, M., T. Fujimoto, and S. Hirata. 1978. Embryotoxic effects of methylmercuric chloride administered to mice and rats during organogenesis. Teratology 18(3):353-366.

Fuyuta, M., T. Fujimoto, and E. Kiyofuji. 1979. Teratogenic effects of a single oral administration of methylmercuric chloride in mice. Acta Anat. (Basel) 104(3):356-62.

Geyer, M.A., R.E. Butcher, and K. Fite. 1985. A study of startle and locomotor activity in rats exposed prenatally to methylmercury. Neurobehav. Toxicol. Teratol. 7(6):759-65.

Gilbert, S.G., and K.S. Grant-Webster. 1995. Neurobehavioral effects of developmental methylmercury exposure. Environ. Health Perspect. 103(Suppl. 6):135-42.

Gilbert, S.G., T.M. Burbacher, and D.C. Rice. 1993. Effects of in utero methylmercury exposure on a spatial delayed alternation task in monkeys. Toxicol. Appl. Pharmacol. 123(1):130-6.

Gilbert, S.G., D.C. Rice, and T.M. Burbacher. 1996. Fixed interval/fixed ratio performance in adult monkeys exposed in utero to methylmercury. Neurotoxicol. Teratol. 18(5):539-46.

Ghosh, A.K., S. Sen, A. Sharma, and G. Talukder. 1991. Effect of chlorophyllin on mercuric chloride-induced clastogenicity in mice. Food Chem. Toxicol. 29(11):777-779.

Grandjean, P., P. Weihe, and R.F. White. 1995. Milestone development in infants exposed to methylmercury from human milk. Neurotoxicology 16(1):27-34.

Grandjean, P., P. Weihe, P.J. Jørgensen, T. Clarkson, E. Cernichiari, and T. Viderø. 1992. Impact of maternal seafood diet on fetal exposure to mercury, selenium, and lead. Arch. Environ. Health 47(3):185-195.

Grandjean, P., P. Weihe, R.F. White, F. Debes, S. Araki, K. Yokoyama, K. Murata, N. Sørensen, R. Dahl, and P.J. Jørgensen. 1997. Cognitive deficit in 7-year-old children with prenatal exposure to methylmercury. Neurotoxicol. Teratol. 19(6):417-428.

Grandjean, P., P. Weihe, R.F. White, and F. Debes. 1998. Cognitive performance of children prenatally exposed to "safe" levels of methylmercury. Environ. Res. 77(2):165-172.

Grandjean, P., R. White, A. Nielsen, D. Cleary, and E.C. de Oliveira Santos. 1999. Methylmercury neurotoxicity in Amazonian children downstream from gold mining. Environ. Health Perspect. 107(7):587-591.

Gunderson, V.M., K.S. Grant, T.M. Burbacher, J.F. Fagan, 3d, and N.K. Mottet. 1986. The effect of low-level prenatal methylmercury exposure on visual recognition memory in infant crab-eating macaques. Child Dev. 57(4):1076-83.

Gunderson, V.M., K.S. Grant-Webster, T.M. Burbacher, and N.K. Mottet. 1988. Visual recognition memory deficits in methylmercury-exposed Macaca fascicularis infants. Neurotoxicol. Teratol. 10(4):373-9.

Hallee, T.J. 1969. Diffuse lung disease caused by inhalation of mercury vapor. Am. Rev. Respir. Dis. 99(3):430-6.

Harada, M. 1995. Minamata Disease: Methylmercury poisoning in Japan caused by environmental pollution. Crit. Rev. Toxicol. 25(1):1-24.

Harada, M. 1997. Neurotoxicity of methylmercury: Minamata and the Amazon. Pp. 177-188 in Mineral and Metal Neurotoxicology, M. Yasui, M.J. Strong, K. Ota, and M.A. Verity, eds. Boca Raton, FL: CRC Press.

Harada, M., H. Akagi, T. Tsuda, T. Kizaki, and H. Ohno. 1999. Methylmercury level in umbilical cords from patients with congenital Minamata disease. Sci. Total Environ. 234(1-3):59-62.

Hirano, M., K. Mitsumori, K. Maita, and Y. Shirasu. 1986. Further carcinogenicity study on methylmercury chloride in ICR mice. Nippon Juigaku Zasshi (Jpn. J. Vet. Sci.) 48(1):127-135.

Höök, O., K.D. Lundgren, and A. Swensson. 1954. On alkyl mercury poisoning: With a description of two cases . Acta Med. Scand. 150(2):131-137.

Hu, H., G. Moller, and M. Abedi-Valugerdi. 1999. Mechanism of mercury-induced autoimmunity: Both T helper 1- and T helper 2-type responses are involved. Immunology 96(3):348-57.

Hua, J., L. Pelletier, M. Berlin, and P. Druet. 1993. Autoimmune glomerulonephritis induced by mercury vapour exposure in the Brown Norway rat. Toxicology 79(2):119-29.

Hughes, J.A., and Z. Annau. 1976. Postnatal behavioral effects in mice after prenatal exposure to methylmercury. Pharmacol. Biochem. Behav. 4(4):385-391.

Hultman, P., and H. Hansson-Georgiadis. 1999. Methyl mercury-induced autoimmunity in mice. Toxicol. Appl. Pharmacol. 154(3):203-11.

Hunter, D., R.R. Bomford, and D.S. Russell. 1940. Poisoning by methyl mercury compounds. Quart. J. Med. 9(July):193-213.

Ilbäck, N.G. 1991. Effects of methyl mercury exposure on spleen and blood natural killer (NK) cell activity in the mouse. Toxicology 67(1):117-124.

Ilbäck, N.G., J. Sundberg, and A. Oskarsson. 1991. Methyl mercury exposure via placenta and milk impairs natural killer (NK) cell function in newborn rats. Toxicol. Lett. 58(2):149-58.

Ilbäck, N.G., L. Wesslen, J. Fohlman, and G. Friman. 1996. Effects of methyl mercury on cytokines, inflammation and virus clearance in a common infection (coxsackie B3 myocarditis). Toxicol. Lett. 89(1):19-28.

Inouye, M., and Y. Kajiwara. 1988. Developmental disturbances of the fetal brain in guinea pigs caused by methylmercury. Arch. Toxicol. 62(1):15-21.

Inouye, M., and U. Murakami. 1975. Teratogenic effect of orally administered methylmercuric chloride in rats and mice. Congenital Anomalies 15(1):1-9.

Inskip, M.J., and J.K. Piotrowski. 1985. Review of the health effects of methylmercury. J. Appl. Toxicol. 5(3):113-33.

IPCS (International Programme on Chemical Safety). 1990. Environmental Health Criteria Document 101 - Methylmercury. Geneva: World Health Organization.

Jacobs, J.M., N. Carmichael, and J.B. Cavanagh. 1977. Ultrastructural changes in the nervous system of rabbits poisoned with methyl mercury. Toxicol. Appl. Pharmacol. 39(2):249-61.

Jalili, H.A., and A.H. Abbasi. 1961. Poisoning by ethyl mercury toluene sulphonanilide. Br. J. Ind. Med. 18:303-308.

Janicki, K., J. Dobrowolski, and K. Krasnicki. 1987. Correlation between contamination of the rural environment with mercury and occurrence of leukemia in men and cattle. Chemosphere 16(1):253-257.

Kanematsu, N., M. Hara, and T. Kada. 1980. Rec assay and mutagenicity studies on metal compounds. Mutat. Res. 77(2):109-116.

Kazantzis, G., K.F. Schiller, A.W. Asscher, and R.G. Drew. 1962. Albuminuria and the nephrotic syndrome following exposure to mercury and its compounds. Quart. J. Med. 31(Oct.):403-418.

Khera, K.S. 1973a. Reproductive capability of male rats and mice treated with methylmercury. Toxicol. Appl. Pharmacol. 24(2):167-77.

Khera, K.S. 1973b. Teratogenic effects of methylmercury in the cat: Note on the use of this species as a model for teratogenicity studies. Teratology 8(3): 293-303.

Khera, K.S., and S.A. Tabacova. 1973. Effects of methylmercuric chloride on the progeny of mice and rats treated before or during gestation. Food Cosmet. Toxicol. 11(2):245-254.

Khera, K.S., F. Iverson, L. Hierlihy, R. Tanner, and G. Trivett. 1974. Toxicity of methylmercury in neonatal cats. Teratology 10(1):69-76.

Kinjo, Y., S. Akiba, N. Yamaguchi, S. Mizuno, S. Watanabe, J. Wakamiya, M. Futatsuka, and H. Kato. 1996. Cancer mortality in Minamata disease pa-

tients exposed to methylmercury through fish diet. J. Epidemiol. 6(3):134-8.
Kinjo, Y., H. Higashi, A. Nakano, M. Sakamoto, and R. Sakai. 1993. Profile of subjective complaints and activities of daily living among current patients with Manamata disease after 3 decades. Environ. Res. 63(2):241-251.
Kjellström, T., P. Kennedy, S. Wallis, and C. Mantell. 1986. Physical and Mental Development of Children with Prenatal Exposure to Mercury from Fish. Stage I: Preliminary tests at age 4. National Swedish Environmental Protection Board Report 3080. Solna, Sweden.
Kjellström, T., P. Kennedy, S. Wallis, A. Stewart, L. Friberg, B. Lind, T. Wutherspoon, and C. Mantell. 1989. Physical and Mental Development of Children with Prenatal Exposure to Mercury from Fish. National Swedish Environmental Protection Board Report No. 3642.
Koller, L.D. 1975. Methylmercury: effect on oncogenic and nononcogenic viruses in mice. Am. J. Vet. Res. 36(10):1501-4.
Koller, L.D., J.H. Exon, and B. Arbogast. 1977. Methylmercury: effect on serum enzymes and humoral antibody. J. Toxicol. Environ. Health 2(5):1115-1123.
Koopman-Esseboom, C., N. Weisglas-Kuperus, M.A. de Ridder, C.G. Van der Paauw, L.G. Tuinstra, and P.J. Sauer. 1996. Effects of polychlorinated biphenyl/dioxin exposure and feeding type on infants mental and psychomotor development. Pediatrics 97(5):700-706.
Kosatsky, T., and P. Foran. 1996. Do historic studies of fish consumers support the widely accepted LOEL for methylmercury in adults. Neurotoxicology 17(1): 177-86.
Kostka, B., M. Michalska, U. Krajewska, and R. Wierzbicki. 1989. Blood coagulation changes in rats poisoned with methylmercuric chloride (MeHg). Pol. J. Pharmacol. Pharm. 41(2):183-9.
Lauwerys, R., H. Roels, P. Genet, G. Toussaint, A. Bouckaert, and S. De Cooman. 1985. Fertility of male workers exposed to mercury vapor or to manganese dust: a questionnaire study. Am. J. Ind. Med. 7(2):171-6.
Lebel, J., D. Mergler, M. Lucotte, M. Amorim, J. Dolbec, D. Miranda, G. Arantes, I. Rheault, and P. Pichet. 1996. Evidence of early nervous system dysfunction in Amazonian populations exposed to low-levels of methylmercury. Neurotoxicology 17(1):157-167.
Lebel, J., D. Mergler, F. Branches, M. Lucotte, M. Amorim, F. Larribe, and J. Dolbec. 1998. Neurotoxic effects of low-level methylmercury contamination in the Amazonian Basin. Environ. Res. 79(1):20-32.
Lee, J.H., and D.H. Han. 1995. Maternal and fetal toxicity of methylmercuric chloride administered to pregnant Fischer 344 rats. J. Toxicol. Environ. Health 45(4):415-425.
Leyshon, K., and A.J. Morgan. 1991. An integrated study of the morphological and gross-elemental consequences of methyl mercury intoxication in rats,

with particular attention on the cerebellum. Scanning Microsc. 5(3):895-904.

MacDonald, J.S., and R.D. Harbison. 1977. Methyl mercury-induced encephalopathy in mice. Toxicol. Appl. Pharmacol. 39(2):195-205.

Magos, L., and W.H. Butler. 1972. Cumulative effects of methylmercury dicyandiamide given orally to rats. Food Cosmet. Toxicol. 10(4):513-7.

Marsh, D.O., T.W. Clarkson, C. Cox, G.J. Myers, L. Amin-Zaki, and S. Al-Tikriti. 1987. Fetal methylmercury poisoning: Relationship between concentration in single strands of maternal hair and child effects. Arch. Neurol. 44(10):1017-1022.

Marsh, D.O., M.D. Turner, J.C. Smith, P. Allen, and N. Richdale. 1995a. Fetal methylmercury study in a Peruvian fish-eating population. Neurotoxicology 16(4):717-726.

Marsh, D.O., T.W. Clarkson, G.J. Myers, P.W. Davidson, C. Cox, E. Cernichiari, M.A. Tanner, W. Ledhar, C. Shamlaye, O. Choisy, C. Hoareau, and M. Berlin. 1995b. The Seychelles study of fetal methylmercury exposure and child development: Introduction. Neurotoxicology 16(4):583-596.

Matsumoto, H., G. Koya, and T. Takeuchi. 1965. Fetal Minamata disease. A neuropathological study of two cases of intrauterine intoxication by a methyl mercury compound. J. Neuropathol. Exp. Neurol. 24(4):563-74.

Matsuo, N., T. Suzuki, and H. Akagi. 1989. Mercury concentration in organs of contemporary Japanese. Arch. Environ. Health 44(5):298-303.

McKeown-Eyssen, G.E., J. Ruedy, and A. Neims. 1983. Methyl mercury exposure in northern Quebec. II. Neurologic findings in children. Am. J. Epidemiol. 118(4): 470-479.

Merigan, W.H., J.P. Maurissen, B. Weiss, T. Eskin, and L.W. Lapham. 1983. Neurotoxic actions of methylmercury on the primate visual system. Neurobehav. Toxicol. Teratol. 5(6):649-58.

Miller, C.T., Z. Zawidska, E. Nagy, and S.M. Charbonneau. 1979. Indicators of genetic toxicity in leukocytes and granulocytic precursors after chronic methylmercury ingestion by cats. Bull. Environ. Contam. Toxicol. 21(3):296-303.

Mitsumori, K., K. Maita, and Y. Shirasu. 1984. Chronic toxicity of methylmercury chloride in rats: Pathological study. Nippon Juigaku Zasshi (Jpn. J. Vet. Sci.) 46(4):549-557.

Mitsumori, K., M. Hirano, H. Ueda, K. Maita, and Y. Shirasu. 1990. Chronic toxicity and carcinogenicity of methylmercury chloride in B6C3F1 mice. Fundam. Appl. Toxicol. 14(1):179-190.

Mitsumori, K., K. Maita, T. Saito, S. Tsuda, and Y. Shirasu. 1981. Carcinogenicity of methylmercury chloride in ICR mice: Preliminary note on renal carcinogenesis. Cancer Lett. 12(4):305-310.

Mitsumori, K., K. Takahashi, O. Matano, S. Goto, and Y. Shirasu. 1983.

Chronic toxicity of methylmercury chloride in rats: Clinical study and chemical analysis. Nippon Juigaku Zasshi (Jpn. J. Vet. Sci.) 45(6):747-757.

Mohamed, M., T. Burbacher, and N. Mottet. 1987. Effects of methylmercury on testicular functions in Macaca fascicularis monkeys. Pharmacol. Toxicol. 60(1):29-36.

Moszczynski, P., S. Slowinski, J. Rutkowski, S. Bem, and D. Jakus-Stoga. 1995. Lymphocytes, T and NK cells, in men occupationally exposed to mercury vapours. Int. J. Occup. Med. Environ. Health 8(1):49-56.

Mottet, N.K., C.M. Shaw, and T.M. Burbacher. 1987. The pathological lesions of methyl mercury intoxication in monkeys. Pp. 73-103 in The Toxicity of Methyl Mercury, C.U. Eccles, and Z. Annau, eds. Baltimore, MD: Johns Hopkins.

Munro, I.C., E.A. Nera, S.M. Charbonneau, B. Junkins, and Z. Zawidzka. 1980. Chronic toxicity of methylmercury in the rat. J. Environ. Pathol. Toxicol. 3(5-6):437-447.

Murata, K., P. Weihe, A. Renzoni, F. Debes, R. Vasconcelos, F. Zino, S. Araki, P.J. Jorgensen, R.F. White, and P. Grandjean. 1999a. Delayed evoked potentials in children exposed to methylmercury from seafood. Neurotoxicol. Teratol. 21(4):343-348.

Murata, K., P. Weihe, S. Araki, E. Budtz-Jorgensen, and P. Grandjean. 1999b. Evoked potentials in Faroese children prenatally exposed to methylmercury. Neurotoxicol. Teratol. 21(4): 471-472.

Murphy, M.J., E.J. Culliford, and V. Parsons. 1979. A case of poisoning with mercuric chloride. Resuscitation 7(1):35-44.

Müsch, H.R., M. Bornhausen, H. Kriegel, and H. Greim. 1978. Methylmercury chloride induces learning deficits in prenatally treated rats. Arch. Toxicol. 40(2):103-108.

Myers, G.J., P.W. Davidson, C.F. Shamlaye, C.D. Axtell, E. Cernichiari, O. Choisy, A. Choi, C. Cox, and T.W. Clarkson. 1997. Effects of prenatal methylmercury exposure from a high fish diet on developmental milestones in the Seychelles Child Development Study. Neurotoxicology 18(3):819-830.

Myers, G.J., D.O. Marsh, C. Cox, P.W. Davidson, C.F. Shamlaye, M.A. Tanner, A. Choi, E. Cernichiari, O. Choisy, and T.W. Clarkson. 1995a. A pilot neurodevelopmental study of Seychellois children following in utero exposure to methylmercury from a maternal fish diet. Neurotoxicology 16(4):629-638.

Myers, G.J., D.O. Marsh, P.W. Davidson, C. Cox, C.F. Shamlaye, M. Tanner, A. Choi, E. Cernichiari, O. Choisy, and T.W. Clarkson. 1995b. Main neurodevelopmental study of Seychellois children following in utero exposure to methylmercury from a maternal fish diet: Outcome at six months. Neurotoxicology 16(4):653-664.

Myers, G.J., P.W. Davidson, C. Cox, C.F. Shamlaye, M.A. Tanner, O. Choisy, J. Sloane-Reeves, D.O. Marsh, E. Cernichiari, A. Choi, M. Berlin, and T.W. Clarkson. 1995c. Neurodevelopmental outcomes of Seychellois children sixty-six months after in utero exposure to methylmercury from a maternal fish diet: Pilot study. Neurotoxicology 16(4):639-652.

Nakatsuru, S., J. Oohashi, H. Nozaki, S. Nakada, and N. Imura. 1985. Effect of mercurials on lymphocyte functions in vitro. Toxicology 36(4):297-306.

Newberne, P.M., O. Glaser, L. Friedman, and B.R. Stillings. 1972. Chronic exposure of rats to methyl mercury in fish protein. Nature 237(5349):40-41.

Newland, M.C., S. Yezhou, B. Logdberg, and M. Berlin. 1994. Prolonged behavioral effects of in utero exposure to lead or methyl mercury: Reduced sensitivity to changes in reinforcement contingencies during behavioral transitions and in steady state. Toxicol. Appl. Pharmacol. 126(1):6-15.

NRC (National Research Council). 1991. Frontiers in Assessing Human Exposures to Environmental Toxicants: Report of the Symposium. Washington, DC: National Academy Press.

NRC (National Research Council). 1997. Environmental Epidemiology, Vol. 2.: Use of the Gray Literature and Other Data in Environmental Epidemiology. Washington, DC: National Academy Press.

O'Kusky, J. 1983. Methylmercury poisoning of the developing nervous system: Morphological changes in neuronal mitochondria. Acta Neuropathol. (Berl) 61(2):116-22.

Olson, K., and G.M. Bousch. 1975. Decreased learning capacity in rats exposed prenatally and postnatally to low doses of mercury. Bull. Environ. Contam. Toxicol. 13(1):73-9.

Ortega, H.G., M. Lopez, A. Takaki, Q.H. Huang, A. Arimura, and J.E. Salvaggio. 1997. Neuroimmunological effects of exposure to methylmercury forms in the Sprague-Dawley rats. Activation of the hypothalamic-pituitary-adrenal axis and lymphocyte responsiveness. Toxicol. Ind. Health 13(1):57-66.

Popescu, H.I., L. Negru, and I. Lancranjan. 1979. Chromosome aberrations induced by occupational exposure to mercury. Arch. Environ. Health 34(6):461-3.

Queiroz, M.L., and D.C. Dantas. 1997. B lymphocytes in mercury-exposed workers. Pharmacol. Toxicol. 81(3):130-3.

Queiroz, M.L., and D.C. Dantas. 1997a. T lymphocytes in mercury-exposed workers. Immunopharmacol. Immunotoxicol. 19(4):499-510.

Queiroz, M.L., C. Bincoletto, M.R. Quadros, and E.M. De Capitani. 1999. Presence of micronuclei in lymphocytes of mercury exposed workers. Immunopharmacol. Immunotoxicol. 21(1):141-50.

Rasmussen, E.B., and M.C. Newland. 1999. Acquisition of a Multiple DRH

Extinction Schedule of Reinforcement in Rats Exposed during Development to Methylmercury. No. 697. Pp. 149. SOT 1999 Annual Meeting.

Reuhl, K.R., L.W. Chang, and J.W. Townsend. 1981a. Pathological effects of in utero methylmercury exposure on the cerebellum of the golden hamster. 1. Early effects upon the neonatal cerebellar cortex. Environ. Res. 26(2):281-306.

Reuhl, K.R., L.W. Chang, and J.W. Townsend. 1981b. Pathological effects of in utero methylmercury exposure on the cerebellum of the golden hamster. II. Residual effects on the adult cerebellum. Environ. Res. 26(2):307-27.

Rice, D.C. 1989. Delayed neurotoxicity in monkeys exposed developmentally to methylmercury. Neurotoxicology 10(4):645-650.

Rice, D.C. 1992. Effects of pre- plus postnatal exposure to methylmercury in the monkey on fixed interval and discrimination reversal performance. Neurotoxicology 13(2):443-52.

Rice, D.C. 1996. Evidence for delayed neurotoxicity produced by methylmercury. Neurotoxicology 17(3-4):583-596.

Rice, D.C. 1998. Age-related increase in auditory imapirment in monkeys exposed in utero plus postnatally to methylmercury. Toxicol. Sci. 44(2):191-196.

Rice, D.C., and S.G. Gilbert. 1982. Early chronic low-level methylmercury poisoning in monkeys impairs spatial vision. Science 216(4547):759-761.

Rice, D.C., and S.G. Gilbert. 1990. Effects of developmental exposure to methyl mercury on spatial and temporal visual function in monkeys. Toxicol. Appl. Pharmacol. 102(1):151-63.

Rice, D.C., and S.G. Gilbert. 1992. Exposure to methyl mercury from birth to adulthood impairs high-frequency hearing in monkeys. Toxicol. Appl. Pharmacol. 115(1):6-10.

Rice, D.C., and S.G. Gilbert. 1995. Effects of developmental methylmercury exposure or lifetime lead exposure on vibration sensitivity function in monkeys. Toxicol. Appl. Pharmacol. 134(1):161-9.

Robison, S.H., O. Cantoni, and M. Costa. 1984. Analysis of metal-induced DNA lesions and DNA-repair replication in mammalian cells. Mutat. Res. 131(3-4): 173-81.

Rogan, W.J., and B.C. Gladen. 1991. PCBs, DDE, and child development at 18 and 24 months. Ann. Epidemiol. 1(5):407-413.

Rowland, A.S., D.D. Baird, C.R. Weinberg, D.L. Shore, C.M. Shy, and A.J. Wilcox. 1994. The effect of occupational exposure to mercury vapour on the fertility of female dental assistants. Occup. Environ. Med. 51(1):28-34.

Salonen, J.T., K. Seppänen, K. Nyyssönen, H. Korpela, J. Kauhanen, M. Kantola, J. Tuomilehto, H. Esterbauer, F. Tatzber, and R. Salonen. 1995. Intake of mercury from fish, lipid peroxidation, and the risk of myocardial infarction and coronary, cardiovascular, and any death in Eastern Finnish men. Circulation 91(3):645-655.

Samuels, E.R., H.M. Heick, P.N. McLaine, and J.P. Farant. 1982. A case of accidental inorganic mercury poisoning. J. Anal. Toxicol. 6(3):120-2.

Schalock, R.L., W.J. Brown, R.A. Kark, and N.K. Menon. 1981. Perinatal methylmercury intoxication: behavioral effects in rats. Dev. Psychobiol. 14(3):213-9.

Schroeder, H., and M. Mitchener. 1975. Life-time effects of mercury, methyl mercury, and nine other trace metals in mice. J. Nutr. 105(4):452-458.

Sekowski, J.W., L.H. Malkas, Y. Wei, and R.J. Hickey. 1997. Mercuric ion inhibits the activity and fidelity of the human cell DNA synthesome. Toxicol. Appl. Pharmacol. 145(2):268-76.

Shaw, C.M., N.K. Mottet, and D.V. Finocchio. 1979. Cerebrovascular lesions in experimental methyl mercurial encephalopathy. Neurotoxicology 1(1):57-74.

Shenker B.J., P. Berthold, C. Rooney, L. Vitale, K. DeBolt, and I.M. Shapiro. 1993. Immunotoxic effects of mercuric compounds on human lymphocytes and monocytes. III. Alterations in B-cell function and viability. Immunopharmacol. Immunotoxicol. 15(1):87-112.

Shenker, B.J., T.L. Guo, and I.M. Shapiro. 1999. Induction of apoptosis in human T-cells by methyl mercury: Temporal relationship between mitochondrial dysfunction and loss of reductive reserve. Toxicol. Appl. Pharmacol. 157(1):23-35.

Siblerud, R.L.. 1990. The relationship between mercury from dental amalgam and the cardiovascular system. Sci. Total Environ. 99(1-2):23-35.

Skerfving, S., K. Hansson, and J. Lindsten. 1970. Chromosome breakage in humans exposed to methyl mercury through fish consumption. Arch. Environ. Health 21(2):133-139.

Skerfving, S., K. Hansson, C. Mangs, J. Lindsten, and N. Ryman. 1974. Methylmercury-induced chromosome damage in man. Environ. Res. 7(1):83-98.

Slotkin, T.A., S. Pachman, J. Bartolome, and R.J. Kavlock. 1985. Biochemical and functional alterations in renal and cardiac development resulting from neonatal methylmercury treatment. Toxicology 36(2-3):231-41.

Solecki, R., L. Hothorn, M. Holzweissig, and V. Heinrich. 1991. Computerised analysis of pathological findings in longterm trials with phenylmercuric acetate in rats. Arch. Toxicol. (Suppl.):14:100-3.

Soni, J.P., R.U. Singhania, A. Bansal, and G. Rathi. 1992. Acute mercury vapor poisoning. Indian Pediatr. 29(3):365-8.

Sørensen, N., K. Murata, E. Budtz-Jørgensen, P. Weihe, and P. Grandjean. 1999.

Prenatal methylmercury exposure as a cardiovascular risk factor at seven years of age. Epidemiology 10(4):370-375.

Spyker, J.M. 1975. Assessing the impact of low level chemicals on development: Behavioral and latent effects. Fed. Proc. 34(9):1835-44.

Spyker, J., S. Sparber, and A.M. Goldberg. 1972. Subtle consequences of methylmercury exposure: Behavioral deviations in offspring of treated mothers. Science 177(49):621-623.

Steuerwald, U., P. Weihe, P. Jorgensen, K. Bjerve, J. Brock, B. Heinzow, E. Budtz-Jørgensen, and P. Grandjean. 2000. Maternal seafood diet, methylmercury exposure, and neonatal neurological function. J. Pediatr. 136(5):599-605.

Su, M.Q., and G.T. Okita. 1976. Behavioral effects on the progeny of mice treated with methylmercury. Toxicol. Appl. Pharmacol. 38(1):195-205.

Tamashiro, H., H. Akagi, M. Arakaki, M. Futatsuka, and L.H. Roht. 1984. Causes of death in Minamata disease: Analysis of death certificates. Int. Arch. Occup. Environ. Health 54(2):135-146.

Tamashiro, H., M. Arakaki, M. Futatsuka, and E.S. Lee. 1986. Methylmercury exposure and mortality in southern Japan: A close look at causes of death. J. Epidemiol. Community Health 40(2):181-185.

Takeuchi, T. 1968. Pathology of Minamate disease. Pp. 141-228 in Minamata Disease. Study Group of Minanata Disease, ed. Kumamoto, Japan: Kumamoto University.

Thompson, S.A., K.L. Roellich, A. Grossmann, S.G. Gilbert, and T.J. Kavanagh. 1998. Alterations in immune parameters associated with low level methylmercury exposure in mice. Immunopharmacol Immunotoxicol 20(2):299-314.

Thuvander, A., J. Sundberg, and A. Oskarsson. 1996. Immunomodulating effects after perinatal exposure to methylmercury in mice. Toxicology 114(2):163-75.

Tsubaki, T., and H. Takahashi, eds. 1986. Clinical aspects of Minamata disease. Neurological aspects of methylmercury poisoning in Minamata. Pp. 41-57 in Recent Advances in Minamata Disease Studies. Tokyo: Kodansha.

Tubbs, R.R., G.N. Gephardt, J.T. McMahon, M.C. Pohl, D.G. Vidt, S.A. Barenberg, and R. Valenzuela. 1982. Membranous glomerulonephritis associated with industrial mercury exposure. Study of pathogenetic mechanisms. Am. J. Clin. Pathol. 77(4):409-13.

Uchino, M., T. Okajima, K. Eto, T. Kumamoto, I. Mishima, and M. Ando. 1995. Neurologic features of chronic Minamata disease (organic mercury poisoning) certified at autopsy. Intern. Med. 34(8):744-7.

Verschaeve, L., M. Kirsch-Volders, C. Susanne, C. Groetenbriel, R.

Haustermans, A. Lecomte, and D. Roossels. 1976. Genetic damage induced by occupationally low mercury exposure. Environ. Res. 12(3):306-16.
Verschuuren, H.G., R. Kroes, E.M. Den Tonkelaar, J.M. Berkvens, P.W. Helleman, A.G. Rauws, P.L. Schuller, and G.J. Van Esch. 1976. Toxicity of methylmercury chloride in rats. III. Long-term toxicity study. Toxicology 6(1):107-123.
Vorhees, C.V. 1985. Behavioral effects of prenatal methylmercury in rats: A parallel trial to the Collaborative Behavioral Teratology Study. Neurobehav. Toxicol. Teratol. 7(6):717-25.
Vroom, F.Q., and M. Greer. 1972. Mercury vapour intoxication. Brain 95(2): 305-18.
Wakita, Y. 1987. Hypertension induced by methyl mercury in rats. Toxicol. Appl. Pharmacol. 89(1):144-7.
Warkany, J., and D.M. Hubbard. 1953. Acrodynia and mercury. J. Pediat. 42(3):365-386.
Wasserman, G., J.H. Graziano, P. Factor-Litvak, D. Popovac, N. Morina, A. Musabegovic, N. Vrenezi, S. Capuni-Paracka, V. Lekic, E. Preteni-Redjepi, S. Hadjialjevic, V. Slavkovich, J. Kline, P. Shrout, and Z. Stein. 1992. Independent effects of lead exposure and iron deficiency anemia on developmental outcome at age 2 years. J. Pediatr. 121(5 Pt. 1):695-703.
Wassick, K.H., and A. Yonovitz. 1985. Methyl mercury ototoxicity in mice determined by auditory brainstem responses. Acta Otolaryngol. 99(1-2): 35-45.
Weiss, B. 1998. A risk assessment perspective on the neurobehavioral toxicity of endocrine disruptors. Toxicol. Ind. Health 14(1-2):341-59.
WHO (World Health Organization). 1976. Mercury. Environmental Health Criteria 1. Geneva, Switzerland: World Health Organization.
Wild, L.G., H.G. Ortega, M. Lopez, and J.E. Salvaggio. 1997. Immune system alteration in the rat after indirect exposure to methylmercury chloride or methylmercury sulfide. Environ. Res. 74(1):34-42.
Williams, M.V., T. Winters, and K.S. Waddell. 1987. In vivo effects of mercury (II) on deoxyuridine triphosphate nucleotidohydrolase, DNA polymerase (alpha, beta), and uracil-DNA glycosylase activities in cultured human cells: relationship to DNA damage, DNA repair, and cytotoxicity. Mol. Pharmacol. 31(2):200-7.
Wössmann, W., M. Kohl, G. Grüning, and P. Bucsky. 1999. Mercury intoxication presenting with hypertension and tachycardia. Arch. Dis. Child. 80(6):556-7.
Wulf, H.C., N. Kromann, N. Kousgaard, J.C. Hansen, E. Niebuhr, and K. Alboge. 1986. Sister chromatid exchange (SCE) in Greenlandic Eskimos.

Dose-response relationship between SCE and seal diet, smoking, and blood cadmium and mercury concentrations. Sci. Total Environ. 48(1-2):81-94.

Yasutake, A., Y. Hirayama, and M. Inouye. 1991. Sex Difference of nephrotoxicity by methylmercury in mice. Pp. 389-396 in Nephrotoxicity: Mechanisms, Early Diagnosis, and Therapeutic Management. Fourth International Symposium of Nephrotoxicity, Guilford, England, UK, 1989. P.H. Bach, and K.J. Ullrich, eds. New York: Marcel Dekker.

Yoshida, Y., H. Kamitsuchibashi, R. Hamada, Y. Kuwano, I. Mishima, and A. Igata. 1992. Truncal hypesthesia in patients with Minamata disease. Intern. Med. 31(2):204-7.

Zenick, H. 1974. Behavioral and biochemical consequences in methylmercury chloride toxicity. Pharmacol. Biochem. Behav. 2(6):709-13.

Zenick, H. 1976. Evoked potential alterations in methylmercury chloride toxicity. Pharmacol. Biochem. Behav. 5(3):253-5.

6

Comparison of Studies for Use in Risk Assessment

UNTIL recently, the data base available for risk assessments of MeHg has been limited to high-dose poisoning episodes in Japan and Iraq. More recently, however, epidemiological studies have been conducted on the health effects of exposure to low doses of MeHg (for details of health effects, see Chapter 5). The low-dose MeHg exposure studies are more relevant to levels of exposures in the United States and, therefore, more appropriate for use in risk assessments. The two largest and most comprehensive studies to address the health effects of MeHg—the Seychelles Child Development Study (SCDS) and the Faroe Islands studies—reached different conclusions. A range of adverse neuropsychological and neurophysiological outcomes were found to be associated with prenatal Hg exposure in the Faroe Islands study, whereas adverse effects were not found in the main Seychelles study. This chapter compares those two studies, as well as data from the pilot phase of the SCDS, and a smaller study carried out on a cohort in New Zealand.

MeHg exposure in the SCDS and Faroe Islands studies were similar; the arithmetic mean maternal hair Hg concentration in the Seychelles cohort (6.8 μg/g) was slightly higher than the geometric mean reported in the Faroe cohort (4.3 μg/g). Several differences in research design and cohort characteristics have been identified that might account for the discrepant findings. Some of those explanations seem less persuasive, however, when the data from the New Zealand study are considered. That study found associations with MeHg exposure in a population whose sources of MeHg exposure were similar to those in the Seychelles and used end points similar to those examined in the Seychelles. Although the New Zealand data have been available for some time, they have not been used extensively for risk assessment, possibly because until recently, they had not been subjected to standard peer-review procedures. A re-analysis of the New Zealand data by Crump et al. (1998), which underwent peer review, reported associations of prenatal MeHg exposure with several end points (when one extreme outlier was excluded), including four that were not found to be related to MeHg in the Seychelles study. The New Zealand study has been criticized for errors in matching exposed children to controls and for testing exposed children and controls at different ages (Myers et al. 1998). Those errors occurred in the 4-year follow-up but were corrected in the 6-year follow-up, which is the data set reviewed in this section. In addition, there is no information that would suggest the presence of differential measurement error across the studies. Any error of that type is likely to be nondifferential (i.e., unbiased), and it would reduce the likelihood of detecting associations between MeHg exposure and neurobehavioral test scores.

Data from the peer-reviewed pilot SCDS of 217 children assessed at 5.5 years (Myers et al. 1995) are also considered in this chapter. (Note that the nonstandard treatment of the data from the Revised Denver Developmental Screening Test (DDST-R) discussed in Chapter 5 was not an issue in the 5.5-year follow-up since the DDST-R was not given at that age.) Two of the four outcomes that were tested in both the pilot and the main Seychelles studies at 5.5 years of age were found to be associated with prenatal Hg exposure in the pilot study. The Seychelles investigators were cautious about drawing inferences from their pilot data, because the effects were substantially weaker when four outliers were excluded from the analyses and because socioenvironmental influences were not adequately assessed and controlled statistically. It is not clear, however, that it is appropriate statistically to exclude influential data points; many statisticians would instead recommend the use of data transformation to reduce their influence. Exclusion is appropriate only where a value appears biologically implausible (see discussion of the New Zealand outlier in the Benchmark Analysis section in Chapter 7). With regard to socioenvironmental influences, T.W. Clarkson (principal investigator in the SCDS, personal commun., January 20, 2000) indicated to the committee that the most heavily contaminated fish consumed in the Seychelles islands—swordfish, shark, and tuna—tend to be among the most expensive fish, so that, if anything, exposure levels might be higher among mothers with *higher* socioeconomic status. There is, therefore, no reason to expect a confounding of exposure with lower socioeconomic status, and low socioeconomic status is not likely to explain the association of Hg exposure with adverse development outcomes in the Seychelles pilot study.

ASSESSMENT OF PRENATAL Hg EXPOSURE: CORD BLOOD VERSUS MATERNAL HAIR AND TIMING OF EXPOSURE

The principal measure of prenatal exposure in the Faroe study was Hg in cord blood; in the Seychelles, it was Hg in maternal hair. The Faroe investigators also analyzed maternal-hair samples, but no cord-blood specimens were obtained in the Seychelles. In a recently published analysis, the Faroe investigators compared the relation of the cord-blood and maternal-hair Hg measures with their 7-year end points (Grandjean et al. 1999). As shown in Table 6-1, cord-blood Hg concentration was significantly associated with a slightly larger number of end points than maternal-hair Hg concentration, and in most cases the associations were slightly stronger. For various pharmacokinetic and neurodevelopmental reasons, cord-blood measurements might be more sensitive indicators of the neurodevelopmental effects of MeHg. However, given that hair Hg concentrations in the Faroe Islands study were only a slightly weaker predictor of Hg effects than cord blood, it would be reasonable to expect that, if children were affected in the main Seychelles study, some indication of an association between child performance and maternal-hair Hg concentration would be apparent in that study. With the possible exception of the Bender Gestalt scores for boys, there is no indication of even a trend in the predicted direction in the data published to date from the main SCDS (e.g., see Figures 5-7 and 5-8).

It should be noted that the maternal-hair samples obtained in the Faroe and Seychelles studies did not necessarily reflect exactly the same period of pregnancy. In part, this is because the Seychelles study ob-

TABLE 6-1 Change in Neuropsychological Test Performance Associated with a Doubling of the Hg Concentration in the Faroese Cohort[a]

	Cord Blood Hg		Maternal Hair Hg	
Test	Change[b]	p[c]	Change[b]	p[c]
NES[d] Finger Tapping				
Preferred hand	-5.37	0.049	-5.99	0.039
Other hand	-1.97	0.460	-4.40	0.120
Both hands	-4.11	0.136	-6.64	0.024
NES Hand-Eye Coordination				
Error score	3.70	0.187	5.40	0.070
NES Continuous Performance Test				
Missed responses	10.08	0.024	5.14	0.241
Reaction Time	15.93	<0.001	8.99	0.035
Wechsler Intelligence Scale for Children-Revised				
Digit Spans	-5.62	0.049	-4.39	0.147
Similarities	-0.37	0.902	-2.07	0.525
Block Designs	-4.36	0.109	-2.86	0.322
Bender Visual Motor Gestalt Test				
Error on copying	3.83	0.154	3.60	0.208
Delayed recall	-4.64	0.104	-1.26	0.679
Boston Naming Test				
No cues	-9.75	<0.001	-6.98	0.016
With cues	-10.47	<0.001	-7.47	0.009
California Verbal Learning Test (Children)				
Learning	-4.33	0.123	-3.96	0.184
Immediate recall	-6.64	0.019	-5.93	0.049
Delayed recall	-5.69	0.047	-5.15	0.092
Recognition	-4.24	0.151	-3.15	0.318

Source: Adapted from Grandjean et al., 1999, Table 2.

[a]All Hg values were log transformed for the analyses presented here.

[b]Expressed as % of standard deviation.

[c]Statistical significance of the effect associated with Hg exposure in a mutiple regression including potential confounders.

[d]Neurobehavioral Evaluation System.

tained Hg concentrations from a 9-cm length of hair reflecting average MeHg exposure during pregnancy and the Faroe study obtained concentrations from hair samples of variable length, some of 3-cm (reflecting late second and third trimester) and some of 9-cm in length. Additionally, the Faroe maternal-hair samples, which were obtained at delivery, did not include the last 3 weeks of gestation, because it takes approximately 20 days after ingestion for the newly formed portion of the hair strand to emerge above the scalp. If the third trimester is particularly important for the development of the neural substrate for cognitive and neuromotor function, it is perhaps not surprising that the maternal-hair sample obtained in the Faroe Islands might be somewhat less sensitive than the cord-blood sample, which primarily reflects third-trimester exposure (see Chapter 3). A maternal-hair sample that reflected the last 20 days of pregnancy might have been more sensitive. In the SCDS, the maternal-hair samples were obtained at delivery and at 6 months postpartum. The portion of the hair strands corresponding to the pregnancy period was analyzed, assuming 1 cm of hair growth per month. If third-trimester exposure is critical for the neurodevelopmental end points, the SCDS measure of Hg exposure averaged across the entire pregnancy might be less sensitive in detecting them, compared with cord blood, which primarily reflects third-trimester exposure. It might be informative for the SCDS group to re-analyze their data using the concentration of Hg in the portions of hair corresponding only to the third trimester as the exposure measure.

It is also of interest that the neurophysiological end points in the Faroe study (e.g., brain-stem auditory-evoked potentials) were associated only with the maternal-hair Hg measure, not with the cord-blood Hg. Because the hair measure presumably reflects an earlier period of gestation, the data suggest an earlier sensitive or critical period for the neurophysiological effects. In summary, it does not appear that the failure of the SCDS to collect cord-blood Hg samples can account for the discrepancies between their findings and those in the Faroe study because, in the latter study, associations were found between neurobehavioral test scores and both cord-blood Hg and maternal-hair Hg concentrations (Table 6-1). Moreover, the findings reported in New Zealand and the pilot SCDS were based solely on maternal-hair-sample data averaged across the entire pregnancy.

DIFFERENCES IN THE NEUROBEHAVIORAL END POINTS ASSESSED AND THE CHILDREN'S AGES AT ASSESSMENT

The Faroe and Seychelles studies used very different neurobehavioral test batteries. For the most part, the tests selected for the SCDS are considered apical or omnibus tests (e.g., the McCarthy Scales of Children's Abilities), which yield global scores that integrate performance over many separate neuropsychological domains. In contrast, because the Faroe investigators hypothesized multifocal domain-specific neuropsychological effects, their test battery largely consisted of highly focused tests selected from those commonly used in clinical neuropsychology (e.g., California Verbal Learning Test—Children and Boston Naming Test). The Faroe test battery does not include an apical test of global function.

The subscales from the McCarthy test (verbal, perceptual-performance, quantitative, memory, and motor) that assess specific domains of function might be expected to be more directly comparable to the tests administered in the Faroe Islands. For instance, given the finding in the Faroe study that memory, as assessed by the California Verbal Learning Test, was significantly associated with prenatal Hg exposure, it would be expected that children's scores on the McCarthy memory scale in the SCDS would be associated with Hg exposure. However, they were not. In fact, prenatal Hg exposure was not associated with scores on any of the McCarthy subscales. It is important to examine in detail the extent to which the individual McCarthy subscales are comparable to the domain-specific tests selected for the Faroe study. Psychometrically, they are different. The California Verbal Learning Test, for example, involves five learning trials of a 12-word list, with free- and cued-recall trials following short and long delays, and a recognition trial. None of the 18 tests that contribute to scores on the McCarthy scales examine rate of learning, and the memory scale combines scores on four tests that involve recall of differing types of information: pictorial (six common objects arrayed on a page), auditory sequence (xylophone notes), word list (ranging from 3 words in a specified sequence to a 13-word sentence with 9 key words that are scored), connected discourse (recall of individual story elements), and numbers (forward and backward recall of strings of numbers up to seven digits long.) Clearly, a child's score on the McCarthy memory scale integrates performance on a much wider variety of memory skills than does either the short- or long-delay free-recall trials of the California Verbal Learning Test. Scores on some of the 18 specific subscales of the McCarthy test might offer greater comparability with the key end points of the California Verbal Learning Test assessed in the Faroe study. Each of the 18 subscales is quite brief, however, and thus less psychometrically sound than the richer California Verbal Learning Test, which assesses only one domain of function but does that in considerable depth.

Similarly, although the Boston Naming Test, which was included in the Faroe Islands test battery, and the preschool language scale and the verbal scale of the McCarthy verbal scale which were included in the SCDS 66-month test battery of the SCDS can be considered tests of language skills, the specific skills they assess are quite different. The Boston Naming Test specifically assesses confrontational naming skills, consisting of line drawings of common objects that a child has to name under time pressure (20 seconds). If the child cannot retrieve the correct name spontaneously, semantic and then phonemic cues are provided. In contrast, the total score on the preschool language scale (PLS) integrates a child's performance on the auditory comprehension and expressive communication subscales, both of which assess a broad range of language skills (eg., comprehension and production of vocabulary; concepts of quantity, quality, space, and time; morphology; syntax; and inference drawing). Like the total score on the PLS, the total score on the McCarthy verbal scale integrates a child's performance across many language-relevant domains in the following tests: pictorial memory (same as test described for memory scale), word knowledge (pointing to the picture of an object named by the examiner, providing the name for four pictured objects, and providing word definitions), verbal memory (same as test described for memory scale), verbal fluency (generating words in 20-sec trials to fit specific semantic constraints, such as things to eat or animals), and opposite analogies (providing antonyms). Thus, although the four items of the word-knowledge test that assess naming could be isolated and considered an index of confrontational naming, similar to the Boston Naming Test, the four items are unlikely to possess the same sensitivity insofar as the latter test consists of 60 items.

Thus, although the Faroe Islands and SCDS test batteries include tests of language and memory, it is not appropriate to view the end points used in the studies to assess each domain to be equivalent either in terms of the specific skills assessed or the test sensitivity.

Although the Bender-Gestalt Test was administered in the Faroe study and SCDS as a measure of visual-spatial abilities, different scoring systems were used (the Gottingen system in the Faroe Islands and the Koppitz system in the Seychelles). The finding of a significant association with Hg in the former but not the latter study is similar to the finding reported by Trillingsgaard et al. (1985) that scores derived using the more-detailed Gottingen system were significantly associated with low-dose lead exposure, and scores on the Koppitz system were not. Thus, the Gottingen system used in the Faroe Islands might be more sensitive. Although the Seychelles data could be rescored using the Gottingen system, the committee was told that the data might still not be comparable, because the more sensitive memory for design conditions was not administered in the Seychelles study.

To help determine the degree to which the discrepant results from the Faroe study and SCDS are attributable to differences in the neurobehavioral tests used, the Seychelles group administered several of the more domain-specific tests from the Faroe battery in their 8-year follow-up. The results of those assessments, however, are not yet available.

A second important difference in the assessment batteries used in the Faroe study and SCDS relates to the age of assessment—7 years in the Faroe Islands and 5.5 years of age in the SCDS. The final assessment in the New Zealand cohort was at 6 years of age. Generally speaking, developmental assessments are likely to be less sensitive in detecting subtle neurotoxic effects when they are administered during a period of rapid developmental change. The period covering ages 60-72 months, when the SCDS and New Zealand cohorts were evaluated, is one such period during which marked individual differences in the rate of cognitive maturation are likely to eclipse subtle differences in function attributable to a teratogenic exposure (Jacobson and Jacobson 1991). The assessments performed in the SCDS during infancy, particularly the 19- and 29-month Bayley scales, were also not administered at optimal age points. Studies of prenatal exposure to alcohol and other substances that have administered the Bayley scales at multiple ages have repeatedly failed to detect effects at 18 months, probably because it too is a period of rapid cognitive maturation, involving the emergence of spoken language. Twenty-nine months is likely to be an insensitive testing point for the Bayley scales because it is at the end of the age range for which the version of this test used in the Seychelles was standardized, leading to a substantial risk of a "ceiling effect" (i.e., too many children receiving the highest possible scores on numerous items). The next round of testing in the Seychelles will be at 8 years of age, a point in development that should be more optimal for detecting neurodevelopmental effects.

Although differences in end points assessed and age of assessment might explain the failure of the SCDS to detect the associations found in the Faroe Islands study, findings from the New Zealand study and the Seychelles pilot study suggest that the discrepancies between the Faroe Islands and the main Seychelles studies are probably not due to differences in the assessments. The New Zealand study found associations between MeHg exposure and scores on the McCarthy Scales of Children's Abilities (the primary outcome measure used in the SCDS) at about the same age of assessment as in the Seychelles study, in a study with full control for potential confounding influences. Associations with prenatal Hg exposure were even seen on the McCarthy scales and the PLS in the 217-member Seychelles pilot study at 5.5 years of age, albeit with only limited control for socioenvironmental influences.

STABLE VERSUS EPISODIC PATTERN OF EXPOSURE

The predominant source of Hg exposure in the Seychelles is daily fish consumption. Maternal fish consumption averages 12 meals per week. Hg exposure in the Faroe Islands, by contrast, is often more episodic. In the Faroe Islands, pilot-whale meals are relatively infrequent (less than once per month on the average), but whale meat has concentrations of MeHg between 10 and 20 times greater than those in many fish consumed in the Faroe Islands (Grandjean et al. 1992); thus, the whale meals might represent toxicologically more significant peak or bolus doses. Laboratory animal experiments on prenatal alcohol exposure have demonstrated that maternal ingestion of a given dose of alcohol over a short time causes greater neuronal (Bonthius and West 1990) and behavioral impairment (Goodlett et al. 1987) than that caused by gradual ingestion of the same total dose over several days. Thus, it is possible that the more episodic exposure pattern in the Faroe Islands, with heavier doses per occasion, has a more adverse impact on neuronal development than the more gradual exposure in the Seychelles. However, it is difficult to compare the 12 fish meals per week reported in the Seychelles with the three fish "dinners" per week in the Faroe Islands, because the types of fish eaten and their Hg concentrations are different. Moreover, the exposure-associated differences in neurobehavior found in the New Zealand cohort and the Seychelles pilot study where no whale meat was eaten suggest that bolus doses are not necessary to generate cognitive deficits at those levels of exposure.

The importance of high episodic ("spiking") exposures is unclear. However, as discussed in Chapter 4, the degree of spiking in the Faroe study is likely to be in the low-to-moderate range (i.e., less than a doubling in hair Hg concentrations, assuming an individual at the Faroe Islands median exposure level consumes three consecutive 4-ounce whale meals). Spiking might also occur in the Seychelles given the availability of fish species with characteristically moderate-to-high concentrations of Hg (e.g. tuna), although the absence of dietary data does not allow this issue to be examined further.

STUDY DIFFERENCES IN CONTROL FOR CONFOUNDERS

A potential confounder is a variable related to both the exposure of interest (e.g., MeHg) and to the outcome of interest (e.g., neurobehavior). If the relation between exposure and outcome is no longer significant after controlling statistically for a potential confounder, it is inferred that the relation between exposure and outcome is spurious and due to confounding by the control variable being examined. Because random assignment to predetermined exposure levels cannot be used to control for confounding in human exposure studies, it is important to assess whether a broad range of control variables confound any associations observed between exposure and outcome. Table 6-2 lists the control variables examined in the Faroe and Seychelles studies. Both studies evaluated most of the variables that are known to be at least moderately related to childhood cognitive outcome, including maternal cognitive competence (e.g. Ravens test), child age, gender, maternal alcohol consumption and smoking during pregnancy, and parental income. A few variables that are sometimes modestly related to those outcomes

TABLE 6-2 Control Variables Assessed in the Faroe Islands and the Seychelles Studies

Covariates	Faroe Islands Study	Seychelles Study
Birth weight		X
Maternal cognitive competence (Ravens)	X	X
Child's age	X	--[a]
Child's sex	X	X
Gestational age		X
Smoking during pregnancy	X	X
Alcohol during pregnancy	X	X
Duration breast feeding	X	X
PCBs	X	--[b]
Education (mother and father)	X	
Employment (family income)	X	X
Obstetric care	X	
Daycare	X	X
Computer acquaintance	X	
Examiner	X	
Birth order		X
Maternal age		X
Child's medical history		X
Language at home	N/A	X
Maternal medical history		X
Maternal hair lead	X[c]	X[d]

[a]Test scores adjusted for age, based on U.S. norms.

[b]No PCBs were detected in any of the 49 serum samples obtained at 66 months postpartum. These data support the assumption that there is virtually no PCB exposure in this population.

[c]Whole pregnancy.

[d]At parturition, i.e., pregnancy minus 45 days.

Source: Adapted from NIEHS 1998.

were assessed in one study but not in the other (e.g., maternal age and birth order in the Seychelles study; obstetrical care in the Faroe study). However, the influences of those variables on cognitive outcome are probably too weak to account for any major inconsistencies between the two studies. Parental education was not assessed in the Seychelles study, but it is likely to have added little information over and above family income and maternal cognitive competence.

At a workshop sponsored by the White House Office of Science and Technology Policy in November 1998, the Faroe investigators noted that, apparently due to social-class differences, the maternal Ravens scores and the child verbal-test scores were generally higher among families residing in one of the three Faroe towns than among those living in the countryside. Because more fish and whale meat are consumed by rural residents, the associations of Hg exposure with child verbal-test scores could be spurious, reflecting those social-class differences. (Although the Ravens scores were controlled statistically in the analyses, that single test might not have fully controlled for social-class confounding.) Data presented at the workshop showed, however, that these associations remained significant, even after controlling for a dichotomous town-country control variable (Table 6-3). Although that analysis is reassuring, it would not be appropriate to control for town routinely in all analyses. Because fish and whale consumption constitute a large proportion of the rural diet, the disappearance of associations after controlling for residence could be due to the fact that residing in a rural area leads to increased Hg exposure which, in turn, causes an adverse outcome. It would not necessarily indicate that the lower social class associated with rural residence is the true cause of the Hg-associated deficit. The disappearance of an association between Hg and neurobehavior under those circumstances would be very difficult to interpret, because the interpretation would depend upon what condition is considered the reason for the association between living in a rural area and poor outcome (i.e., lower social class or greater Hg exposure).

Because the rural residents had to travel relatively long distances to the testing site in Torshavn, there has been concern that fatigue and the strangeness of the urban setting might have caused the rural children to perform more poorly. As noted above, however, the data in Table 6-3 make clear that the regression coefficients for prenatal Hg exposure remain significant even after controlling for child's residence.

The neuropsychological test examiner is one potentially important factor that was routinely controlled for in the Faroe Islands study (see NIEHS 1998, section 3.5), but was not controlled for in the SCDS. The

TABLE 6-3 Effects of Residence (Town vs. Country) and Prenatal Hg Exposure on Developmental Outcomes in the Faroe Islands Study

	Residence[a]		Hg Without Controlling for Residence[b]		Hg Controlling for Residence[c]	
Test	b[d]	p	b	p	b	p
NES Finger Tapping						
Preferred Hand	0.03	0.95	-1.10	0.05	-1.14	0.04
Other Hand	-0.47	0.26	-0.39	0.46	-0.55	0.31
Both Hands	-1.41	0.11	-1.67	0.14	-2.04	0.07
NES Hand-Eye Coordination						
Error Score	0.001	0.94	0.03	0.19	0.04	0.18
NES Continuous Performance						
Missed Responses	-0.12	0.16	0.27	0.02	0.24	0.04
Reaction Time	-14.55	0.06	40.30	<0.001	35.34	0.002
WISC-R						
Digit Spans	0.06	0.58	-0.27	0.05	-0.26	0.07
Similarities	0.04	0.91	-0.05	0.90	-0.08	0.84
Block Designs	0.19	0.02	-0.17	0.11	-0.12	0.26
Bender Visual Motor Gestalt Test						
Error on Copying	-1.03	0.005	0.67	0.15	0.45	0.35
Delayed Recall	0.35	0.004	-0.25	0.10	-0.17	0.28
Boston Naming Test						
No cues	1.10	0.005	-1.77	<0.001	-1.51	0.003
With cues	1.24	0.001	-1.91	<0.001	-1.60	0.001
CVLT						
Learning	0.89	0.16	-1.25	0.12	-1.10	0.18
Immediate recall	0.37	0.06	-0.57	0.02	-0.49	0.05
Delayed recall	0.02	0.92	-0.55	0.05	-0.56	0.05
Recognition	0.15	0.92	-0.29	0.15	-0.30	0.14

[a]Controlling for all independent variables except Hg; 0 = country, 1 = town.
[b]Controlling for all independent variables except residence.
[c]Controlling for all independent variables.
[d]Raw (unstandardized) regression coefficient.

Abbreviations: NES, Neurobehavioral Evaluation System; CVLT, California Verbal Learning Test.

Source: Adapted from Appendix III-B, Table 3, NIEHS (1998).

omission of that control variable might not seem important in light of the lack of observed effects in that study. However, if one examiner who is less adept at eliciting optimal performance from the subjects tested a large proportion of less exposed children, the results could be affected. If those children performed more poorly than they otherwise would have on the test, an association between Hg concentration and test scores might be obscured by failure to control for the examiner. That result could also occur if an adept tester tested a large proportion of the more heavily exposed children, leading them to achieve higher scores than they would have if tested by other examiners.

The SCDS controlled for age by converting the raw test scores to age-corrected standard scores with conversion tables based on U.S. norms (NIEHS 1998). In contrast, the Faroe study analyzed the raw scores by adjusting statistically for the child's age (measured in days since birth). The latter approach is preferable for three reasons. First, the applicability of U.S. norms to these study populations is uncertain. Second, the use of age-corrected standard scores can reduce the sensitivity of the test, because several adjacent raw scores are treated as equivalent in converting to standard scores. Moreover, because age-corrected standard scores use 3-month intervals, for the purposes of conversion of raw to standard scores, a child whose age is 4 years, 3 months, and 31 days is considered to be the same age as a child who is 4 years, 0 months, and 1 day. However, that same child is considered to be different in age from a child who is only 1 day older (i.e., 4 years, 4 months, and 1 day). Finally, the Faroe approach of controlling statistically for age by multiple regression seems appropriate, because the effect of age is likely to be linear across the relatively short age period (3 months in both studies). Although it seems unlikely that the difference in approach to controlling for age could account for the discrepancies in the findings of those two studies, it would be of interest to see a re-analysis of the SCDS data using the approach that was used in the Faroe study.

There appears to have been no need to control for PCB exposure in the Seychelles, because PCB body burdens in that population are exceptionally low. In contrast to North America and Europe, where these contaminants are routinely detected in serum samples, 28 samples obtained from Seychelles study children showed no detectable concentrations of any PCB congeners. In the Faroe study, prenatal PCB exposure was measured in umbilical cord tissue rather than cord blood or maternal blood or milk, as in most previous studies, and specimens were obtained for only half the newborns. Cord-tissue PCB concentration has never been validated in relation to blood or milk concentration, and because cord tissue is lean, it might provide a less reliable indication of total PCB body burden. Although PCBs are measured most accurately on a lipid-adjusted basis in most tissue, the wet-weight measure used in the Faroes was probably more valid for cord tissues, because the gravimetric approach used to measure fat content is not sufficiently reliable in a lean medium.

With respect to confounding by PCBs, prenatal PCB exposure was associated with four of the eight outcomes whose relation to cord Hg concentration was statistically significant. Those outcomes related primarily to verbal and memory performance, domains found in previous studies to be associated with prenatal PCB exposure (Jacobson and Jacobson 1996; Patandin et al. 1999). When PCBs and Hg were included together in the model, one outcome—continuous performance test (CPT) reaction time—was independently related to Hg exposure (Grandjean et al. 1997, Table 5). For the other three outcomes, however, the associations with both PCB and Hg fell short of conventional levels of statistical significance. One likely explanation is that both of those contaminants adversely affect those outcomes, but their relative contributions cannot be determined given their co-occurrence in the Faroe population (r = 0.41). It is unfortunate that cord specimens were not obtained from a greater proportion of the children.

In a second set of analyses (Budtz-Jørgensen et al. 1999), potential confounding by prenatal PCB exposure was reduced by dividing the sample into tertiles in terms of the infants' cord PCB concentrations. Regressions assessing the associations between Hg exposure and the five principal 7-year outcomes were then run separately for each of the groups. The regression coefficients for Hg in the lowest PCB tertile were no weaker than those among the infants exposed to moderate or heavy PCB doses, lending additional support to the conclusion that the associations between Hg and these outcomes are not attributable to confounding by prenatal PCB exposure.

POPULATION DIFFERENCES IN VULNERABILITY

Vulnerability to prenatal Hg exposure might be enhanced or attenuated by differences in genetic susceptibility, diet, or exposure to other contaminants. The SCDS cohort is predominantly African in descent; the Faroe cohort is Caucasian. Moreover, the Faroe population is thought to be descendant from a small number of "founders," which could increase genetic vulnerability to toxic insult. Although racial differences in vulnerability are possible, it should be noted that such differences have not been seen for environmental exposure to lead, which has been studied in racially diverse samples. Moreover, evidence of MeHg neurotoxicity was found in the genetically heterogeneous and racially diverse sample assessed in New Zealand, a sample that was predominantly non-Caucasian.

In principle, poor nutrition might also make a population particularly vulnerable to teratogenic insult. However, the data on birth weight and gestation length in the Faroe and Seychelles studies suggest that energy and macronutrient (protein and carbohydrate) deficiency is unlikely. Nevertheless, micronutrient deficiency in association with low intakes of fortified or unrefined grains or fruits and vegetables is possible. It is also possible that children in one or both samples might have been weaned to breast-milk substitutes or milks of other species that provide inadequate amounts of iron, other minerals, or vitamins. Conversely, certain nutrients found in fish eaten by the Seychelles residents (e.g., omega-3 fatty acids and selenium) could attenuate adverse effects of Hg exposure. The general health status of a population might also enhance or attenuate vulnerability to teratogenic exposure. Because the Faroe and Seychelles populations apparently both receive excellent health care, however, health status seems unlikely to explain the differences in the study findings.

As stated above, one substantial difference between the Faroe and the Seychelles populations relates to their PCB exposure. Whereas PCB concentrations in the Seychelles population are among the lowest observed anywhere in the world, the portion of the Faroe population that eats whale blubber accumulates unusually high PCB body burdens. Although it is conceivable that PCB exposure in the Faroe Islands might enhance fetal vulnerability to Hg, that hypothesis is speculative at present; experimental animal studies would be needed to test its plausibility. The possibility of effect modification by PCB exposure was examined in regression analyses that, in addition to confounders, also included the Hg and PCB exposure variables and their product (Hg ×

PCBs; Budtz-Jørgensen et al. 1999). All five *p*-values for the Hg × PCBs interaction terms exceeded 0.20, suggesting an absence of potentiation of the effects of one of the contaminants by the other. Thus, it seems unlikely that differences in vulnerability due to PCB exposure can explain the differences between the Faroe Islands and the Seychelles findings.

The sample in the main Seychelles study appears to have been developmentally robust. There was an exceptionally low number of abnormal scores on the Denver Developmental Screening Test, an unusually high mean Psychomotor Development Index score, and a very low rate of referral for mental retardation. On the other hand, the means and standard deviations of the cognitive measures administered at later ages were similar to U.S. norms. It is unclear to what extent the developmental robustness of that particular sample might have buffered it from any adverse effects of prenatal Hg exposure.

RANDOM VARIATION IN THE DETECTABILITY OF EFFECTS AT LOW EXPOSURES

The magnitude of the associations found in the MeHg studies resembles that reported with respect to other environmental contaminants, such as low-dose lead and PCBs. When the magnitude of an association is subtle, it is not surprising that it is not detected in every cohort studied. With respect to lead exposure, a strong scientific consensus has developed that blood lead concentrations in excess of 10 µg/dL place a child at increased risk of poor developmental outcomes. However, not all lead studies have found an association, and substantial variability exists in the magnitudes of the reported effects (Bellinger 1995). If two studies from the lead literature were chosen randomly, it is likely that the results of the two would not be entirely concordant. The uncertainties inherent in such studies (e.g., the assessment of exposure histories, the measurement of critical population characteristics, the idiosyncratic pattern of potential confounding factors, and the measurement of neurodevelopmental outcomes) render it likely that evidence of neurotoxicity will not be detected in some of the study cohorts assessed. With respect to the SCDS, the evidence consistent with such effects found in the pilot phase, coupled with the suggestion of unusual developmental robustness in the main study, suggest that the failure to detect apparent adverse effects in the main study could be due to the substantial sample-to-sample variation expected when trying to identify weak associations in an inherently "noisy" system of complex, multi-determined neurobehavioral end points.

Given the large sample size in the main Seychelles cohort, it might seem surprising that that study could lack the power needed to detect an association between increased MeHg exposure and neurobehavioral impairments. However, power analyses that are based on total sample size can be misleading if adverse effects occur primarily among the most heavily exposed children, who typically comprise a very small proportion of the sample. Although the sample size of 700 children in the SCDS would seem to be more than adequate, only about 35 children were exposed at 15 μg/g or higher. Because multiple regression analysis examines associations that are averaged across the entire distribution of exposure, associations that hold only for the most highly exposed children can be difficult to detect. Thus, if adverse effects of prenatal MeHg exposure occur primarily in the upper range, the power to detect them will be limited, and it would not be surprising if associations found in one Seychelles cohort (the pilot study) were not detected in the next cohort (the main study) (see Chapter 7 for further discussion on the issue of statistical power).

CONCLUSIONS

- Three well-designed, prospective, longitudinal studies have examined the relation of prenatal MeHg exposure to neuropsychological function in childhood. MeHg was associated with poorer performance in the Faroe Islands study but not in the SCDS. Little attention has been paid to the New Zealand study because, until recently, it had not been subjected to peer review. Differences in the primary biomarker of Hg exposure (cord blood versus maternal hair), type of neuropsychological tests administered (domain specific versus global), age of testing (7 versus 5.5 years), and sources of exposure (whale meat versus fish) between the Faroe study and the SCDS have been suggested to account for the differences in the findings of the two studies. When the New Zealand data are considered, however, those differences no longer seem determinative, because the New Zealand study, in which the exposure and research design were very similar to the SCDS, also found associations between higher MeHg levels and worse neurobehavioral test scores, as did the pilot SCDS.

- There is no empirical evidence or hypothesized mechanism to support the suggestion that PCB exposure might enhance vulnerability to MeHg. The lack of any evidence of statistical interaction between Hg and PCB exposure in the Faroes data also makes it unlikely that a difference in PCB exposure can explain the differences between the Faroe Islands and the Seychelles findings.
- The lack of statistical control for examiner in the SCDS, population differences in susceptibility among the study populations, and dietary factors might explain some of the differences among the study findings.
- It is possible that the differences are primarily due to between-sample variability in the expression of neurotoxicity at low doses. Even large sample studies can lack adequate power to detect adverse associations if a relatively small number of children are exposed in the upper ranges of the exposure distributions, where the adverse effects are most likely to be found.
- Although none of the between-study differences noted above appears to be determinative, the combined influence of two or more of these factors is difficult to predict. For example, it is possible that slightly reduced vulnerability in the Seychelles population combined with the use in that study of a biomarker of exposure that averages across pregnancy could make it difficult to detect neurocognitive effects that might be specific to third trimester exposure.
- When the two studies reporting associations between MeHg and neurobehavior are compared, the strengths of the New Zealand study include an ethnically heterogeneous sample and the use of developmental end points with greater predictive validity. The advantages of the Faroe study include a larger sample size, the use of two different biomarkers of exposure, and extensive scrutiny in the epidemiological literature. The Faroe data have also undergone extensive re-analyses in response to questions raised by panelists in the NIEHS (1998) workshop and by this committee in the course of its deliberations.

RECOMMENDATIONS

- It would be helpful to obtain more comprehensive nutritional data from all three populations as well as single-strand hair analyses to address more effectively the issue of spiking or bolus dose. A reanalysis of the 5.5-year SCDS data controlling statistically for examiner might also be useful.
- Most of the MeHg exposure standards currently in effect are based on extrapolations from the Iraqi MeHg poisoning episode, in which exposure was due to the consumption of highly contaminated grain and resulted in body burdens that greatly exceeded those found in the general population of fish consumers. Given the availability of data from three well-designed epidemiological studies in which prenatal MeHg exposures were in the range of general-population exposures, exposure standards should be based on data from these newer studies.

REFERENCES

Bellinger, D. 1995. Interpreting the literature on lead and child development: The neglected role of the "experimental system". Neurotoxicol. Teratol. 17(3):201-212.

Bonthius, D.J., and J.R. West. 1990. Alcohol-induced neuronal loss in developing rats: Increased brain damage with binge exposure. Alcohol Clin. Exp. Res. 14(1):107-118.

Budtz-Jørgensen, E., N. Keiding, P. Grandjean, and R. White. 1999. Methylmercury neurotoxicity independent of PCB exposure. [Letter]. Environ. Health Perspect. 107(5):A236-A237.

Crump, K.S., T. Kjellström, A.M. Shipp, A. Silvers, and A. Stewart. 1998. Influence of prenatal mercury exposure upon scholastic and psychological test performance: benchmark analysis of a New Zealand cohort. Risk Anal. 18(6):701-713.

Goodlett, C.R., S.J. Kelly, and J.R. West. 1987. Early postnatal alcohol exposure that produces high blood alcohol levels impairs development of spatial navigation learning. Psychobiology 15(1):64-74.

Grandjean, P., P. Weihe, P.J. Jørgensen, T. Clarkson, E. Cernichiari, and T. Videro. 1992. Impact of maternal seafood diet on fetal exposure to mercury, selenium, and lead. Arch. Environ. Health 47:185-195

Grandjean, P., E. Budtz-Jørgensen, R.F. White, P.J. Jørgensen, P. Weihe, F. Debes, and N. Keiding. 1999. Methylmercury exposure biomarkers as indicators of neurotoxicity in children aged 7 years. Am. J. Epidemiol. 150(3):301-305.

Grandjean, P., P. Weihe, R.F. White, F. Debes, S. Araki, K. Yokoyama, K. Murata, N. Sørensen, R. Dahl, and P.J. Jørgensen. 1997. Cognitive deficit in 7-year-old children with prenatal exposure to methylmercury. Neurotoxicol. Teratol. 19(6):417-428.

Jacobson, J.L., and S.W. Jacobson. 1991. Assessment of teratogenic effects on cognitive and behavioral development in infancy and childhood. Pp. 248-261 in Methodological Issues in Controlled Studies on Effects of Prenatal Exposure to Drugs of Abuse, Research Monograph 114, M.M. Kilbey, and K. Asghar, eds. Rockville, MD: National Institute on Drug Abuse.

Jacobson, J.L., and S.W. Jacobson. 1996. Intellectual impairment in children exposed to polychlorinated biphenyls in utero. N. Engl. J. Med. 335(11):783-789.

Myers, G.J., P.W. Davidson, and C.F. Shamlaye. 1998. A review of methylmercury and child development. Neurotoxicology 19(2):313-28.

Myers, G.J., P.W. Davidson, C. Cox, C.F. Shamlaye, M.A. Tanner, O. Choisy, J. Sloane-Reeves, D.O. Marsh, E. Cernichiari, A. Choi, M. Berlin, and T.W. Clarkson. 1995. Neurodevelopmental outcomes of Seychellois children sixty-six months after in utero exposure to methylmercury from a maternal fish diet: Pilot study. Neurotoxicology 16(4):639-652.

NIEHS (National Institute of Environmental Health Sciences). 1998. Scientific Issues Relevant to Assessment of Health Effects from Exposure to Methylmercury. Workshop organized by Committee on Environmental and Natural Resources(CENR) Office of Science and Technology Policy (OSTP) The White House, November 18-20, 1998, Raleigh, NC.

Patandin, S., C.I. Lanting, P.G. Mulder, E.R. Boersma, P.J. Sauer, and N. Weisglas-Kuperus. 1999. Effects of environmental exposure to polychlorinated biphenyls and dioxins on cognitive abilities in Dutch children at 42 months of age. J. Pediatr. 134(1):33-41.

Trillingsgaard, A., O.N. Hansen, and I. Beese. 1985. The Bender-Gestalt Test as a neurobehavioral measure of preclinical visual-motor integration deficits in children with low-level lead exposure. Pp. 189-193 in WHO Environmental Health, Document 3. Neurobehavioral Methods in Occupational and Environmental Health, Second International Symposium, Copenhagen, Denmark, Aug.6-9, 1985. Copenhagen, Denmark: World Health Organization.

7

DOSE-RESPONSE ASSESSMENT

THIS chapter focuses on dose-response analysis and its role in choosing a point of departure to be used in the risk assessment for MeHg. The chapter begins with a brief review of risk assessment for noncancer end points. Problems with the traditional approach that is based on no-observed-effect levels (NOAELs) will be discussed, along with advantages of more recent approaches that are based on dose-response modeling and benchmark-dose calculations. The chapter then reviews some of the specific challenges that arise when considering benchmark-dose calculations for MeHg. Foremost among these challenges is the fact that three studies of similar quality yield different results regarding the association between low-level exposures to MeHg and adverse developmental outcomes. After exploring the possibility that differences in power might explain some of this discrepancy, the committee presents and discusses several approaches that could be used to provide a benchmark dose based on data from the three studies. Among the methods considered are traditional approaches based on selecting a single outcome from a single study and an integrative analysis that combines information from different studies and outcomes. Results are found to be sensitive to model choice and recommendations are made for model choice.

RISK ASSESSMENT FOR NONCANCER END POINTS

Quantitative risk assessment for noncancer effects is commonly based on determination of a NOAEL from a controlled study in animals. In this context, the NOAEL is defined as the highest experimental dose that does not produce a statistically or biologically significant increase in adverse effects over those of controls. An "acceptably safe" daily dose for humans is then derived by dividing the NOAEL by a safety factor, usually 10 to 1,000, to account for sensitive subgroups of the population, data insufficiency, and extrapolation from animals to humans. The U.S. Environmental Protection Agency (EPA) refers to the resulting quantity as the reference dose (RfD), the Food and Drug Administration (FDA) uses the term allowable daily intake (ADI) and the Agency for Toxic Substances and Disease Registry (ATSDR) uses minimum risk level (MRL). The concept is also similar to the upper limits (ULs) recently introduced by the National Academy of Sciences for nutrient recommendations. In the event that the lowest experimental dose shows a significant difference from the control, it is termed a LOAEL (lowest-observed-adverse-effect level), and an extra factor of 10 is used in the determination of the RfD, ADI, or MRL (see, for example, EPA 1998). Various reports have provided RfDs for MeHg that are derived from animal studies (Rice 1992; Gilbert et al. 1993; Zelikoff et al. 1995; Rice 1996). Typically, these calculations have used the results from a series of nonhuman primate studies, which indicate that adverse developmental effects in several outcomes occur at 50 μg/kg per day maternal dose. Uncertainty factors of 10 were used for LOAEL to NOAEL, species differences, and individual variation in response for an RfD of 0.05 μg/kg per day.

In recent years, use of the NOAEL has become controversial among risk assessors and regulators because of several serious statistical drawbacks (Gaylor 1983; Crump 1984; Kimmel and Gaylor 1988; Kimmel 1990; Leisenring and Ryan 1992). For instance, because the NOAEL must, by definition, correspond to one of the experimental doses, it can vary considerably across different experiments, yet this statistical variation is usually ignored when computing RfD values. Furthermore, estimation of the NOAEL is sensitive to sample size: because the NOAEL is based on statistical comparisons between exposed and unexposed dose groups, larger studies have higher power to detect small changes and therefore tend to produce lower NOAELs. In contrast, smaller studies tend to yield higher NOAELs due to their lower power to detect real effects. Because NOAEL calculations are traditionally based on pairwise comparisons of exposed groups and controls, there is no widely accepted procedure for calculating a NOAEL in settings where exposure is measured on a relatively continuous scale. Indeed, the current definition of NOAEL involves an implicit assumption that the dose levels are grouped in some way. Grouping is common in the context of controlled animal studies, but most epidemiological studies, including the available MeHg studies, measure exposure on a continuous scale.

Problems with the NOAEL and LOAEL approach have led to increasing interest in the development of alternative approaches based on dose-response modeling techniques. The benchmark dose was defined by Crump (1984) as a lower 95% confidence limit on the dose corresponding to a moderate increase (e.g., 1%, 5%, or 10%) over the background rate. Because the benchmark dose generally occurs within the range of experimental data, Crump and others have argued that its estimation is relatively robust to model choice. In an extensive empirical comparison of NOAEL and benchmark-dose calculations, Allen et al. (1994) found that the NOAEL in a typically sized developmental toxicity study was, on average, 6 times larger than the BMDL corresponding to a 5% risk. The NOAEL was higher than even a 10% BMD, on average, by a factor of 3. Leisenring and Ryan (1992) came to somewhat similar conclusions based on analytical considerations. Crump (1984) used the abbreviation BMD to refer to the benchmark dose. Other authors, including Crump (1995), use BMD to denote the estimated dose that corresponds to a specified risk above the background risk and BMDL to denote the corresponding lower limit. This latter notation has become standard usage now and will be used throughout the remainder of this chapter.

BENCHMARK-DOSE CALCULATIONS FOR CONTINUOUS OUTCOMES

Benchmark-dose calculations for quantitative outcomes (e.g., birth weight or IQ) are more complicated than those for quantal responses, such as presence or absence of a defect. Although Crump (1984) discussed how to calculate a BMD for a quantitative outcome, Gaylor and Slikker (1992) were the first to develop the approach in any detail. Their first step is to fit a regression model characterizing the mean of the outcome of interest as a function of dose and assuming that the data are normally distributed. The next step is to specify a cutoff to define values for which the outcome can be considered abnormal. For example, a weight lower than 0.8 g might be considered abnormal for a teratology study in mice. Using the fitted model, one then calculates the dose-specific probability of falling into the abnormal region. The BMD is estimated as the dose corresponding to a specified increase in that probability, compared with the background probability. The BMDL is the corresponding 95% lower limit on that dose. Figure 7-1 illustrates the ideas behind the approach. The curve in the top panel represents the distribution of IQs in an unexposed population, and the curve in the lower panel has been shifted to the left in response to an exposure. Note that the mean IQ in the unexposed population is 100, and the standard deviation (SD) is 15. The shaded areas in the left tails of each distribution represent the proportion of the exposed and unexposed populations that fall below a specified cutoff point (we will refer to cutoff point as C), designated as the IQ level that indicates an adverse response. In Figure 7-1, we have used a value of C = 75, which represents the lower 5% of the control population. From the figure, it is easy to see that the further left we move the curve corresponding to the exposed group, the higher the percentage of the exposed population that falls below the cutoff point. Gaylor and Slikker's suggestion simply involves finding the exposure level that leads to a specific increase in the proportion of the population falling below the cutoff point. To be more precise, let Y_i represent the outcome for the ith study subject, and suppose that a lowered outcome is considered to be adverse (e.g., as for IQ). Then, let P_0 denote the probability that an unexposed individual falls below the value (C) that defines an adverse effect. The BMD is then defined as the dose, x, such that

$$\Pr(Y < C \mid \text{dose} = x) - P_0 = \text{BMR},$$

where BMR denotes the "benchmark response" and refers to a specific risk increase above background risk. As in the quantal-response setting, BMR values of 0.1, 0.05, or possibly 0.01 are generally chosen. Later in the chapter, the committee focuses mostly on the case where $P_0 = 0.05$ BMR = 10% of the children experiencing an adverse effect. Thus, these choices of P_0 and BMR will result in a BMD that represents a doubling of the proportion of the population that falls into the adverse effect

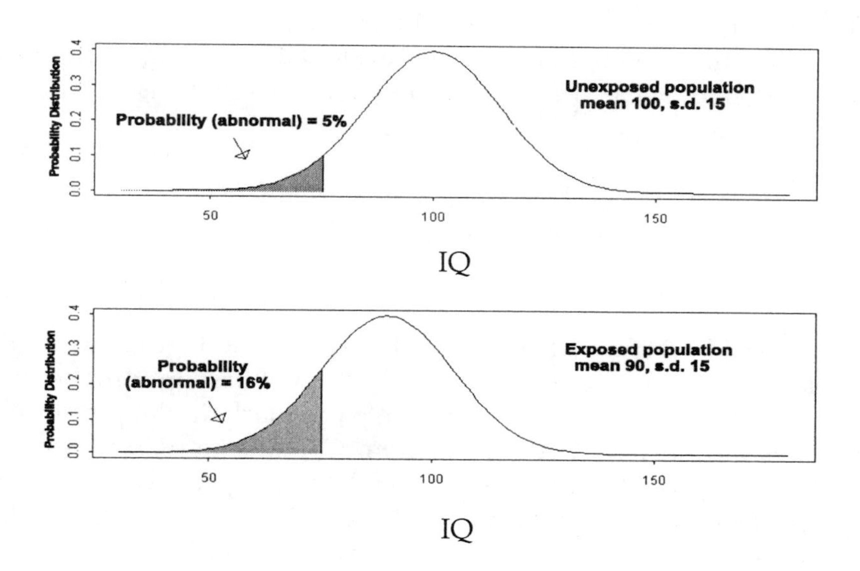

FIGURE 7-1 Hypothetical IQ distribution in an exposed and unexposed population.

region. Two broad approaches are available for BMD calculations that are based on continuous outcomes. As described above, one option is to fix P_0 at a specified percentile of performance in the unexposed population (e.g., 0.05 or 0.10). Assuming that the data follow a linear model ($Y_i = a_0 + a_1X_i + \varepsilon_i$, where X_i represents the exposure level for the ith subject, a_o and a_1 are unknown regression coefficients, and ε_i is random error, assumed to be normally distributed with variance σ^2), specifying a fixed P_0 is equivalent to setting the cutoff value at a specified number of SDs below the mean in the unexposed group:

$$C = a_0 + \sigma\Phi^{-1}(P_0),$$

where Φ (x) represents the normal cumulative distribution function (i.e., the area under a standard normal curve up to and including the value x). When $P_0 = 0.05$, for example, the value of C is $a_0 - 1.645\ \sigma$ (i.e., 1.64 SDs below the control mean). Alternatively, one can choose the cutoff value directly on the basis of clinical considerations or other information. For example, it might be appropriate to define 2,500 g as the cutoff in an epidemiological study of birth weight. In that case, P_0 can be expressed as a function of C as follows:

$$P_0 = \Phi\left(\frac{C - a_0}{\sigma}\right).$$

As discussed by Crump et al. (1998), there are advantages to using the approach based on specifying a fixed P_0 (i.e., the first option), because the calculations simplify in this setting, particularly in the presence of covariates (see also E. Budtz-Jørgensen, Copenhagen University, N. Keiding, Copenhagen University, and P. Grandjean, University of Southern Denmark, unpublished material, May 5, 2000). Under the assumption that the error terms follow a normal distribution, it follows that the benchmark dose will be the solution, BMD, to

$$\Phi\left(\frac{C - a_0 - a_1 \mathrm{BMD}}{\sigma}\right) = \mathrm{BMR} + P_0,$$

which is

$$\mathrm{BMD} = [\sigma\Phi^{-1}(P_0) - \sigma\Phi^{-1}(\mathrm{BMR} + P_0)] / a_1. \qquad (7\text{-}1)$$

Notice that the estimated BMD simply corresponds to a constant divided by the dose-response slope from the regression model. That concept is important, because it provides some theoretical justification for some analyses (presented later) that are based on the inverse of the estimated benchmark doses from several MeHg studies.

Several authors have suggested variations on how to calculate BMDs for continuous outcomes. For example, Kodell and West (1993) and West and Kodell (1993) extended the Gaylor and Slikker approach to allow the model variance to depend on dose level (the calculations above assume a constant σ^2). Crump (1995) developed a more general approach that relaxed the normality assumption required by previous approaches. Bosch et al. (1996) proposed a nonparametric approach that avoided the need for specifying a distribution altogether. In general, the different approaches are unlikely to yield dramatically different results when the data are approximately normally distributed with constant variance, which is the case in most of the MeHg epidemiological studies.

SOME SPECIFIC CONSIDERATIONS FOR MeHg

Aside from the general issues discussed above, several specific issues further complicate the application of benchmark-dose methods for MeHg. Foremost among these issues is the existence of three studies of comparable quality that lead to seemingly conflicting results in terms of the association between MeHg and adverse developmental or neurological outcomes. Previous chapters have discussed in-depth some of the possible explanations for this conflict (e.g., unmeasured confounders, co-exposures, and variations in population sensitivity). Another possibility is that the differences are due to random chance. Indeed, study results have been presented and summarized largely in terms of *p* values based on statistical tests of the association between exposure and outcome. Only recently have several papers focused on dose-response modeling and benchmark-dose calculations. When the focus is on statistical testing rather than modeling, it is common to encounter apparent contradictions, wherein one study will yield a statistically significant association at $p < 0.05$, and another one does not. To assess study concordance more fully, it is useful to consider the statistical power[1] that each has to detect effects of the magnitude observed.

For simplicity here, suppose that all confounders have already been accounted for, so that we can consider the power that a study will have to detect a true non-zero slope based on a simple linear regression ($Y_i = a_0 + a_1 X_i + \varepsilon_i$, where Y_i, X_i, a_0, a_1, and ε_i are as defined above). It is straightforward to compute the power to detect specific values of the dose-response parameter a_1, but comparing such calculations across studies is complicated, because the computed power depends on the distributions of exposure levels and outcomes within each study (see Zar

[1]Statistical power refers to the probability of correctly rejecting the null hypothesis of no association when, in fact, a true association exists (see Zar 1998).

1998). Cohen (1988, p. 75) argues that standardized regression coefficients provide a useful way to discuss power for the linear regression setting. The standardized regression coefficient corresponds to the raw slope (in our case, a_1) multiplied by the standard deviation of exposure and divided by the standard deviation of the error term. In the simple linear regression setting, the standardized regression coefficient corresponds precisely to the Pearson correlation between X and Y. Because the standardized regression coefficient is a unitless quantity, power calculations are simplified considerably and involve only sample size. According to Table 3.4 of Cohen (1988), the New Zealand study would have had high power (85% or greater) to detect correlations of approximately ±0.2 or larger, and the Seychelles and the Faroe Islands studies would have had power to detect smaller correlations of approximately ±0.1 or more. Figure 7-2 graphs the power that each study would have had to detect various values of the standardized regression coefficients. The Faroe Islands study, being the largest study, has the highest power, and the New Zealand study has the lowest.

To further aid in interpreting the power calculations summarized in Figure 7-2, we have computed the standardized regression observed in the studies. Table 7-1 shows the standardized regression coefficients for the significant outcomes in the Faroe Islands and New Zealand studies. Five of the eight effects reported in the Faroe Islands study were very small, ranging from −0.05 to 0.08. The power to detect such small effects in the Seychelles study—even with a sample of 700 children—was only about 50% (see Figure 7-2). Thus, some of the inconsistency between the findings of the Faroe Islands and the Seychelles studies could be due to limited power to detect very small effects, even in these very large samples. On the other hand, these analyses cannot explain the failure of the main Seychelles study to detect the neuropsychological effects of the magnitude reported in the New Zealand study, because the Seychelles study should have had adequate power to detect those effects.

There is at least one important caveat to the power considerations discussed above. Standard power calculations for the linear model setting are based on the assumption of a true linear relationship between exposure and outcome. In a real world dose-response setting, such as encountered for MeHg, there is likely to be some nonlinearity. That means that the observed level of statistical significance in a study might depend less on the total sample size than on the spread of the exposure

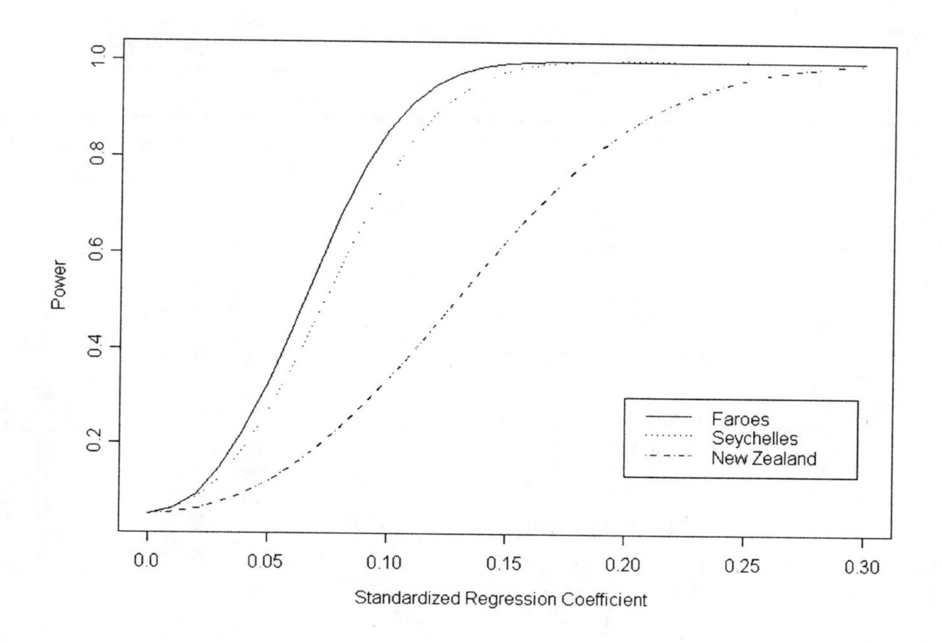

FIGURE 7-2 The power that each study has to detect a given standardized regression coefficient.

levels and, in particular, on whether there are sufficient observations at high exposure levels to characterize the true shape of the dose response in that region. In fact, all three studies had fairly skewed exposure distributions, with a large number of subjects clustered at low exposure levels, along with a few subjects exposed at moderate to high levels. Such skewness in the observed exposure levels can be associated with other problems as well. For example, extreme observations have the potential to exert a strong influence on the results in such settings. Indeed, Crump et al. (1998) reported nonsignificant results from a regression analysis on all the children in the New Zealand cohort, but significant results after omission of a single child whose mother's hair Hg concentration was 86 ppm (4 times higher than that of the next highest exposure level in the study). We will see presently that dose-

TABLE 7-1 Estimates of Standardized Regression Coefficients Based on Reported Study Results

Study	Exposure SD	Outcome	Outcome SD	Raw Regression Coefficient	Standardized Regression Coefficient
Faroe Islands [a]	0.375	Finger Tapping	6.15	-1.1	-0.07
		CPT-Errors [b]	0.54	0.12	0.08
		CPT-Reaction Time	80	40.3	0.18
		Digit Span	1.5	-.27	-0.06
		Boston Naming Test–no cues	5.3	-1.77	-0.12
		Boston Naming Test–cues	5.3	-1.91	-0.13
		CVLT-Short-term	3.1	-0.57	-0.06
		CVLT-Long-term	3.8	-0.55	-0.05
New Zealand [c]	3.31	TOLD-Language Development	16	-0.6	-0.12
		WISC-R:PIQ	16	-0.54	-0.11
		WISC-R:FSIQ	16	-0.55	-0.11
		McCarthy Perceptual Performance	10	-0.53	-0.17
		McCarthy Motor Test[b]	0.15	-0.007	-0.15

[a]Exposures measured on the log-scale. Exposure SD and regression coefficients provided by study investigators (Grandjean et al. 1997). Outcome SDs estimated by dividing the interquartile range by 1.3.

[b]Log transformed.

[c]Data from Crump et al. 1998.

Abbreviations: CPT, Continuous Performance Test; CVLT, California Verbal Learning Test; TOLD, Test of Language Development; WISC-R:PIQ, Wechsler Intelligence Scale for Children-Revised Performance IQ; WISC-R:FSIQ, Wechsler Intelligence Scale for Children-Revised Full-Scale IQ.

response estimates for the Faroe Islands study were also sensitive to omission of some of the observations with very high exposure levels.

COMPARING BENCHMARK DOSES

From a statistical perspective, reconciling differences among the various studies is more appropriately accomplished by a comparison of dose-response estimates (i.e., regression slopes) rather than *p* values resulting from the application of hypothesis tests. In this section, we present and compare benchmark-dose calculations from the New Zealand, Faroe Islands, and Seychelles studies. These have been reported individually for the New Zealand study by Crump et al. (1998) and also for the Seychelles study (Crump et al. 2000). BMDs for the Faroe Islands study have been calculated in a report prepared for EPA (Budtz-Jørgensen et al. 1999) and in an unpublished technical report (E. Budtz-Jørgensen, Copenhagen University, N. Keiding, Copenhagen University, and P. Grandjean, University of Southern Denmark, unpublished material, May 5, 2000), both of which were made available to the committee. The committee also requested and obtained some additional calculations to be discussed presently.

It might seem counterintuitive to present benchmark-dose calculations for the Seychelles study which did not show statistically significant associations between exposure and outcome. However, the idea makes more sense if we think of a benchmark dose as simply a transformation of the estimated dose response (as we saw in Eq. 7-1). Just as it can make sense to compare slope estimates from several studies, some of which are significantly different from zero and some of which are not, so can it make sense to compare benchmark doses. In the following example, we assume that lower values of the outcomes in question are adverse. Crump et al. (2000) argued that it is possible to calculate a BMDL even for studies where the estimated dose response goes in the 'wrong' direction. In such settings, the estimated BMD will not even exist. That is, when the estimated regression line suggests a beneficial effect of exposure (as was the case for several outcomes in the Seychelles study), a linear regression model predicts that there will be no exposure level resulting in a 10% adverse response. However, even in such a setting, the BMDL will be finite so long as the estimated regression coefficient is not statistically significantly different from zero, so that its upper or lower confidence limit (depending on whether a larger or smaller response is considered adverse) still goes in the expected direction. Indeed, Crump et al. (2000) presented BMDLs for five outcomes from the Seychelles study. An important caveat to this discussion is that BMDLs based on negative studies should be interpreted very cautiously. Although such calculations can be useful in a setting like ours where we are interested in comparing results over several MeHg studies, the committee advises that particular care be applied in using this approach in settings involving a single negative study as the basis for a risk assessment. Further research on this topic would be useful.

Crump et al. (1998) calculated BMDs for the New Zealand cohort at age 6, using the K-power model[2] and assuming $P_0 = 0.05$ and BMR = 0.10. Five outcomes were considered: TOLD Language Development, WISC-R Performance IQ, WISC-R Full-Scale IQ, McCarthy Perceptual Performance Scale, and McCarthy Motor Scale. It is important to note again that the results of the analysis reported here are based on omitting the highest exposed individual (86 ppm). A hair Hg concentration of 86 ppm is more than 4 times the next highest hair Hg concentration in the study. If the one-compartment pharmacokinetic model and EPA's standard default input assumption are used, it can be estimated that a 60-kg woman would have to eat an average of 0.5 pounds (227 g) of fish containing 2.2 ppm of Hg to reach a hair Hg concentration of 86 ppm. Consistent exposure at such a dose seems unlikely when the mean Hg concentration in fish from fish-and-chips shops, a principal source of exposure in New Zealand (Kjellström et al. 1986), is 0.72 ppm (Mitchell et al. 1982). On the basis of those considerations, the committee concluded that analyzing the New Zealand data without the data from that individual is appropriate.

Budtz-Jørgensen et al. (1999) presented BMDs and BMDLs for five outcomes measured in the Faroe Islands study: motor speed (finger tapping), attention (CPT reaction time), visuospatial performance (Bender), language (Boston Naming Test), and short-term memory (California Verbal Learning Test). The calculations for each outcome were done using the K-power model, as well as standard linear models applied to the untransformed exposure and square-root and log-transformed exposures. Calculations for the Faroe Islands study were per-

---

[2]The K-power model assumes the following mean: $a_0 + a_1x^k$, where K is a parameter to be estimated along with a_0 and a_1, thus allowing for a nonlinear dose-response relationship. The estimated value of K is typically constrained to be greater than or equal to 1.

formed for both maternal-hair and cord-blood Hg. Methods similar to those reported by Crump et al. (1998) were used, fixed P_0 values being 0.05 and 0.16 and excess risks (BMRs) being 5% and 10%. For comparability, the committee requested that the Crump analyses on the Seychelles and New Zealand data be expanded to include BMD and BMDL calculations for P_0 = 0.05 and 0.16 and for BMR = 0.05 and 0.10. The committee requested these analyses only for the outcomes measured when the children were 5 to 7 years old, because that age period was the only one available from the Faroe Islands study, and data from that age group have better predictive ability than data from earlier ages (E. Budtz-Jørgensen, Copenhagen University, N. Keiding, Copenhagen University, and P. Grandjean, University of Southern Denmark, unpublished material, May 5, 2000). For the Seychelles study, six 66-month end points were considered: Bender Gestalt errors, Child Behavior Checklist-Total, McCarthy-General Cognitive Index, Preschool Language-Total Score, Woodcock-Johnson Applied Problems, and Woodcock-Johnson Letter-Word recognition. For reasons to be discussed in more detail presently, the focus here is on calculations derived from the K-power model applied with P_0 = 0.05 and BMD = 0.05. The results of these analyses are summarized in Table 7-2 and graphically represented in Figure 7-3.

Table 7-2 and Figure 7-3 reveal some interesting patterns. First, we see that although study-to-study variability is substantial, within-study consistency (i.e., outcome to outcome) is relatively high. BMDs tended to be lowest for the New Zealand study. BMD and BMDL estimates for the Seychelles study tended to be either nonexistent or quite large (nonexistent values or values greater than 100 are indicated by asterisks in the table). Despite the substantial variability, however, the analyses yield a range of BMD values that are moderately consistent across the three studies. The next section discusses how those data might be used as the basis for a risk assessment.

CHOOSING A CRITICAL DOSE FOR A POINT OF DEPARTURE

An important step in the risk-assessment process is choosing an appropriate dose to be used as the "point of departure" (i.e., choosing the dose to which uncertainty factors will be applied to estimate an

TABLE 7-2 Benchmark Dose Calculations (ppm MeHg in maternal hair) from Various Studies and for Various End Points

Study	End Point	BMD[a]	BMDL
Seychelles[b]	Bender Copying Errors	*** [e]	25
	Child Behavior Checklist	21	17
	McCarthy General Cognitive	***	23
	Preschool Language Scale	***	23
	WJ Applied Problems	***	22
	WJ Letter/Word Recognition	***	22
Faroe Islands[c]	Finger Tapping	20	12
	CPT Reaction Time	17	10
	Bender Copying Errors	28	15
	Boston Naming Test	15	10
	CVLT: Delayed Recall	27	14
New Zealand[d]	TOLD Language Development	12	6
	WISC-R:PIQ	12	6
	WISC-R:FSIQ	13	6
	McCarthy Perceptual Performance	8	4
	McCarthy Motor Test	13	6

[a]BMDs are calculated from the K-power model under the assumption that 5% of the responses will be abnormal in unexposed subjects (P_0 = 0.05), assuming a 5% excess risk (BMR = 0.05).

[b]Data from Crump et al. 1998, 2000. "Extended" covariates.

[c]Data from Budtz-Jørgensen et al. 1999.

[d]Data from Crump et al. 1998, 2000.

[e] *** indicates value exceeds 100.

Abbreviations: WJ, Woodcock-Johnson Tests of Achievement; CPT, Continuous Performance Test; CVLT, California Verbal Learning Test; TOLD, Test of Language Development; WISC-R:PIQ, Wechsler Intelligence Scale for Children-Revised Performance IQ; WISC-R:FSIQ, Wechsler Intelligence Scale for Children-Revised Full-Scale IQ.

RfD). In a traditional setting, NOAELs are computed for each end point in each study, and the point of departure would most likely be chosen to correspond to the highest observed NOAEL. In the present setting, however, where BMDs and BMDLs are calculated instead of NOAELs, the appropriate choice is not immediately clear. It is not necessarily

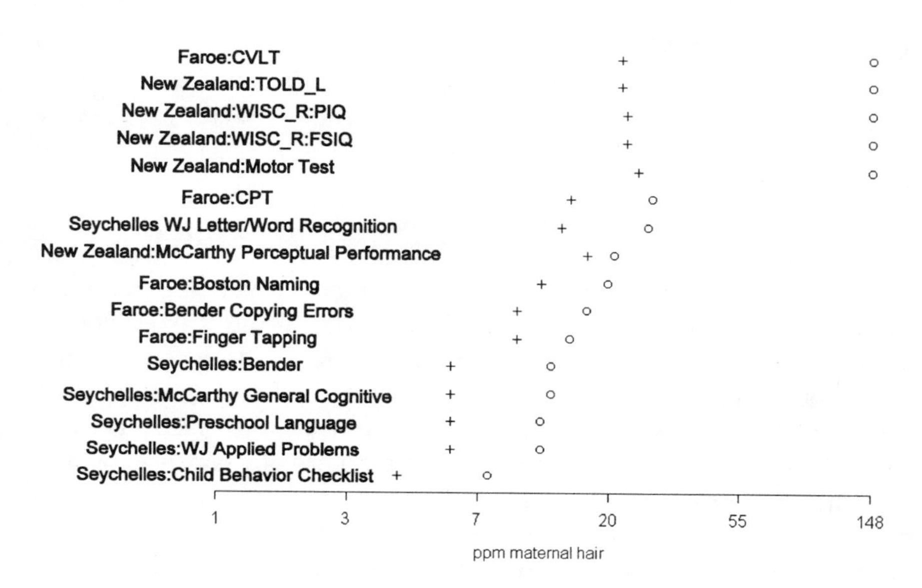

FIGURE 7-3 Benchmark dose (○) and lower 95% confidence limit on the benchmark dose (+): P_0 = BMR = 0.05, K-power model. Abbreviations: WJ, Woodcock-Johnson Tests of Achievement; CPT, Continuous Performance Test; CVLT, California Verbal Learning Test; TOLD, Test of Language Development; WISC-R:PIQ, Wechsler Intelligence Scale for Children-Revised Performance IQ; WISC-R:FSIQ, Wechsler Intelligence Scale for Children-Revised Full-Scale IQ.

appropriate to choose the lowest BMDL as the point of departure because that could easily result in choosing an unreliable end point (i.e., one with great uncertainty). One suitable option might be to choose the BMDL corresponding to the end point with the lowest BMD, selected from among those end points that show a statistically significant effect of exposure. That would lead to choosing the BMD or BMDL based on McCarthy Perceptual Performance from the New Zealand study. The BMD for this end point estimated from the K-power model is 8 ppm; the corresponding BMDL is 4.

The committee had some reservations about choosing the New Zealand study as the basis for risk assessment. First, it is a relatively small study, with only 237 children in contrast to over 700 studied in both the

Seychelles and the Faroe Islands studies. The original study results were not reported in the peer-reviewed literature but in a technical report to the Swedish Government. Although Crump et al. (2000) reported some statistical modeling of the data, the study has not been as comprehensively evaluated nor subject to the same level of scrutiny and re-analysis as the Seychelles and Faroe Islands studies.

The committee concluded that it would be inappropriate to pick the Seychelles study as the basis for risk assessment, given the available evidence for positive effects in the New Zealand and Faroe Islands studies, as well as in the Seychelles own pilot study. The committee felt that a good argument can be made for choosing the Faroe Islands study as the basis for the risk assessment. The Faroe Islands study is large (over 900 children) and measured two biomarkers of exposure. Moreover, it has been extensively analyzed and re-analyzed to explore the extent of confounding and the impact of outliers.

As discussed in Chapter 4, both hair and blood are reliable biomarkers for MeHg exposure. Given the lack of knowledge of differential effects of MeHg at different periods of gestation, there is currently no compelling reason to consider one biomarker of fetal exposure more appropriate than the other. Comparison of the analyses based on hair and cord blood in the Faroe Islands study suggests that the cord-blood measure explains more of the variability in more of the outcomes than hair Hg (see Table 6-1). On that basis, the committee recommends that cord blood be used as the biomaker for a risk assessment of the Faroe Islands data.

Table 7-3 shows the BMDs and BMDLs for the five principal Faroe Islands outcomes based on cord-blood measurements. Normally, the most sensitive adverse end point is selected as the basis for risk assessment. However, in the context of a neuropsychological test battery, the reliability of the individual end points can be highly variable. Therefore, it might not be appropriate in all cases to select the one most sensitive end point as a point of departure for the BMDS. In the Faroe Islands cord-blood analyses, the CPT reaction time measure had the smallest BMD and BMDL. However, because of difficulties in test administration, the data from the second half of the cohort were discarded for the analysis of this end point. Under the circumstances, the committee felt that it would be more appropriate to select the second most sensitive end point, the Boston Naming Test, for which no administration difficulties were encountered. It is noteworthy that the Boston Naming Test

TABLE 7-3 Benchmark Dose Calculations (ppb MeHg in cord blood) from the Faroe Islands Study for Various End Points

End Point	BMD[a]	BMDL
Finger Tapping	140	79
CPT Reaction Time	72	46
Bender Copying Errors	242	104
Boston Naming Test	85	58
CVLT: Delayed Recall	246	103

[a]BMDs are calculated from the K-power model under the assumption that 5% of the responses will be abnormal in unexposed subjects (P_0 = 0.05), assuming a 5% excess risk (BMR = 0.05).

Abbreviations: CPT, Continuous Performance Test; CVLT, California Verbal Learning Test.

Source: Budtz-Jørgensen et al. 1999.

was the most sensitive end point in the analyses based on maternal-hair Hg concentration.

The results of the Faroe Islands cord-blood Hg analysis for the Boston Naming Test provide a BMD of 85 ppb and a BMDL of 58 ppb. Corresponding values for hair Hg can be calculated by dividing the cord-blood concentration by a factor of 5 ppb of blood per ppm hair (Grandjean et al. 1992). Such a calculation results in a BMD of 17 ppm and a BMDL of 12 ppm for hair. It is interesting to note that those values are in fact quite close to the BMD and BMDL calculated for the Boston Naming Test based directly on the hair Hg concentration (i.e., BMD, 15 ppm; BMDL, 10 ppm).

Despite several strong arguments in favor of choosing a point of departure based on the Faroe Islands study, there was some concern that the estimated BMDs and corresponding BMDLs could be confounded by PCB exposure, which was not adjusted statistically in the benchmark analysis. To address this question, the committee requested that the Faroe Islands research group provide some additional calculations. Accordingly, Budtz-Jørgensen and colleagues[3] provided estimates of the

---

[3]E. Budtz-Jørgensen, Copenhagen University, N. Keiding, Copenhagen University, and P. Grandjean, University of Southern Denmark, unpublished material, June 21, 2000.

BMDs and BMDLs for four end points (Finger Tapping, CPT Reaction Time, Boston Naming, and CVLT Delayed Recall) based on (1) models that include log(PCB) as an additional covariate and (2) the subset of subjects in the lowest tertile of PCB exposures. Because PCBs were measured only for children examined in 1993, only about half of the full cohort (approximately 450 children) are used for analysis 1, and only one-sixth (approximately 150 children) are used for analysis 2. Results were provided for Hg measured in both maternal hair and cord blood (see Table 7-4). The reduced sample sizes in these additional analyses increased the variability among the results. There was no clear pattern with respect to how the PCB-adjusted analyses differed from the original results.

Because of the potential for measurement error to cause additional bias with respect to estimating the Hg effects in the PCB-adjusted models, the committee gave greater weight to interpreting the results of analyses performed in the low PCB subset. Comparing the low PCB subset with the full cohort results, for example, the BMDs for Finger Tapping and CPT Reaction Time were 5-13 ppm lower for maternal hair and 19-99 ppb lower for cord blood, and the BMDs for Boston Naming and Delayed Recall were 5-6 ppm higher for maternal hair and 42-147 ppb higher for cord blood. Thus, the BMDs for the low-PCB-exposed subset for the two end points that were related to PCB exposure—Boston Naming and California Verbal Learning—did not differ from the BMDs for the total sample by any more than the BMDs for the two end points that did not relate to PCB exposure. It should also be noted that the variability seen in Table 7-4 is well within that expected by chance; note that, in all cases, the BMDs and BMDLs for both the PCB-adjusted and the low-PCB subset analyses lie well within the intervals defined by the BMDs and corresponding BMDLs derived for the full cohort. For example, the BMD for the Boston Naming Test based on cord blood in the full cohort (85 ppb) is smaller than the BMD based on the low PCB subset (127 ppb). In fact, the difference between the BMDs based on the full cohort and the low PCB subset is less than one standard error of the BMD based on the low PCB cohort. In weighing all these considerations, the committee concludes that results based on the full cohort provide a reliable basis for establishing a point of departure for a risk assessment for MeHg. Because cord blood explains more of the ob-

TABLE 7-4 BMD (BMDL) Estimates from the Faroe Islands Study with and Without Adjustment for PCBs and in the Subset of Low PCB-Exposed Children (Results are reported separately for MeHg measured in hair and cord blood and are calculated using the K-power model.)

		Full Cohort	Adjusted for PCBs	Low PCB subset
Exposure	End Point	BMD (BMDL)[a]	BMD (BMDL)	BMD (BMDL)
Hair	Finger Tapping	20 (12)	17 (9)	7 (4)
	CPT Reaction Time	18 (10)	27 (11)	13 (5)
	Boston Naming Test	15 (10)	24 (10)	21 (6)
	CVLT: Delayed Recall	27 (14)	39 (12)	32 (7)
Cord Blood	Finger Tapping	140 (79)	149 (66)	41 (24)
	CPT Reaction Time	72 (46)	83 (49)	53 (28)
	Boston Naming Test	85 (58)	184 (71)	127 (40)
	CVLT: Delayed Recall	246 (103)	224 (78)	393 (52)

[a]BMDs are calculated under the assumption that 5% of the responses will be abnormal in unexposed subjects ($P_0 = 0.05$), assuming a 5% excess risk (BMR = 0.05).

Abbreviations: CPT, Continuous Performance Test; CVLT, California Verbal Learning Test.

Source: E. Budtz-Jørgensen, Copenhagen University, N. Keiding, Copenhagen University, and P. Grandjean, University of Southern Denmark, unpublished material, April 28, 2000.

served variability than Hg in maternal hair, the committee believes that it provides a more suitable biomarker for determining the point of departure.

AN INTEGRATIVE ANALYSIS

Although the committee felt comfortable with recommending that risk assessment be based on the Boston Naming Test from the Faroe Islands study, it also explored a weight-of-evidence approach based on an integrative analysis that allows a quantitative synthesis of information available across studies (see Hedges and Olkin 1985 for a more general discussion). Indeed, the recent draft EPA guidelines for carcinogen risk assessment suggest that such approaches can be useful in settings where it is difficult to chose a single study to serve as the basis for a risk assessment (EPA 1999). When well conducted, integrative analysis can provide valuable information to bolster or support a weight-of-evidence argument. Of course, synthesizing data across studies requires a careful statistical analysis that takes proper account of appropriate study-to-study and, in this case, outcome-to-outcome heterogeneity. One approach is to use a hierarchical random effects model. The committee conducted such an analysis using an extension of a method discussed by Dominici et al. (in press). Although the technical details can be complicated, the hierarchical modeling basically serves to smooth away some of the random variation that complicates the interpretation of the data presented in Table 7-2.[4] The approach also provides a way to quantify study-to-study and outcome-to-outcome variability. To motivate the approach, it is useful to consider the graphical presentation of our data in Figure 7-3. The figure displays estimated BMDs and corresponding BMDLs for the outcomes listed in Table 7-2. Results are presented from the K-power model, the parameters P_0 and BMR both taking the value 0.05. The plot is organized in order of increasing BMD values. The circles indicate BMD and the crosses indicate BMDLs. To allow the eye to distinguish more easily between the values associated with the Faroe Islands and New Zealand studies, the plot is drawn on the log scale. As discussed earlier, several of the BMDs did not exist for the Seychelles study. The committee has arbitrarily assigned those a value of 150 for the purpose of plotting. The figure illustrates the large study-to-study variability relative to the outcome-to-outcome variability. The figure suggests that it might make sense to borrow strength from the different studies and outcomes to gain increased precision. That is what the hierarchical model achieves. The results allow us to do several things. First, we can obtain a revised, smooth estimate of the BMDs in each study. Table 7-5 provides these

---

[4]To handle nonexistent BMDs, the hierarchical model was applied to the inverse of the BMDs reported in Table 7-2. Nonexistent BMDs were assigned an inverse value of 0. See appendix for more detail.

TABLE 7-5 Results of Applying the Hierarchical Analysis to BMDs (ppm of MeHg in Hair) Calculated Using the K-power Model

Study	End Point	Original	Smoothed
		BMD (BMDL)[a]	BMD (BMDL)
Seychelles[b]	Bender Copying Errors	*** [e] (25)	*** (26)
	McCarthy General Cognitive	*** (23)	*** (24)
	WJ Applied Problems	*** (22)	*** (24)
	Child Behavior Checklist	21 (17)	22 (18)
	Preschool Language Scale	*** (23)	*** (25)
	WJ Letter/Word Recognition	*** (22)	*** (24)
Faroe Islands[c]	Finger Tapping	20 (12)	20 (13)
	CPT Reaction Time	18 (10)	19 (12)
	Bender Copying Errors	28 (15)	24 (15)
	Boston Naming Test	15 (10)	17 (12)
	CVLT: Delayed Recall	27 (14)	24 (15)
New Zealand[d]	TOLD Language Development	12 (6)	13 (8)
	WISC-R:PIQ	12 (6)	13 (8)
	WISC-R:FSIQ	13 (6)	13 (8)
	McCarthy Perceived Performance	8 (4)	12 (7)
	McCarthy Motor Test	13 (6)	13 (8)

[a]BMDs are calculated under the assumption that 5% of the responses will be abnormal in unexposed subjects ($P_0 = 0.05$), assuming a 5% excess risk (BMR = 0.05).

[b]Data from Crump et al. (1998) and Crump et al. 2000. "Extended" covariates.

[c]Data from Budtz-Jørgensen et al. 1999.

[d]Data from Crump et al. 1998, 2000.

[e]*** indicates nonexistent values or values greater than 50.

Abbreviations: WJ, Woodcock-Johnson Tests of Achievement; CPT, Continuous Performance Test; CVLT, California Verbal Learning Test; TOLD, Test of Language Development; WISC-R:PIQ, Wechsler Intelligence Scale for Children-Revised Performance IQ; WISC-R:FSIQ, Wechsler Intelligence Scale for Children-Revised Full-Scale IQ.

smoothed BMDs along with corresponding BMDLs. For comparison, the original unsmoothed values are also included. Results are shown for the K-power model with $P_0 = 0.05$ and BMR = 0.05. Note that the effect of the hierarchical modeling is to smooth away much of the random variability observed in the original data. That is especially true of the more extreme values. Estimated BMDs are relatively unchanged for the Faroe Islands study, although even in that study, the outcome-to-outcome variability is reduced. Smoothing increases the BMDs slightly for the New Zealand study. BMDs for the Seychelles study remain high and most are still indicated with asterisks. Another interesting thing to notice from the table is that all the BMDLs tend to move closer to their respective BMDs. That is because the hierarchical model is able to reduce the variability inherent to each individual BMD estimate by drawing strength from the other end points. An important feature of the table is that although much of the outcome-to-outcome variability seems to be smoothed away through the hierarchical modeling, substantial study-to-study variability remains.

Although the hierarchical modeling provides a useful tool for separating random versus systematic variation and provides more stable estimates of study-specific and outcome-specific BMDs, the question remains regarding how the results might be used for risk assessment. There are several possible approaches. One would be to repeat the exercise described in the previous section, basing the risk assessment on either the Faroe Islands or the New Zealand studies but replacing the original BMD and BMDL estimates with the smoothed values from Table 7-5. The argument in favor of this approach is that it will have removed some of the bias associated with selecting an extreme value, and also it will have reduced some of the statistical variability. One could also argue for using the estimate of central tendency derived from the hierarchical modeling approach. The committee's analysis based on the K-power model suggests a mean BMD of 21 ppm, which coincidentally corresponds precisely to the mean of the smoothed BMDs from the Faroe Islands study. (The mean of the unsmoothed BMDs from the Faroe Islands study is 22 ppm). A third approach would be to produce a theoretical estimate of the BMDL on the basis of the lower 5th percentile point from the estimated distribution of BMDs obtained from the hierarchical modeling exercise. Applying this approach to the results for the K-power model yields an estimate of 7 ppm. The various approaches discussed in this section are summarized in Table 7-6.

MODEL CHOICE ISSUES

As mentioned earlier, the calculations provided by the Faroe Islands research group to the EPA included BMD and BMDL calculations under square-root and log transformations as well as calculations for the K-power model (Budtz-Jørgensen et al. 1999). To enable a full comparison with the results of other studies, the committee requested that the Crump analyses be expanded to include results based on the square-root and log transformations for the New Zealand and Seychelles studies. At first inspection, the results were troubling. Although standard statistical assessments of model adequacy could not distinguish between models based on the K-power model applied to untransformed data, or linear models based on square-root or log dose, the corresponding BMDs and BMDLs differed fairly dramatically. In general, BMDs and BMDLs were lowest for the log model and highest for the linear model. Budtz-Jørgensen and colleagues provided some extended discussion on this issue in the context of the Faroe Islands study (E. Budtz-Jørgensen, Copenhagen University, N. Keiding, Copenhagen University, and P. Grandjean, University of Southern Denmark, unpublished material, May 5, 2000). Because of the profound importance of model choice on estimation of the BMD, the committee requested that the Faroe Islands research group[5] provide some additional calculations to aid the committee's deliberations. The committee wondered, for example, if the influence of a few highly exposed individuals on the estimated dose response might explain the large model-to-model variations. The Faroe Islands study research group conducted sensitivity analyses repeating the regression models after omitting some of the highest observations. The results suggested that the influence of the extreme observations did not explain the model-to-model variability.

---

[5]E. Budtz-Jørgensen, Copenhagen University, N. Keiding, Copenhagen University, and P. Grandjean, University of Southern Denmark, unpublished material, April 28, 2000.

TABLE 7-6 Approaches to Benchmark Dose Calculation (ppm MeHg in Hair)

Approach	BMD	BMDL
Most sensitive end point from New Zealand	8	4
Median end point from New Zealand	12	6
Most sensitive end point from Faroe Islands study	15	10
Median end point from Faroe Islands study	20	12
Integrative analysis	21[a]	8[b]

[a]Logically equivalent to a BMD.

[b]Logically equivalent to a BMDL. Lower 5th percentile from the estimated distribution of BMDs.

After extensive discussion, the committee concluded that the most reliable and defensible results for the purpose of risk assessment are those based on the K-power model. The argument for this conclusion is as follows. In dose-response settings like those with MeHg, when there are no internal controls (i.e., no unexposed individuals) and where the dose response is relatively flat, the data will often be fit equally well by linear, square-root and log models. The models can yield very different results for BMD calculations, however, because these calculations necessitate extrapolating to estimate the mean response at zero exposure level. Both the square-root and the log models take on a supralinear shape at low doses, that is, they postulate a steeper slope at low doses. Thus, they tend to lead to lower estimates of the BMD than linear or K-power models. From a toxicological perspective, the K-power model has greater biological plausibility, because it allows for the dose response to take on a sublinear form, if appropriate. Sublinear models would be appropriate, for instance, in the presence of a threshold. The K-power model is typically fit under the constraint that $K \geq 1$, so that supralinear models are ruled out. Figure 7-4 contrasts several classes of dose-response models.

The model sensitivity described here might seem in conflict with the concept, put forward by Crump and others, that by estimating risks at moderate levels, such as 5% or 10%, the BMD should be relatively robust to model specification. As discussed by Budtz-Jørgensen and colleagues, key to understanding this apparent contradiction is that the Faroe Islands study does not include any true controls (i.e., subjects with

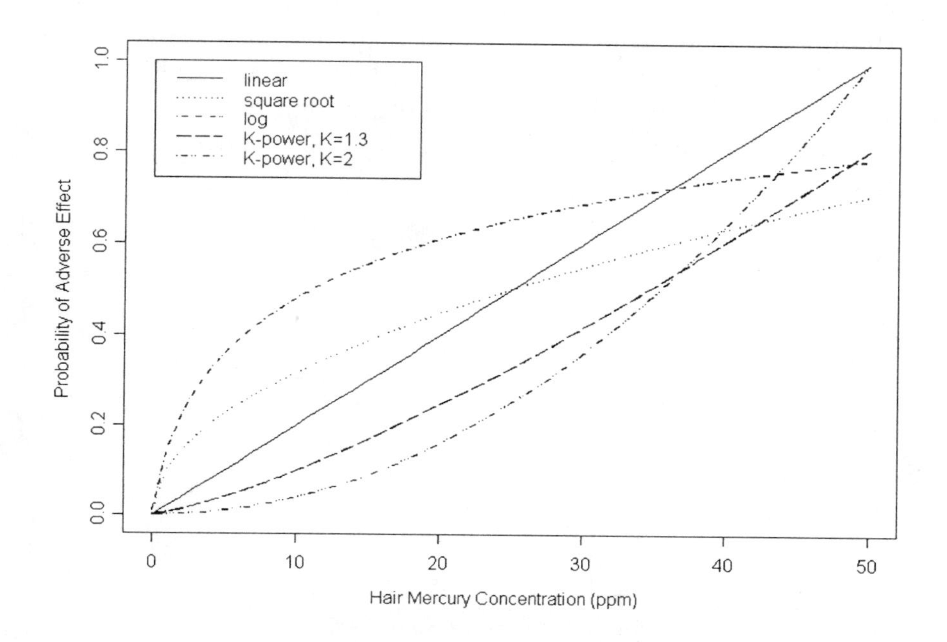

Figure 7-4 The estimated expected excess response due to Hg exposure as a function of the Hg concentration calculated using the linear, square root, and logarithmic (log) model. Source: E. Budtz-Jørgensen, Copenhagen University, N. Keiding, Copenhagen University, and P. Grandjean, University of Southern Denmark, unpublished material, May 5, 2000.

zero exposure) (E. Budtz-Jørgensen, Copenhagen University, N. Keiding, Copenhagen University, and P. Grandjean, University of Southern Denmark, unpublished material, May 5, 2000). The majority of exposures resulted in hair Hg concentrations that exceeded 5 ppm (24 ppb cord blood). The interquartile range for hair Hg concentration was 3 to 8 ppm (13 to 40 ppb for cord blood) (Grandjean et al. 1992). When models are fitted to the data, they are really capturing the shape of the dose response in this middle range of exposure, as illustrated in Figure 7-5. The figure shows dose-response curves fitted to hair Hg data for the linear, square-root and log transformations. The data and fitted curves are plotted on the log scale, so that the fitted log model appears linear and the linear model shows the highest degree of curvature. What becomes clear from Figure 7-5 is that variations in estimated BMDs are not explained by differences in how well the models fit the bulk of the data but rather what the models predict for the mean response for *unexposed* individuals.

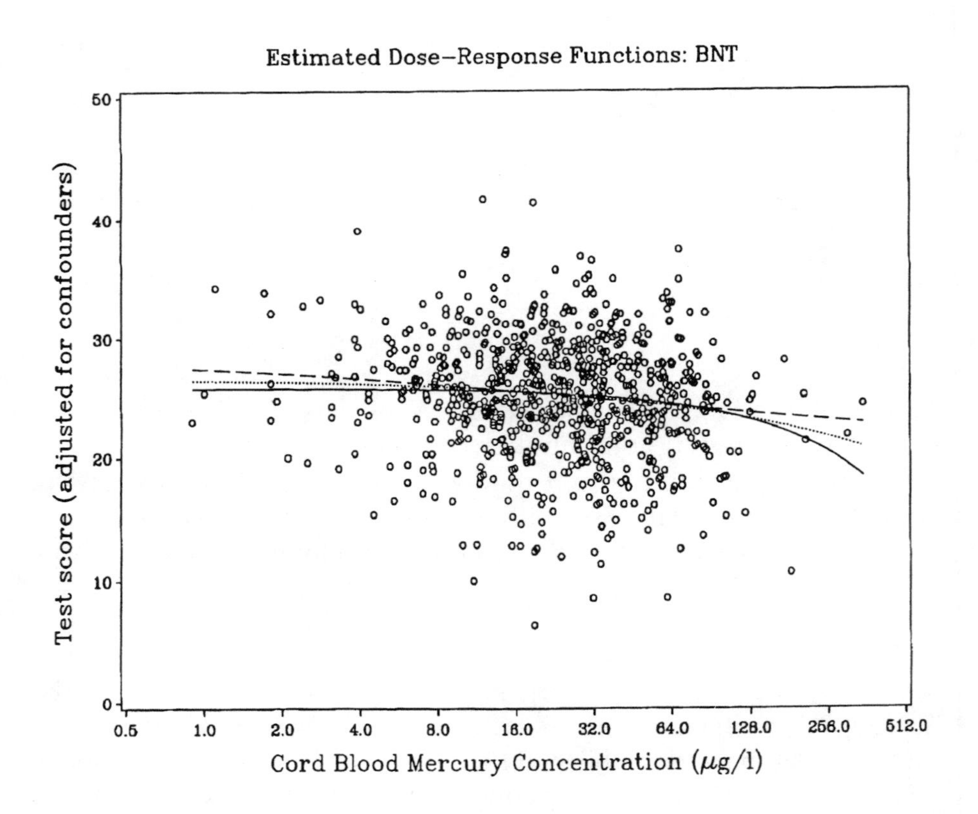

FIGURE 7-5 Dose-response curves fitted to cord-blood Hg data for the linear, square root, and log transformations. Source: E. Budtz-Jørgensen, University of Copenhagen, unpublished material, November 12, 1999.

Because the BMD estimation is based essentially on extrapolation to zero exposure from three models that fit equally well, the committee concludes that the choice regarding which model to use cannot be based on statistical grounds alone. Instead, a more biologically based argument is needed. One useful way to think of differences between the various models is that the linear model implicitly assumes an additive effect of Hg exposure, the log model assumes a multiplicative effect, and the square root lies somewhere in between. All three models fit essentially equally well to data that for the most part correspond to concentrations between 2 and 20 ppm in hair. However, the models differ fairly dramatically with regard to how they extrapolate to values below those levels. The linear model would predict that the change in mean outcome as MeHg concentration goes from 0 to 10 ppm in hair should be the same as the change observed in the mean outcome as concentration increases from 10 to 20 ppm. In contrast, the log-model would predict that the change in mean outcome associated with any doubling of MeHg concentration should be the same as the change observed in the mean outcome as concentration increases from 10 to 20 ppm. Thus, the log model would predict that the same magnitude change in outcome would be expected as the concentration goes from 1 to 2 ppm or from 4 to 8 ppm as that observed for the concentration going from 10 to 20 ppm—that is, the extrapolation down to zero exposure will predict a very steep slope at low doses. Given the relative absence of exposures at very low levels, a decision should be made on biological grounds regarding which model makes the most sense for risk assessment. The committee believes that an additive (linear) or perhaps sublinear model is the most justifiable from a biological perspective, thus ruling out square-root and log-transformed models. For MeHg, the committee believes that a good argument can be made for the use of a K-power model with K constrained to be greater than or equal to 1. That rules out square-root ($K = 0.5$) and log models (the limiting case as K approaches 0). More generally, the committee concludes that considerable caution should be used in fitting models based on log or square-root transformations of exposure, which might not be appropriate in dose-response settings, such as those for MeHg, where there are no internal controls and where the dose response is relatively flat. In such settings, linear models based on log- or square-root-transformed data are likely to yield results very similar to those based on untransformed exposures. The supralinear shape of the log and square-root models at low doses will tend to result in smaller BMD estimates than those based on untransformed models.

SUMMARY AND CONCLUSIONS

- Benchmark-dose calculations are available for the Seychelles, New Zealand, and Faroe Islands studies. The calculations reveal fairly high within-study consistency (i.e., outcome to outcome) and high study-to-study variability.
- In general, risk assessments for various toxicants based on animal studies have used a BMR of 0.1, because it usually represents the low range of the observed exposure data. Crump et al. (2000) used a BMR of 0.1 (i.e., 10% of the population is at risk) in their analyses of the New Zealand and Seychelles studies. For the end points studied, the baseline rate in the population is 0.05 (P_0). Selection of a BMR of 0.1, therefore, could result in as much as a tripling of the percentage of the population falling into the abnormal range of neurological performance. The committee, to be more protective of public health, has used a P_0 of 0.05, a BMR of 0.05 (i.e., 5% of the population is at risk). Specification of P_0 in the context of the MeHg studies, however, is somewhat problematic because of the absence of subjects with true zero exposure. The mean response rate at zero is not actually based on observed data but is extrapolated from the fitted model. Because of that, extra caution is needed in choosing a dose-response model as the basis of a risk assessment. Choosing to have P_0 and BMR both equal to 0.05 could lead to a doubling of the proportion of the population falling into the abnormal range. The committee recognizes, however, that the choice of P_0 and the BMR is at the interface of science and policy and should be a science-informed policy judgment. That decision involves choosing a level of risk (i.e., 10%, 5%, or 1%) that is considered to be an "acceptable risk," similar to the choice of an acceptable risk for a carcinogen (i.e., 1×10^{-6} cases). That decision should be guided by the full body of evidence and based on the best available and most relevant data. In choosing the P_0 and the BMR for MeHg, it is

preferable that the range of MeHg concentrations and outcomes observed in the Faroe Islands study be considered.

- Basing the risk assessment on the single most sensitive end point from the most sensitive study would lead to the use of McCarthy Perceptual Performance Scale from the New Zealand study.
- The Faroe Islands study provides the strongest basis for choosing the critical dose for defining a point of departure. This large, high-quality study has been extensively analyzed and re-analyzed to explore the possibility of confounding, outliers, differential sensitivity, and other factors. The Boston Naming Test scores are the most sensitive, reliable end point.
- The potential for confounding by PCB exposure is of some concern for the Faroe Islands study. However, on the basis of a series of sensitivity analyses provided by the Faroe Islands research group, the committee concluded that the PCB exposures were unlikely to be causing serious bias in BMD estimates. Although BMD and BMDL estimates were available in the subset of low PCB exposed subjects, the committee decided to use the estimates based on the full cohort because the considerably larger sample size was felt to result in more reliable estimates.
- It would not be appropriate to base risk-assessment decisions on the Seychelles study because it did not find an association between MeHg and adverse neurodevelopment effects. That finding is not consistent with the weight of evidence demonstrating such an association in the Faroe Islands and New Zealand studies.
- A risk assessment could also be based on an integrative analysis that combines the results of all three studies. One advantage of this approach is that it increases the precision of critical-dose estimates. One could choose either a measure of central tendency (leading to a BMD of 21 ppm in hair) or a lower 5% limit based on the estimated theoretical distribution of benchmark doses (leading to an estimate of 7 ppm in hair). Because the integrative analysis is exploratory, it would be premature to use this approach as the basis for risk assessment for MeHg. However, the approach was useful for facilitating a weight-of-evidence assessment. Furthermore, it is reassuring that the results based on this approach are consistent with those based on the more classic approaches that select a single study.

- Model choice is an important source of uncertainty for the purpose of quantitative risk assessment. Changing the underlying modeling assumptions can have a dramatic effect on the estimated benchmark dose. The committee suggests that the K-power ($K \geq 1$) model results be used.
- Even when such modeling decisions have been made, benchmark-dose calculations require specification of the cutoff point used to define an adverse effect (P_0) and the risk level (BMR) of the benchmark dose. Those are, in part, policy decisions.
- The committee concludes that, given these considerations, the results from the Boston Naming Test in the Faroe Islands study should be used. For that end point, dose-response data based on Hg concentrations in cord blood should be modeled. For that data set, the K-power model ($K \geq 1$) is the model of choice. This analysis estimates a BMD of 85 ppb and a BMDL of 58 ppb. Using a conversion factor of 5 ppb of blood per ppm of hair, that point of departure approximately corresponds to a BMD based on a hair Hg concentration of 17 ppm and a BMDL of 12 ppm. Those values are very close to the values estimated directly from the analysis based on hair Hg concentrations.

RECOMMENDATIONS

- Until better statistical methods become available, risk assessment for MeHg should be based on benchmark dose calculations rather than NOAELs or LOAELs.
- Given the available data, risk assessment should be based on the Boston Naming Test from the Faroe Islands study using MeHg measured in cord blood.
- Despite some potential for PCB exposures to bias BMD estimates based on the Faroe Islands study, the committee recommends using estimates based on the full cohort and not adjusting for PCB exposure, mostly because the larger sample size is believed to result in more reliable estimates.
- Benchmark doses should be based on the K-power model with K constrained to take a value of 1 or greater.
- Because the integrative analysis is exploratory, it would be premature to recommend it for use now. However, the approach should

be considered in context of a weight-of-evidence argument. Further research on the use of integrative models for risk assessment would be useful.

- Further research is generally needed on statistical issues related to risk assessment that is based on epidemiological data. In particular, further research to develop more appropriate methods for handling model uncertainty (e.g., the Bayesian technique of *model averaging* (Carlin and Louis 1998)) would be useful. Further work is also needed to develop risk assessment methods for a setting like MeHg where the study population contains no true controls.

REFERENCES

Allen, B.C., R.J. Kavlock, C.A. Kimmel, and E.M. Faustman. 1994. Dose-response assessment for developmental toxicity. II. Comparison of generic benchmark dose estimates with no observed adverse effect levels. Fundam. Appl. Toxicol. 23(4):487-95.

Bosch, R.J., D. Wypij, and L.M. Ryan. 1996. A semiparametric approach to risk assessment for quantitative outcomes. Risk Anal. 16(5):657-665.

Budtz-Jørgensen, E., N. Keiding, and P. Grandjean. 1999. Benchmark Modeling of the Faroese Methylmercury Data. Research Report 99/5. Prepared at the University of Copenhagen, Denmark for the U.S. Environmental Protection Agency.

Carlin, B.P., and T.A. Louis. 1998. Bayes and Emperical Bayes Methods for Data Analysis. New York: Chapman Hall.

Cohen, J. 1988. Statistical Power Analysis for the Behavioral Sciences, 2nd Ed. Hillsdale, NJ: Lawrence Erblaum Associates.

Crump, K.S. 1984. A new method for determining allowable daily intakes. Fundam. Appl. Toxicol. 4(5):854-871.

Crump, K.S. 1995. Calculation of benchmark doses from continuous data. Risk Anal. 15(1):79-89.

Crump, K.S., C. Van Landingham, C. Shamlaye, C. Cox, P.W. Davidson, G.J. Myers, and T.W. Clarkson. 2000. Benchmark dose concentrations for methylmercury obtained from the Seychelles Child Development Study. Environ. Health Perspect. 108:257-263.

Crump, K.S., T. Kjellström, A.M. Shipp, A. Silvers, and A. Stewart. 1998. Influence of prenatal mercury exposure upon scholastic and psychological test performance: Benchmark analysis of a New Zealand cohort. Risk Anal. 18(6):701-713.

Dominici, F., J.M. Samet, and S.L. Zeger. In press. Combining evidence on air pollution and daily mortality from the 20 largest US cities: A hierarchical modelling strategy (with discussion). Royal Statistical Society: Series A.

EPA (U.S. Environmental Protection Agency). 1998. Guidelines for Neurotoxicity Risk Assessment. Fed. Regist. 63(93):26925-26954.

EPA (U.S. Environmental Protection Agency). 1999. Guidelines for Carcinogen Risk Assessment. Review Draft. NCEA-F-0644, Risk Assessment Forum, U.S. Environmental Protection Agency, Washington, DC, (July 1999). [Online]. Available: http://www.epa.gov/nceawww1/raf/crasab.htm

Gaylor, D.W. 1983. The use of safety factors for controlling risk. J. Toxicol. Environ. Health 11(3):329-36.

Gaylor, D.W., and W. Slikker. 1992. Risk assessment for neurotoxicants. Pp. 331-343 in Neurotoxicology, H. Tilson, and C. Mitchell, eds. New York: Raven Press.

Gilbert, S.G., T.M. Burbacher, and D.C. Rice. 1993. Effects of in utero methylmercury exposure on a spatial delayed alternation task in monkeys. Toxicol. Appl. Pharmacol. 123(1):130-6.

Grandjean, P., P. Weihe, P.J. Jørgensen, T. Clarkson, E. Cernichiari, and T. Viderø. 1992. Impact of maternal seafood diet on fetal exposure to mercury, selenium, and lead. Arch. Environ. Health 47(3):185-195.

Grandjean, P., P. Weihe, R.F. White, F. Debes, S. Araki, K. Yokoyama, K. Murata, N. Sørensen, R. Dahl, and P.J. Jørgensen. 1997. Cognitive deficit in 7-year-old children with prenatal exposure to methylmercury. Neurotoxicol. Teratol. 19(6):417-428.

Hedges, L.V., and I. Olkin. 1985. Statistical Methods for Meta-Analysis. Orlando, FL: Academic Press.

Kimmel, C.A. 1990. Quantitative approaches to human risk assessment for noncancer health effects. Neurotoxicology 11(2):189-98.

Kimmel, C.A., and D.W. Gaylor. 1988. Issues in qualitative and quantitative risk analysis for developmental toxicology. Risk Anal. 8(1):15-20.

Kjellström, T., P. Kennedy, S. Wallis, and C. Mantell. 1986. Physical and Mental Development of Children with Prenatal Exposure to Mercury from Fish. Stage I: Preliminary tests at age 4. National Swedish Environmental Protection Board Report 3080. Solna, Sweden.

Kodell, R.L., and R.W. West. 1993. Upper confidence limits on excess risk for quantitative responses. Risk Anal. 13(2):177-82.

Leisenring, W., and L. Ryan. 1992. Statistical properties of the NOAEL. Regul. Toxicol. Pharmacol. 15(2 Pt. 1):161-171.

Mitchell, J.W., T.E.U. Kjellström, and .L. Reeves. 1982. Mercury in takeaway fish in New Zealand. N. Z. Med. J. 95(702):112-4.

Rice, D.C. 1992. Effects of pre-plus postnatal exposure to methylmercury in the monkey on fixed interval and discrimination reversal performance. Neurotoxicology 13(2):443-452.

Rice, D.C. 1996. Sensory and cognitive effects of developmental methylmercury exposure in monkeys, and a comparison to effects of rodents. Neurotoxicology 17(1):139-154.

West, R.W., and R.L. Kodell. 1993. Statistical methods of risk assessment for continuous variables. Communications in Statistics: Theory and Methods 22(12):3363-3376.

Zar, J.H. 1998. Biostatistical Analysis, 4th Ed. Englewood Cliffs, NJ: Prentice-Hall.

Zelikoff, J.T., J.E. Bertin, T.M. Burbacher, E.S. Hunter, R.K. Miller, E.K. Silbergeld, S. Tabacova, and J.M. Rogers. 1995. Health risks associated with prenatal metal exposure. Fundam. Appl. Toxicol. 25(2):161-170.

8

RISK CHARACTERIZATION AND PUBLIC HEALTH IMPLICATIONS

THE purpose of this chapter is to present a summary of the findings of the committee concerning the health effects of methylmercury (MeHg), end points of toxicity, the critical studies, exposure and dose metrics, and sources of uncertainty that should be considered by EPA in deriving the reference dose (RfD). It includes a discussion of the relevant health end points and the scientific basis and public-health rationale for selecting neurotoxicity in children exposed in utero as the critical end point for the EPA RfD.

The committee was directed to investigate the toxicological effects of MeHg and to evaluate research relevant to EPA's MeHg RfD. The activities of the committee included the following:

1. An evaluation of the available human epidemiological and animal toxicity data.
2. An examination of the critical studies, end points of toxicity, and uncertainty factors used in the derivation of the RfD.
3. A review of exposure data from the available epidemiological studies focusing on consumption of MeHg in fish.
4. Consideration of new and emerging health-effects data.
5. Identification of knowledge gaps and recommendations for future research.

The committee evaluated the body of evidence that has provided the scientific basis for the risk assessments conducted by EPA and other regulatory and health agencies. The committee also reviewed new findings that have emerged since the development of the current RfD and met with the investigators of major ongoing epidemiological studies to examine and compare the methods and results.

Mercury (Hg) is pervasive and persistent in the environment, released from a large variety of natural and anthropogenic sources. The serious health impacts of high-level exposures have long been recognized. Between 1950 and 1975, major poisoning episodes in Japan and Iraq resulted in outbreaks of serious neurotoxic effects, including death, and led to the identification of developmental neurotoxicity as the health effect of greatest concern following high-level episodic exposure. As a result of its well-recognized toxicity, widespread industrial use, and environmental persistence, Hg has been extensively studied. Compared with data bases on many other pollutants, there is a robust data base on Hg, which includes environmental fate and transport; examination of toxicokinetics and toxicodynamics; biological and environmental measures of exposure and dose; and in vitro, animal, and human studies for a broad range of toxicity end points.

Historically, epidemiological investigations have focused on high exposures and related health impacts. More recently, large prospective epidemiological studies have been conducted to examine chronic low-level MeHg exposure. These studies examined the association between subtle end points of neurotoxicity and prenatal exposure measured by maternal markers of prior exposure. These markers are presumed to reflect maternal MeHg exposure from fish consumption. The committee focused on these studies because they provide the most comprehensive evidence of low-dose MeHg toxicity and they examine the exposure pathway most relevant to U.S. population exposures, including the sensitive population of children who were exposed to MeHg in utero.

THE CURRENT EPA REFERENCE DOSE

EPA defines an RfD as an estimate (with uncertainty spanning perhaps an order of magnitude) of a daily exposure to the human population (including sensitive subgroups) that is likely to be without an appreciable risk of deleterious effects during a lifetime (EPA 1997a). The current EPA RfD for MeHg is 0.1 μg/kg-day. The RfD is an important risk-characterization tool that is broadly used as a measure of the "acceptability" of population exposure levels. It is used to guide risk-management decisions and regulatory policies ranging from fish-consumption advisories to air-emission permits. This section provides an overview of the development of the MeHg RfD.

Neurotoxicity in children exposed in utero is the health outcome selected by EPA for the current MeHg RfD. The RfD is based on data from the Iraqi poisoning episode, where the population consumed high levels of MeHg from treated seed grain. The critical study for the RfD conducted by Marsh et al. (1987) identified 81 children who had been in utero during the episode and examined their neurodevelopmental outcomes. Maternal-child pairs were selected from one of five Hg-hair-concentration groups, and the combined incidence of developmental effects (late walking, late talking, mental symptoms, seizures, or increased neurological score) was determined for each group. Exposure levels measured by maternal-hair concentration and combined developmental effects were used to estimate a benchmark dose. The benchmark dose of 11 ppm of Hg in hair was calculated as the 95% lower confidence limit on the maternal-hair concentration corresponding to a 10% extra risk level (Crump et al. 1995). In this report, the lower confidence limit is referred to as the BMDL. The following section describes how EPA derived the current RfD from that value.

A ratio of 250:1 was used to convert hair Hg concentration (mg of Hg/kg of hair) to blood Hg concentration (mg of Hg/L of blood) to derive the RfD critical dose (EPA 1997c):

11 mg/kg of hair would correspond to 11/250 = 44 μg/L of blood.

The following equation was used to obtain a daily dietary intake of MeHg that results in a blood Hg concentration of 44 μg/L:

$$d = \frac{C \times b \times V}{A \times f \times bw},$$

where

d = daily dietary intake (micrograms of MeHg per kilogram of body weight per day),

C = concentration in blood (44 μg/L),

b = elimination constant (0.014 days^{-1}),

V = volume of blood in the body (5 L),
A = absorption factor (expressed as a unitless decimal fraction of 0.95),
f = fraction of daily intake taken up by blood (unitless, 0.05), and
bw = body-weight default value of 60 kg for an adult female.

Using that equation, the total daily quantity of MeHg ingested by a 60-kg female to maintain a blood Hg concentration of 44 µg/L or a hair Hg concentration of 11 ppm would be

$$d = \frac{44\ \mu g/L \times 0.014\ days^{-1} \times 5\ L}{0.95 \times 0.05 \times 60\ kg}$$

$$d = 1.1\ \mu g/kg\text{-}day.$$

A composite uncertainty factor (UF) of 10 was used in the derivation of the RfD to account for human-population variability, lack of a two-generation reproductive study, and lack of data on sequelae resulting from longer durations of exposure (EPA 1997c):

$$RfD = \frac{BMDL}{UF}$$

$$= \frac{1.1\ \mu g/kg\text{-}day}{10}.$$

$$\text{Current RfD} = 0.1\ \mu g/kg\text{-}day.$$

As the calculation shows, the application of UFs has a major influence on the quantification of the final RfD. Although the scientific rationale for the application of these factors is strong, it must be recognized that choosing the ultimate magnitude of the UFs is a policy decision, which is influenced by professional judgment, public-health goals, and the regulatory mandates of EPA.

EVALUATING THE RfD–END POINTS OF MeHg TOXICITY

The committee reviewed human epidemiological results and animal toxicity data to examine potential human health effects and evaluate the use of neurotoxicity in children exposed in utero as the health end point for the derivation of the RfD. Other end points evaluated are carcinogenicity and immunological, reproductive, renal, and cardiovascular toxicity. Chapter 5 presents an in-depth presentation of the health effects of MeHg. The following is a summary of major findings.

Carcinogenicity

Studies in humans of the carcinogenic effects of MeHg are inconclusive. Although no studies have found an association between MeHg and overall cancer death rates in humans, two studies (Kinjo et al. 1996; Janicki et al. 1987) have found associations between Hg exposure and acute leukemia. Interpretation of these findings is limited because of small study populations and lack of control for other risk factors. Renal tumors have been observed in male mice (Mitsumori et al. 1981; Hirano et al. 1986) but only at or above the maximum tolerated dose. Hg has also been shown to cause chromosomal damage and aneuploidy in a number of in vivo and in vitro systems. On the basis of the available human, animal, and in vitro data, the International Agency for Research on Cancer (IARC) and EPA have classified MeHg as a "possible" (EPA Class C) human carcinogen (EPA 2000).

Immunotoxicity

Occupational studies suggest that Hg exposure can affect the immune system in humans (Dantas and Queiroz 1997; Moszczynski et al. 1999). In vitro and animal studies have shown that Hg can be immunotoxic. They suggest that exposure to MeHg can increase human susceptibility to infectious diseases and autoimmune disorders by damaging the immune system (Ilbäck et al. 1996). Animal studies have also shown that prenatal and perinatal exposure to MeHg produce long-term effects on the developing immune system (Wild et al. 1997). Immunological studies in animals are summarized in Table 5-3.

Reproductive Effects

The reproductive effects of MeHg exposure have not been evaluated in humans. However, an evaluation of the clinical symptoms and outcomes of over 6,000 MeHg-exposed Iraqi citizens found a low rate of pregnancies (79% reduction) among the exposed population (Bakir et al. 1973). That provides suggestive evidence of an effect of MeHg on human fertility. Animal studies, including work in nonhuman primates, have found reproductive problems, including decreased conception rates, early fetal losses, and stillbirths (Burbacher et al. 1988).

Renal Toxicity

The kidney is sensitive to inorganic Hg exposure, and renal damage has been observed following human ingestion of organic forms of Hg. Renal effects from organic Hg exposure have been observed only at exposure levels that also cause neurological effects. Renal damage was observed in the victims of the Iraqi poisoning, and an evaluation of death rates in an area of Minamata City, which had the highest prevalence of Minamata disease, found an increase in deaths from renal disease among women but not men (Tamashiro et al. 1986). Several reports of animal studies have also described MeHg-induced renal toxicity.

Cardiovascular Effects

The cardiovascular system appears to be a target for MeHg toxicity in humans and animals. Blood-pressure elevations have been observed in occupationally exposed men (Höök et al. 1954) and in children treated with mercurous chloride for medical conditions. More recently, there is evidence that suggests effects at low levels of exposure. A recent study of 1,000 children from the Faroe Islands found a positive association between prenatal exposure to MeHg, and blood pressure and heart rate variability at age 7 (Sørensen et al. 1999). A Finnish cohort study of 1,833 men linked dietary intake of fish and Hg concentrations in hair and urine with increased risk of acute myocardial infarction (AMI) and coronary heart disease and cardiovascular disease (Salonen et al. 1995). Men who consumed at least 30 g of fish a day had a 2.1 higher risk of AMI. Cardiovascular effects have also been observed in several animal models of MeHg toxicity.

Central-Nervous-System Toxicity

The toxic effects of MeHg in the brain have been well documented in human and animal studies. Although both the adult and fetal brains are susceptible, the developing nervous system is more sensitive to the toxic effects of MeHg than is the developed nervous system. It should be pointed out however, that few studies of MeHg effects in adults have investigated the sensitive and subtle types of neurologic endpoints recently examined in children exposed in utero. Studies of Minamata victims indicate that prenatal exposure caused diffuse damage in the brain and adult exposure caused focal lesions. About 10% of the total body burden of MeHg is found in the brain. After ingestion, MeHg accumulates in the brain where it is slowly converted to inorganic Hg. On the basis of available studies, neurodevelopmental effects appear to be a sensitive end point for MeHg toxicity. There is an extensive human data base on neurodevelopmental effects, including studies of populations following high-dose poisonings and chronic low-level Hg exposure. In general, experimental animal studies have reported a continuum of neurodevelopmental effects similar to those reported in studies of humans exposed to MeHg. Of the three major long-term prospective studies, the Faroe Islands study reported an effect of low-level prenatal exposure on children's performance on neurobehavioral tests particularly in the domains of attention, fine-motor function, confrontational naming, visual-spatial abilities, and verbal memory. Similar effects were not found in the main Seychelles study; however, the smaller New Zealand study found effects on standardized tests of cognitive and neuromotor function that were similar to those administered in the main Seychelles study, and there was preliminary evidence of similar effects in the Seychelles pilot study.

SELECTION OF THE END POINT FOR THE RfD

The findings of the committee regarding the end points of MeHg toxicity support the selection of neurotoxicity in children exposed in utero as a suitable end point for the development of the RfD based on the available data. These effects have been well documented in a number of investigations, including prospective epidemiological studies examining low-dose chronic exposure through consumption of contaminated fish and seafood. Evidence from animal studies is consistent with the neurotoxicity findings in humans.

Given the limits of the available data, developmental neurotoxicity is the most sensitive, well-documented health end point. Therefore, its use as the basis for the RfD should be protective for other adverse effects that occur at higher doses of exposure. However, there is emerging evidence of potential effects on both the immune and cardiovascular systems at low doses of exposure. Although these effects are not well understood, emerging data underscore the need for continued research and raise the possibility of adverse effects to other organ systems at or below the current levels of concern for developmental neurotoxicity.

EXAMINATION OF THE CRITICAL STUDIES FOR THE RfD

The traditional approach to development of an RfD and other public-health-based risk guidance numbers is to select a critical study that is well conducted and provides the most sensitive, or lowest, no-observed-adverse-effect level (NOAEL), lowest-observed-adverse-effect level (LOAEL), or a lower 95% confidence limit on the benchmark dose (BMDL). The relevance of the study exposure levels and pathways to the population of concern should also be considered.

The current EPA RfD is based on developmental neurotoxic effects in children exposed in utero to high-level episodic exposure from bread made with grain treated with MeHg as a pesticide (Marsh et al. 1987). Although that study was judged the most appropriate at the time of the development of the current RfD, a number of recognized sources of uncertainty, including possible selection bias in the cohort, cannot be controlled. In addition, the exposure scenario in Iraq is not comparable to the low-level chronic exposure that the general population of North America might experience through the consumption of fish.

Since the establishment of the current RfD, results from the prospective studies in the Faroe Islands (Grandjean et al. 1997, 1998, 1999a) and the Seychelles (Davidson et al. 1995a,b, 1998), as well as a peer-reviewed re-analysis of the New Zealand study (Crump et al. 1998), have added substantially to the body of knowledge concerning the developmental neurotoxic effects of chronic low-level exposure to MeHg. Each of these studies was well designed and carefully conducted. They examined the relation of prenatal MeHg exposure to neuropsychological function in childhood. MeHg was significantly associated with poorer performance in the Faroe Islands and New Zealand studies, but not in the main Seychelles study.

Much of the debate over the adverse effects of MeHg and the selection of a critical study for the RfD and other guidance has focused on the similarities and differences between the Faroe Islands and the Seychelles studies. The levels of maternal exposure are similar in both studies, but a number of differences in design and cohort characteristics might contribute to the disparate findings. They used different primary biomarkers of Hg exposure (cord blood versus maternal hair), different types of neurological tests (domain specific versus global), and different ages at testing (7 years versus 5.5 years). In addition, the studies had different patterns of exposure (due to whale consumption in the Faroe Islands). When the New Zealand study is considered, those research design differences seem less determinative. In New Zealand, adverse effects were found with exposure measures and a research design similar to the Seychelles study. These studies are contrasted and discussed in detail in Chapter 6.

The Faroe Islands population was also exposed to PCBs. The initial statistical analyses published by the investigators of the Faroe Islands study suggest that the associations of prenatal exposure with language, memory, and verbal-learning deficits might be attributable to prenatal PCB exposure, although the associations with attention and neuromotor-function deficits were not. However, prenatal Hg exposure was associated with deficits in language development in the Seychelles pilot and New Zealand studies, in which there was no evidence of increased PCB exposure. A re-analysis of the Faroe Islands data showed that the association of Hg exposure with language and verbal deficits was as strong among children with low PCB exposure as among those with high exposure. Furthermore, a series of sensitivity analyses provided by the Faroe Islands research group (E. Budtz-Jørgensen, Copenhagen University, N. Keiding, Copenhagen University, and P. Grandjean, University of Southern Denmark, unpublished material, June 21, 2000) indicated that the PCB exposures were unlikely to be causing serious bias in BMD estimates. On the basis of these considerations, the committee concluded that the neurodevelopmental sequelae found in the Faroe Islands study were not attributable to PCB exposure and that PCB exposure did not invalidate the use of the Faroe Islands study as the basis of risk assessment.

The committee explored the possibility that differences in power might explain the discrepancies in the findings of the major studies. Five of the eight effects observed in the Faroe Islands study were very small. Despite the large sample size of the Seychelles study, its power to detect such small effects was poor. The Seychelles study had adequate power to detect the effects seen in the New Zealand study; therefore, such power considerations cannot fully explain its failure to detect any adverse effects at 5.5 years of age.

Despite their differences, the Faroe Islands, Seychelles, and New Zealand studies represent exposure scenarios that are more consistent than the Iraqi study with the North American experience. However, their conflicting results present a vexing choice for the development of a revised RfD. A conservative approach would be to derive as a point of departure the lowest BMD from the positive end points in the Faroe Islands study or the New Zealand study. It is possible to derive a lower limit approximation of a NOAEL or BMD from the Seychelles results, as was done by the Agency for Toxic Substances and Disease Registry for its minimal risk level (MRL). However, the choice of a negative study to derive guidance numbers when well-designed, plausible positive studies are available is difficult to defend. The committee recommends a more inclusive approach to developing any future RfD or exposure guidance. Given the availability of well-designed epidemiological studies in which prenatal MeHg levels were within the range of general-population exposures, contemporary exposure standards should consider the findings of all three studies—the New Zealand, Faroe Islands, and Seychelles studies.

To synthesize information from the different studies and outcomes, the committee conducted an integrative analysis to derive and compare estimates of BMDLs. This analysis is described in Chapter 7. The committee debated whether to include the Seychelles study in the BMD evaluation. It concluded that it would be inappropriate to exclude the data from any well-designed study and that the inclusion of the Seychelles study was important to ensure that the analysis would reflect the full range of effects of MeHg exposure.

BMD CONSIDERATIONS: SELECTING A POINT OF DEPARTURE

The current MeHg RfD is based on a BMD estimation. The selection of a particular BMD for the derivation of the RfD represents a critical decision, influenced by both scientific and policy considerations. The BMDL is defined as a lower confidence limit on the dose corresponding to a given increase in response (e.g., 1%, 5%, or 10%) over the background rate (Crump 1984), the benchmark response (BMR). It is intended to be applied as an alternative to the NOAEL to provide a point of departure for low-dose extrapolation. The BMD represents a refinement over the traditional NOAEL or LOAEL, since it is not constrained to be one of the observed or experimental doses, and uses the full-range of dose-response information inherent in the data. Various terms are used for BMD estimates. In this report, the term BMDL denotes the lower confidence limit on the dose corresponding to the BMR of interest, and BMD is used to denote the point estimate of the dose.

The critical studies of MeHg examined a range of neurodevelopmental outcomes. Selection of the most appropriate BMD requires consideration of the biological significance of the effects, including the sensitivity and severity of the outcomes, consideration of the ability to detect both exposure and effects, and selection of an appropriate dose-response model. To examine and compare the results of the critical studies, BMD calculations were conducted and compared for various end points. These results are presented and discussed in detail in Chapter 7.

Various analyses were conducted as part of the committee's consideration of the overall weight of the evidence of developmental neurotoxic effects from low-level MeHg exposure. It is intended as a bounding exercise to evaluate and present the range of effects, BMDs, and BMDLs for each of the major epidemiological studies. The results provide a range of BMDLs, which should be considered in selecting the critical

BMD for development of a revised RfD. The methods considered included (1) approaches based on selecting a single outcome from a single study, and (2) an integrative analysis that synthesizes information over different studies and outcomes. Because the integrative analysis is exploratory, it would be premature to use this approach as the basis for risk assessment for MeHg. However, the approach was useful for facilitating a weight-of-evidence assessment.

The BMDLs derived from the various end points of the critical studies (with a P_0 of 0.05, where P_0 denotes the probability that an unexposed individual falls below the cutoff value that defines an adverse effect, and a BMR of 0.05) range from 4 (New Zealand McCarthy Perceptual Performance Test) to 23 (Seychelles Preschool Language Scale Test) in parts per million (ppm) Hg in maternal hair. It should be noted that the choice of P_0 and the BMR are, in part, policy decisions. The full range of findings is presented in Table 7-2. Table 8-1 lists the BMDLs derived using the K-power model from Table 7-2. The K-power model was suggested because from a toxicological perspective, it has greater biological plausibility, since it allows the dose response to take on a sublinear form, if appropriate. The K-power model is typically fit under the constraint that $K \geq 1$, so that supralinear models are ruled out. As shown in Table 8-1, the data suggest fairly high within-study consistency but high study-to-study variability. However, the ratio between the highest and lowest BMDLs was only 6.

The integrative analysis used a hierarchical model to quantify study-to-study and outcome-to-outcome variability, while smoothing away much of the random variability observed in the original data. Smoothed estimates of the BMDs and BMDL for each study were derived (with a P_0 of 0.05 and a BMR of 0.05), and the distribution was examined. Outcome-to-outcome variability is reduced, but substantial study-to-study variability remains. The smoothed mean of the distribution of the various BMDs is 21 ppm, with a lower 5th percentile of 7 ppm.

The mean of the distribution is in close agreement with the unsmoothed mean of the BMDs from the Faroe Islands study (22 ppm). The integrative analysis does not permit the direct calculation of a BMDL. However, the lower 5th percentile of the theoretical distribution of true BMD values is analogous to a BMDL; that value is 8 ppm.

The examination of the BMDs suggests a number of ways to select a point of departure for the derivation of the RfD. The most sensitive end

TABLE 8-1 BMDLs for Study End Points (ppm Hg in maternal hair, BMR = 0.05)

Study	Age	End Point	BMDL (K power)
Seychelles[a]	66 months	Bender Copying Errors	25
		Child Behavior Checklist	17
		McCarthy General Cognitive	23
		Preschool Language Scale	23
		WJ Applied Problems	22
		WJ letter/word Recognition	22
Faroe Islands[b]	7 years	Finger Tapping	12
		CPT Reaction Time	10
		Bender Copying Errors	15
		Boston Naming Test	10
		CVLT: Delayed Recall	14
New Zealand[c]	6-7 years	TOLD Language Development	6
		WISC-R:PIQ	6
		WISC-R:FSIQ	6
		McCarthy Perceptual Performance	4
		McCarthy Motor Test	6

[a]Data from Crump et al. 1998, 2000.
[b]Data from Budtz-Jorgensen et al. 1999.
[c]Data from Crump et al. 1998, 2000.
Abbreviations: BMDL, lower 95% confidence limit on the benchmark dose; BMR, benchmark response; WJ, Woodcock-Johnson Tests of Achievement; CPT, Continuous Performance Test; CVLT, California Verbal Learning Test; TOLD, Test of Language Development; WISC-R:PIQ, Wechsler Intelligence Scale for Children-Revised performance IQ; WISC-R:FSIQ, Wechsler Intelligence Scale for Children-Revised Full-Scale IQ.

point from the most sensitive study is the McCarthy Perceptual Performance from New Zealand (BMD, 8 ppm; BMDL, 4 ppm). The Faroe Islands study represents the central tendency of the three studies, and a central BMD from this study could provide a reasonable point of departure (median BMDL value, 12 ppm). The central tendency of the integrative analysis (BMD, 21 ppm) or lower 5% limit (7 ppm) might also be considered. The Seychelles study, because of the lack of positive findings, does not provide an appropriate point of departure for risk assessment. Although the Seychelles study is a well-conducted study, the cohort appeared to be less sensitive than those of the New Zealand and Faroe Islands studies for reasons that are still not understood.

SELECTION OF THE CRITICAL STUDY AND POINT OF DEPARTURE FOR THE REVISED RfD

The committee conducted an in-depth examination of the methods, strengths, uncertainties, and outcomes of each of the major studies. It included an examination of findings and comparison of BMDs and BMDLs. On the basis of its consideration of the body of evidence, the committee concluded that a well-designed study with positive effects provides the most appropriate public-health basis for the RfD. When the two studies with positive effects are compared, the strengths of the New Zealand study include an ethnically heterogeneous sample, in which the observed effects cannot be attributed to the particular vulnerability of a genetically homogenous ethnic group, and the use of developmental end points with greater predictive validity for school performance than that of the discrete neuropsychological tests used in the Faroe Islands study. The advantages of the Faroe Islands study over the New Zealand study include a larger sample size, the use of two biomarkers of exposure, and more extensive scrutiny in the epidemiological literature. In addition, the Faroe Islands data have undergone extensive re-analysis in response to questions raised by panelists at the NIEHS (1998) workshop and by this committee in the course of its deliberations. Therefore, the committee recommends the Faroe Islands study as the critical study for the revision of the RfD. For that study, dose-response data based on Hg concentrations in cord blood should be used to estimate the BMD. Because the data on the most sensitive end point—the Continuous Performance Test—were analyzed for only half the sample, the committee recommends that the BMDL based on the next most sensitive end point—the Boston Naming Test—be considered as a reasonable and representative point of departure for a revised RfD.

SOURCES OF UNCERTAINTY: CONSIDERATION FOR UNCERTAINTY FACTORS

Evaluation of the sources of uncertainty is essential for the development of a RfD. Some uncertainty is inherent in any experimental or epidemiological study. To address these uncertainties in the derivation of the RfD, the NOAEL or BMDL may be divided by one or more uncertainty factors. Uncertainty factors were originally termed safety factors and were used in the derivation of acceptable daily intakes (ADIs) to account for recognized uncertainties by incorporating an additional margin of safety on the NOAEL (NRC 1994). Traditionally, uncertainty factors and modifying factors of 10 or 3 have been applied to address well-recognized issues, which reflect potentials for additional sensitivities or adverse effects not addressed in the dose-response analysis. These issues include variation in sensitivity among humans, animal-to-human extrapolation, extrapolation from subchronic-to-chronic exposure, LOAEL-to-NOAEL extrapolation, and incomplete data to address all possible outcomes. A modifying factor, based on professional judgment, may also be applied to address uncertainties in the data base or critical study (Dourson et al. 1996). Traditional default uncertainty factors of 10 have been acknowledged to be somewhat arbitrary since they were proposed by O.G. Fitzhugh and A. Lehman in the early 1950s (NRC 1994).

At the present time, there is no consistent approach in the application of uncertainty factors across the various regulatory and public-health agencies. The selection and application of uncertainty factors represents a scientific policy judgment that has a major influence on the determination of the RfD or other risk-management guidance numbers. For example, application of large uncertainty factors might overshadow the moderate differences among the various study findings and their NOAELS or BMDs in determining the magnitude of the RfD. That possibility is particularly relevant to the MeHg RfD, because the body of evidence examined by the committee indicates a general convergence of the lower doses associated with neurodevelopmental effects. Given the relatively small differences in BMDLs, a more consistent approach to the application of uncertainty factors could reduce the current inconsistencies between the EPA RfD and other risk guidance numbers.

To identify sources of uncertainty in deriving the current MeHg RfD, EPA conducted an analysis of uncertainties (EPA 1997b, Vol. VI Appendix) in relation to the critical study of neurodevelopmental effects from the 1971 Iraqi MeHg poisoning incident (Marsh et al. 1987). Major sources of uncertainty were identified as the variability in susceptibilities within the Iraqi cohort, population variability in the pharmacokinetic processes, and response classification error. An additional concern was the applicability of a risk assessment based on a grain-consuming population to the U.S. population for which fish consumption is the primary source of MeHg exposure. A composite uncertainty factor of 10 was applied in the derivation of the RfD to account for several uncertainties, including human-population variability, lack of a two-generation reproductive study, and lack of data on sequelae resulting from longer durations of exposure (EPA 1997c). Although the rationale for the composite uncertainty factor applied in the current RfD is well described, it is not possible to quantitatively validate that it adequately addresses the combined uncertainties in the Iraqi data because some of them have been described only in qualitative terms.

Any refinement of the current RfD will require consideration of sources of uncertainty. The committee has evaluated the body of evidence, focusing on the prospective epidemiological studies of neurotoxicity in children exposed in utero. Refinement of the current RfD based on results from these studies will require both quantitative and qualitative analysis of uncertainties to guide the application of uncertainty factors.

Not all sources of uncertainty require the addition of uncertainty factors in the derivation of the RfD. When the MeHg prospective epidemiological studies provide the basis for the RfD, uncertainty factor adjustments are potentially required only for the following reasons:

- If the uncertainty could result in underestimation of the adverse effects of MeHg exposure on human health.
- If there is reason to suspect that the U.S. population is more sensitive than the study populations to the adverse effects of MeHg.

Although there are multiple sources of uncertainty in the quantitative derivation of the RfD, not all result in an RfD that is insufficiently protective. Table 8-2 lists sources of uncertainty identified by the committee.

Individual responses to MeHg exposure are variable and a key source of uncertainty. Factors that might influence susceptibility include age,

TABLE 8-2 Sources of Uncertainty in Key Epidemiological Studies

Susceptible subpopulations
- Interindividual toxicokinetic variability in dose reconstruction
- Toxicodynamic variability
- Nutritional deficits

Measures of exposure
- Lack of dietary-intake data
- Extrapolation from biomarker Hg content to MeHg intake
- Nutritional and dietary confounders and effect modifiers
- Co-exposure to other neurotoxicants (e.g., PCBs)
- Co-exposure to other forms of Hg
- Inability to measure peak exposures
- Temporal matching of exposure to critical periods of susceptibility for the developing fetal brain

Lack of consideration of other key or most-sensitive health end points
- Potential cardiovascular or immune-system effects
- Neurological sequelae (i.e., late emerging effects)

gender, genetics, health status, nutritional influences including dietary interactions, and interindividual toxicokinetic and toxicodynamic variability. For example, data from Iraq indicate that although some individuals were sensitive to low levels of exposure, some members of the cohort were not sensitive to extremely high levels of exposure. That finding suggests a wide interindividual variability in sensitivity. Development of the RfD must consider this individual variation; in particular, any biomarker-based measure should account for the toxicokinetic variability in the population. At present, there is no clear evidence that the U.S. population is more sensitive than any of the key study populations. However, in any given population, there might be sensitive subpopulations whose sensitivity to MeHg is not adequately represented in the dose-response assessment. That possibility could represent an additional source of uncertainty.

Limitations in the evaluation of exposures also represent a source of uncertainty. Of particular concern is the uncertainty in the linkage between the time and the intensity of exposure to critical periods of brain development. Each dose metric provides different information about exposure. Dietary-recall data might be useful in stratifying exposure levels, but appropriate dietary data were not collected in the key studies. Measurement of cord blood does not detect temporal variability in exposure and reflects exposure during a period late in gestation. Therefore cord-blood concentrations might not correspond to the periods of greatest fetal sensitivity to Hg neurotoxicity. Similarly, average concentrations of Hg in hair provide no information on peak exposures and, because of variation in length and growth rate, might not reflect comparable periods of gestation.

In any experimental or epidemiological data, there is also some uncertainty on whether the measured effects represent the true most sensitive or critical effects. Neurodevelopmental effects are the most extensively studied sensitive end point for MeHg exposure, but there remains some uncertainty about the possibility of other health effects at low levels of exposure. In particular, there are indications of immune and cardiovascular effects, as well as neurological effects emerging later in life, that have not been adequately studied.

A number of additional sources of uncertainty are not possible to quantify but might contribute to the differences in study findings and BMDLs for the various outcomes. Those might include differences in nutritional and dietary confounders and effect modifiers such as beneficial effects from eating fish. Differences in population susceptibilities and unmeasured coexposure to other pollutants, including other forms of Hg, might introduce uncertainty.

On the basis of an evaluation of the sources of uncertainty in the key epidemiological studies, the committee identified two major categories of uncertainty, which should be considered in the determination of uncertainty factors for the revision of the RfD:

- Interindividual toxicokinetic variability in dose reconstruction (see Chapter 3).
- Data-base insufficiency (i.e., because of consideration of possible low-dose sequelae and latent effects, and immunotoxicity and cardiovascular effects) (see Chapter 5).

On the basis of the analysis presented in Chapter 3, the committee believes that an uncertainty factor of 2-3 for dose reconstruction from hair Hg concentrations or an uncertainty factor of about 2 for dose reconstruction from blood Hg concentrations is objective and appropriate. Despite ongoing work to provide a data-based and probabilistic basis for uncertainty-factor adjustments in the derivation of the RfD (e.g., Hattis et al. 1999), the choice of values for most categories of uncertainty other than toxicokinetics, and for the aggregate uncertainty-factor adjustment remains, in part, a policy decision. That is particularly the case for the uncertainty factor category of data-base insufficiency. The choice of values for most uncertainty-factor categories (e.g., animal to human) can be related to extant (although limited) analyses of empirical data. In the case of data-base insufficiency, however, the uncertainty-factor value is intended to address the possibility that more accurate or complete information might result in a lower NOAEL or BMD or might result in a more sensitive end point. If data were available to assess such a possibility adequately and quantitatively, such data might well lead to a more appropriate RfD rather than to an uncertainty-factor adjustment. Thus, the selection of an appropriate uncertainty-factor value for data-base insufficiency is inherently uncertain. Nonetheless, the committee believes that there is a reasonable possibility that significant immunotoxicity and cardiovascular effects, as well as neurotoxic sequelae, might occur at exposure levels below the dose corresponding to the neurodevelopmental BMD identified by the committee. Therefore, given the relatively unambiguous starting point for variability in dose reconstruction, the committee believes that an overall uncertainty-factor adjustment of no less than 10 is necessary and appropriate to provide an adequate margin of protection.

IMPLICATIONS FOR PUBLIC HEALTH AND RISK MANAGEMENT

The RfD provides critical guidance for a broad range of public-health and regulatory initiatives aimed at reducing Hg exposures and preventing adverse health impacts. The goal of the RfD is to estimate a level of daily exposure without adverse public health impacts even for sensitive individuals.

EPA has estimated from food consumption surveys that 7% of women nationwide exceed the RfD. From a food consumption survey in New Jersey it was estimated that 21% of women of childbearing age exceed the current RfD (Stern et al. 1996). EPA has calculated that a hair Hg concentration of 1.0 ppm would approximately result from an intake of MeHg at the current EPA RfD (see calculations in the Current EPA Reference Dose section in this chapter). Although estimates of hair and blood concentrations in the U.S. population are sparse, when that hair Hg concentration (1.0 ppm) is compared with available data, it is again seen that more highly exposed subpopulations frequently exceed the current RfD (EPA 1997c; Stern et al. 2000).

The committee conducted a margin-of-exposure (MOE) analysis to examine the margin of safety between available estimates of population exposure and BMDLs derived from the major epidemiological studies. The MOE approach provides a method of characterizing risks and is being used increasingly to examine potential population risks, particularly for noncancer end points. The MOE approach has been recommended by The Presidential/Congressional Commission on Risk Assessment and Risk Management (1997) as a common metric to be used by both environmental-protection and public-health agencies for assessing and comparing health risks. The MOE is the ratio of the critical dose (NOAEL or BMDL) to the estimated population exposure level. The smaller the ratio, the greater the cause for concern. Because the BMDLs are not adjusted by uncertainty factors, MOEs less than 10 indicate that population exposures might be approaching levels of public-health concern. Table 8-3 presents the results of the MOE analysis. The analysis compared available estimates of the range of population Hg concentration in hair to BMDLS from the major studies: the cord-blood-derived BMDL for the lowest reliable end point (the Boston Naming Test) from the Faroe Islands study; the 5% lower bound BMD from the committee's integrative analysis; and the Iraq study BMDL, which is the point of departure for the current RfD.

MOEs for the estimates of mean population levels range from 7.5 (New Zealand, most sensitive end point) to 77.3 (Seychelles, median end point). Those results indicate that the risk of adverse health impacts from the current exposure level in the majority of the population is low. However, for those at the high end of the population exposure distribution (95th percentile), the MOEs indicate that the margin of safety for the

TABLE 8-3 Population Margins of Exposure (MOE)[a] for Selected BMDLs and Exposure Estimates (ppm of Hg in Maternal Hair or Estimated Equivalent to Maternal Hair)

		MOE					
		Estimated MeHg Exposure in Selected Populations					
		New Jersey Pregnant Women[b]		EPA Region V Population[c]		U.S. Women of Childbearing Age[d]	
Study	Selected BMDL (value, ppm)	Mean (0.53)	95th Percentile (2.0)	Mean (0.29)	95th Percentile (1)	Mean (0.36)	95th Percentile (2.4)
New Zealand	Most sensitive (4)	7.5	2.0	13.8	4	11.1	1.7
Faroe Islands	Most sensitive (10)	18.9	5.0	34.5	10	27.8	4.2
Faroe Islands	Most-sensitive-reliable, cord-blood derived (12)	22.6	6.0	41.4	12	33.3	5.0
Seychelles	Median (22)	41.5	11	77.3	22	61.1	9.2
New Zealand, Faroe Islands, and Seychelles (integrative analysis)	Lower 5% (7)	13.2	3.5	24.1	7	19.4	2.9
Iraq	(11)[e]	20.8	5.5	37.9	11	30.6	4.6

[a]MOE, BMDL/exposure estimate.
[b]Data from Stern et al. 2000.
[c]Data from Pellizzari et al. 1999.
[d]Data from Smith et al. 1997.
[e]Current RfD basis.
Abbreviations: BMDL, lower 95% confidence limit on the benchmark dose; RfD, reference dose.

most highly exposed is consistently below 10. That indicates that the exposure levels of high-end consumers are close to those at which there are observable adverse neurodevelopmental impacts.

To further characterize the risks of MeHg, the committee developed an estimate of the number of children born annually to women most likely to be highly exposed through high fish consumption (highest 5% estimated to consume 100 g per day). Available consumption data and current population and fertility rates indicate that over 60,000 newborns annually might be at risk for adverse neurodevelopmental effects from in utero exposure to MeHg.

The MeHg-associated performance decrements on the neuropsychological tests administered in the Faroe Islands and New Zealand studies suggest that prenatal MeHg exposure is likely to be associated with poorer school performance. In the Faroe Islands sample, MeHg-related deficits were seen across a broad range of specific domains, including vocabulary, verbal learning, visuospacial attention, and neuromotor function. Deficits of the magnitude reported in these studies are likely to be associated with increases in the number of children who have to struggle to keep up in a normal classroom or who might require remedial classes or special education.

Revision of the RfD for MeHg can have far-reaching implications for public health and environmental protection. Currently, 40 states have issued advisories concerning consumption of certain freshwater fish. Any revision of the RfD will have implications for the market for fish and seafood and the dietary choices of Americans. Regulatory impacts might also be substantial, because federal and state agencies use the RfD to develop water-quality criteria and set limits on Hg releases in air and water. Additionally, there are implications for industrial use of Hg and Hg-containing materials, as well as decisions about disposal methods and recycling options.

Ideally, the application of the RfD in risk management should provide a margin of safety for all of the population. The application of the RfD to guide regulatory and risk-management policies must also consider risk tradeoffs, economic and technological limitations, as well as cultural and political influences. It must be recognized that the refinement of the RfD might not eliminate agency differences in risk management. However, improving the scientific basis for decision-making represents an important step forward in developing a cohesive strategy to prevent adverse effects from MeHg.

COMMITTEE FINDINGS AND RECOMMENDATIONS

- Hg is pervasive and persistent in the environment. Its use in products and emission from industrial processes and combustion have resulted in global circulation and atmospheric deposition. There have been well-documented instances of population poisonings, highly exposed occupational groups, and worldwide chronic low-level environmental exposures. The bioaccumulation of MeHg can lead to high concentrations in many species of fish and result in unacceptable levels of exposure and risk to highly exposed or susceptible subpopulations.
- The weight of the evidence of developmental neurotoxic effects from exposure to MeHg is strong. There is a strong data base, which includes multiple human studies and experimental evidence in animals and in vitro tests. Human studies include both high-exposure scenarios and evaluations of effects of chronic low-level exposure. The epidemiological studies also include well-established biomarkers to evaluate exposure levels in study populations.
- The weight of evidence from multiple epidemiological studies supports the selection of neurotoxicity in children exposed in utero as the most sensitive well-documented effect and a suitable end point for the derivation of the BMD. However, emerging evidence of other potential effects should also be considered in the calculation and the implementation of the EPA RfD.
- Given the availability of results from large prospective epidemiological studies, the Iraq study results should no longer be considered the critical study for the EPA RfD. The exposure scenarios in Iraq are not comparable to the low-level chronic exposures in North America. In addition, there are well-recognized uncertainties concerning exposure and response classification in the Iraq study.
- The New Zealand, Faroe Islands, and Seychelles studies are well-designed epidemiological investigations in which prenatal MeHg exposures were within the range of at least some U.S. population exposures. Any revision of the RfD or other exposure standards should consider the findings of these studies.
- After considering the weight of evidence and range of results from the three major epidemiological studies, the committee concludes

that a positive study will provide the strongest public-health basis for the RfD and recommends the Faroe Islands study as the critical study. Within that study, the lowest BMD for a neurobehavioral end point considered to be sufficiently reliable is the Boston Naming Test. The BMDL estimated from that test is 58 ppb Hg in cord blood (approximately corresponding to 12 ppm Hg in hair). That value should be considered a reasonable point of departure for the development of the revised RfD.

- An MOE analysis using available estimates of population exposure levels indicates that average U.S. population risks from MeHg exposure are low. However, those with high exposures from frequent fish consumption might have little or no margin of safety.
- The population at highest risk is the offspring of women of childbearing age who consume large amounts of fish and seafood. The committee estimates that over 60,000 children are born each year at risk for adverse neurodevelopmental effects due to in utero exposure to MeHg.
- There is a critical need for improved characterization of population exposure levels to improve estimates of current exposure, track trends, and identify high-risk subpopulations. Characterization should include improved nutritional and dietary exposure assessment and improved biomonitoring for all population groups. Exposure to other chemical forms of Hg, including exposure to elemental Hg from dental amalgams, should also be investigated.
- The application of uncertainty factors in the revision of the RfD should be based on a thorough quantitative and qualitative evaluation of the full range of uncertainties and limitations of the critical studies. Uncertainty factors applied in the development of a revised RfD should include data-base insufficiency and interindividual toxicokinetic variability in dose reconstruction. As a starting point, an uncertainty factor of 2-3 should be applied to a central tendency estimate of dose derived from maternal hair, or a factor of about 2 should be applied to a central tendency estimate of dose derived from cord blood to account for interindividual pharmacokinetic variability in dose reconstruction. The choice of an uncertainty factor for data-base insufficiency is, in part, a policy decision; however, given the data indicating possible low-dose sequelae and latent effects and immunotoxicity and cardiovascular effects, the

committee concludes that an overall composite uncertainty factor of no less than 10 is needed.

- Concurrent with the revision of the RfD, harmonization efforts should be undertaken to establish a common scientific basis for the establishment of exposure guidance and reduce current differences among agencies. Harmonization efforts should address the risk-assessment process and recognize that risk-management efforts reflect the differing mandates and responsibilities of these agencies.
- Recent studies have found associations between exposure to MeHg and impairments of the immune, reproductive, and cardiovascular systems. Immune and cardiovascular effects have been observed following both prenatal and adult exposures. MeHg exposure levels associated with those effects are comparable to and in some cases lower than those known to cause neurodevelopmental problems. Additional research should be done using animal models and human populations that have chronic, low-dose exposure to MeHg. Effects of exposure during fetal development through the entire life span is needed. Further research is also needed to evaluate MeHg-induced chromosomal aberrations and cancer.
- The committee recommends that results from the Boston Naming Test in the Faroe Islands study be used in the calculation of the RfD. For that study, dose- response data based on Hg concentrations in cord blood should be modeled using the K-power model ($K \geq 1$). On the basis of that study, that test, and that model, the committee's preferred estimate of the BMDL is 58 parts per billion (ppb)[1] of Hg in cord blood (approximately corresponding to 12 ppm Hg in hair). To estimate this BMDL, the committee's calculations involved a series of steps, each involving one or more assumptions and related uncertainties. Alternative assumptions could have an impact on the estimated BMDL value. In selecting a single point of departure, the committee followed established public-health practice of using the lowest value for the most sensitive, relevant end point.

---

[1]The BMDL of 58 ppb is calculated statistically and represents the lower 95% confidence limit on the dose (or biomarker concentration) that is estimated to result in an increased probability that 5% of the population will have an abnormal score on the Boston Naming Test.

- The BMDL of 12 ppm is nearly identical to the BMDL currently used by EPA (11 ppm). Given the toxicokinetic variability and uncertainties in the data, an uncertainty factor of at least 10 is supported by the committee. Therefore, on the basis of its analysis of the available data, the committee finds that the value of EPA's current RfD for MeHg (0.1 μg/kg per day) is scientifically justifiable for the protection of public health.

REFERENCES

Bakir, F., S.F. Damluji, L. Amin-Zaki, M. Murtadha, A. Khalidi, N.Y. al-Rawi, S. Tikriti, H.I. Dhahir, T.W. Clarkson, J.C. Smith, and R.A. Doherty. 1973. Methylmercury poisoning in Iraq. Science 181(96):230-241.

Budtz-Jørgensen, E., N. Keiding, and P. Grandjean. 1999. Benchmark Modeling of the Faroese Methylmercury Data. Research Report 99/5. Prepared at the University of Copenhagen, Denmark, for the U.S. Environmental Protection Agency.

Burbacher T.M., M.K. Mohamed, and N.K. Mottett. 1988. Methylmercury effects on reproduction and offspring size at birth. Reprod. Toxicol 1(4):267-278.

Crump, K.S. 1984. A new method for determining allowable daily intakes. Fundam. Appl. Toxicol. 4(5):854-871.

Crump, K.S., T. Kjellström, A.M. Shipp, A. Silvers, and A. Stewart. 1998. Influence of prenatal mercury exposure upon scholastic and psychological test performance: benchmark analysis of a New Zealand cohort. Risk Anal. 18(6):701-713.

Crump, K.S., C. Van Landingham, C. Shamlaye, C. Cox, P.W. Davidson, G.J. Myers, and T.W. Clarkson. 2000. Benchmark concentrations for methylmercury obtained from the Seychelles child development study. Environ. Health Perspect. 108(3):257-63.

Crump K, J. Viren, A. Silvers, H. Clewell 3rd , J. Gearhart, and A. Shipp. 1995. Reanalysis of dose-response data from the Iraqi methylmercury poisoning episode. Risk Anal. 15(4):523-532.

Dantas, D.C., and M.L. Queiroz. 1997. Immunoglobulin E and autoantibodies in mercury-exposed workers. Immunopharmacol. Immunotoxicol. 19(3): 383-92.

Davidson, P.W., G.J. Myers, C. Cox, C. Shamlaye, O. Choisy, J. Sloane-Reeves, E. Cernchiari, D.O. Marsh, M. Berlin, M. Tanner, and T.W. Clarkson. 1995a. Neurodevelopmental test selection, administration, and performance in the main Seychelles child development study. Neurotoxicology 16(4):665-676.

Davidson, P.W., G.J. Myers, C. Cox, C.F. Shamlaye, D.O. Marsh, M.A. Tanner, M. Berlin, J. Sloane-Reeves, E. Cernichiari, O. Choisy, A. Choi, and T.W. Clarkson. 1995b. Longitudinal neurodevelopmental study of Seychellois children following in utero exposure to methylmercury from maternal fish ingestion: outcomes at 19 and 29 months. Neurotoxicology 16(4):677-688.

Davidson, P.W., G.J. Myers, C. Cox, C. Axtell, C. Shamlaye, J. Sloane-Reeves, E. Cernichiari, L. Needham, A. Choi, Y. Wang, M. Berlin, and T.W. Clarkson. 1998. Effects of prenatal and postnatal methylmercury exposure from fish consumption on neurodevelopment:outcomes at 66 monts of age in the Seychelles child development study. JAMA 280(8):701-707.

Dourson, M.L., S.P. Felter, and D. Robinson. 1996. Evolution of science-based uncertainty factors in noncancer risk assessment. Regul. Toxicol. Pharmacol. 24(2):108-20.

EPA (U.S. Environmental Protection Agency). 1997a. Mercury Study Report to Congress. Vol. I.: Executive Summary. EPA-452/R-97-003. U.S. Environmental Protection Agency, Office of Air Quality Planning and Standards, and Office of Research and Development.

EPA (U.S. Environmental Protection Agency). 1997b. Mercury Study for Congress. Volume VI: Characterization of Human Health and Wildlife Risks from Anthropogenic Mercury Emissions in the United States. EPA-452/R-97-008b. U.S. Environmental Protection Agency, Office of Air Quality Planning and Standards, and Office of Research and Development.

EPA (U.S. Environmental Protection Agency). 1997c. Mercury Study for Congress. Volume VII: Characterization of Human Health and Wildlife Risks from Mercury Exposure in the United States. EPA-452/R-97-009. U.S. Environmental Protection Agency, Office of Air Quality Planning and Standards, and Office of Research and Development.

EPA (U.S. Environmental Protection Agency). 2000. Methylmercury (MeHg) CASRN 22967-92-6. U.S. Environmental Protection Agency IRIS Substance file. [Online]. Available: http://www.epa.gov/iris/subst/0073.htm. Last Updated: 5 May 1998.

Grandjean, P., P. Weihe, R.F. White, and F. Debes. 1998. Cognitive performance of children prenatally exposed to "safe" levels of methylmercury. Environ. Res. 77(2):165-172.

Grandjean, P., E. Budtz-Jørgensen, R.F. White, P.J. Jørgensen, P. Weihe, F. Debes, and N. Keiding. 1999a. Methylmercury exposure biomarkers as indicators of neurotoxicity in children aged 7 years. Am. J. Epidemiol. 150(3):301-305.

Grandjean, P., P. Weihe, R.F. White, F. Debes, S. Araki, K. Yokoyama, K. Murata, N. Sørensen, R. Dahl, and P.J. Jørgensen. 1997. Cognitive deficit in 7-year-old children with prenatal exposure to methylmercury. Neurotoxicol. Teratol. 19(6):417-428.

Hattis, D., P. Banati, R. Goble, and D.E. Burmaster. 1999. Human interindividual variability in parameters related to health risks. Risk Anal. 19(4):711-26.

Hirano, M., K. Mitsumori, K. Maita, and Y. Shirasu. 1986. Further carcinogenicity study on methylmercury chloride in ICR mice. Nippon Juigaku Zasshi (Jpn. J. Vet. Sci.). 48(1):127-135.

Höök, O., K.D. Lundgren, and A. Swensson. 1954. On alkyl mercury poisoning: with a description of two cases . Acta Med. Scand. 150(2):131-137.

Ilbäck, N.G., L. Wesslen, J. Fohlman, and G. Friman. 1996. Effects of methyl mercury on cytokines, inflammation and virus clearance in a common infection (coxsackie B3 myocarditis). Toxicol. Lett. 89(1):19-28.

Janicki, K., J. Dobrowolski, and K. Krasnicki. 1987. Correlation between contamination of the rural environment with mercury and occurrence of leukemia in men and cattle. Chemosphere 16(1):253-257.

Kinjo, Y., S. Akiba, N. Yamaguchi, S. Mizuno, S. Watanabe, J. Wakamiya, M. Futatsuka, and H. Kato. 1996. Cancer mortality in Minamata disease patients exposed to methylmercury through fish diet. J. Epidemiol. 6(3):134-8.

Marsh, D.O., T.W. Clarkson, C. Cox, G.J. Myers, L. Amin-Zaki, and S. Al-Tikriti. 1987. Fetal methylmercury poisoning: Relationship between concentration in single strands of maternal hair and child effects. Arch. Neurol. 44(10):1017-1022.

Mitsumori, K., K. Maita, T. Saito, S. Tsuda, and Y. Shirasu. 1981. Carcinogenicity of methylmercury chloride in ICR mice: Preliminary note on renal carcinogenesis. Cancer Lett. 12(4):305-310.

Moszczynski, P., S. Slowinski, J. Rutkowski, S. Bem, and D. Jakus-Stoga. 1995. Lymphocytes, T and NK cells, in men occupationally exposed to mercury vapours. Int. J. Occup. Med. Environ. Health 8(1):49-56.

NIEHS (National Institute of Environmental Health Sciences). 1998. Scientific Issues Relevant to Assessment of Health Effects from Exposure to Methylmercury. Workshop organized by Committee on Environmental and Natural Resources (CENR) Office of Science and Technology Policy (OSTP) The White House, November 18-20, 1998, Raleigh, NC.

NRC (National Research Council). 1994. Science and Judgment in Risk Assessment. Washington, DC: National Academy Press.

Pellizzari, E.D., R. Fernando, G.M. Cramer, G.M. Meaburn, and K. Bangerter. 1999. Analysis of mercury in hair of EPA region V population. J. Expo. Anal. Environ. Epidemiol. 9(5):393-401.

Presidential/Congressional Commission on Risk Assessment and Risk Management. 1997. Risk Assessment and Risk Management in Regulatory Decision-Making. Final Report. Vol.2. Washington, DC: GPO.

Salonen, J.T., K. Seppänen, K. Nyyssönen, H. Korpela, J. Kauhanen, M. Kantola, J. Tuomilehto, H. Esterbauer, F. Tatzber, and R. Salonen. 1995. Intake of mercury from fish, lipid peroxidation, and the risk of myocardial infarction and coronary, cardiovascular, and any death in Eastern Finnish men. Circulation 91(3):645-655.

Smith, J.C., P.V. Allen, and R. Von Burg. 1997. Hair methylmercury levels in U.S. women. Arch. Environ. Health 52(6):476-80.

Sørensen, N., K. Murata, E. Budtz-Jørgensen, P. Weihe, and P. Grandjean. 1999. Prenatal methylmercury exposure as a cardiovascular risk factor at seven years of age. Epidemiology 10(4):370-375.

Stern, A.H., M. Gochfeld, C. Weisel, and J. Burger. 2000. Mercury and methylmercury exposure in the New Jersey pregnant population. Arch. Environ. Health. In press.

Stern, A.H., L.R. Korn, and B.E. Ruppel. 1996. Estimation of fish consumption and methylmercury intake in the New Jersey population. J. Expo. Anal. Environ. Epidemiol. 6(4):503-525.

Tamashiro, H., M. Arakaki, H. Akagi, K. Hirayama, K. Murao, and M.H. Smolensky. 1986. Sex differential of methylmercury toxicity in spontaneously hypertensive rats (SHR). Bull. Environ. Contam. Toxicol. 37(6):916-24.

Wild, L.G., H.G. Ortega, M. Lopez, and J.E. Salvaggio. 1997. Immune system alteration in the rat after indirect exposure to methylmercury chloride or methylmercury sulfide. Environ. Res. 74(1):34-42.

APPENDIX TO CHAPTER 7

DOMINICI et al. (2000) used a two-stage Bayesian model to pool dose-response information across a relatively large number of studies. The first stage of their analysis estimated dose-response slopes from each study, adjusting for various confounding factors measured for each study. The second stage involved fitting a hierarchical Bayesian model to the estimates obtained at the first stage. The approach is heuristically appealing and is in fact similar to the ad-hoc two-stage algorithm that was often used to fit linear growth curve models before the advent of programs such as SAS PROC MIXED (see, for example, Laird 1990). As noted by Dominici et al., the approach approximates a fully Bayesian analysis on the original data. The authors justified this approximation by (1) empirically checking this approximation for their particular application and (2) pointing to theoretical justification for the approximation given by Daniels and Dass (1998). A two-stage analysis along the same lines is attractive in the context of MeHg for several reasons. First, the approach is natural in settings in which the original study-specific data are unavailable. That is, one can simply fit the second stage of the model to published summary measures (i.e., dose-response slopes and corresponding standard errors) from each study. Second, the approach easily extends to the case of multiple outcomes, because outcome within a study simply represents an additional level in the hierarchical model. As discussed in the chapter, data available to the committee included estimated BMDs and BMDLs computed for each of the individual outcomes assessed for the Faroe Islands, Seychelles, and

New Zealand studies (see Table 7-3). One approach might be to apply the hierarchical analysis directly to the estimated BMDs, although the committee felt it appropriate to apply the analysis to the inverse BMDs instead. One advantage of working with the inverse BMDs is that very large and undefined values are transformed to zero. Working with the inverse BMDs also has some theoretical justification, because in the context of a linear model, the estimated BMD is simply a constant divided by the estimated dose-response slope (see Equation 7-1).

To describe the committee's approach in more detail, it is useful to define some notation. Let $\hat{\beta}_{ij}$ be the inverse of the BMD estimated for the jth outcome, $j = 1, \ldots J_i$, within study, $i = 1, \ldots I$. The corresponding standard errors, $\hat{\sigma}_{ij}$, can be estimated by subtracting $\hat{\beta}_{ij}$ from the inverse of the BMDL and then dividing by 1.64. The hierarchical model can be expressed as

$$\hat{\beta}_{ij} \big| \beta_{ij} \approx N(\beta_{ij}, \hat{\sigma}^2_{ij})$$

$$\beta_{ij} \big| \beta_i \approx N(\beta_i, \gamma^2)$$

$$\beta_i \big| \beta \approx N(\beta, \tau^2)$$

$$\beta \approx N(m,n),\ \gamma^2 \approx \text{InvGamma}(a,b),\ \tau^2 \approx \text{InvGamma}(c,d),$$

where a, b, c, d, m, and n are chosen so that the priors are all relatively noninformative. In other words, we assume that the true inverse BMDs for each outcome are normally distributed around a study-specific mean value and that these study-specific values are in turn normally distributed around an overall mean. We fit the hierarchical model using the BUGS (Bayesian inference Using Gibbs Sampling) software package (Spiegelhalter et al. 1996). The product of the analysis is a series of simulated distributions of the various random variables defined in the model. Applying an inverse transformation again converts those results to yield estimates of the distribution of the quantities of interest, namely, BMDs. In addition to providing an estimate of distribution of true BMDs corresponding to different outcomes from different studies, the output from the program allows computation of so-called posterior estimates of the true BMDs, given the observed values. The advantage of working with the posterior estimates instead of the original values is that they have removed some of the random variation inherent in the observed estimates. The "smoothed BMDs" referred to in Chapter 7 and also in Figure 7-3 are posterior estimates.

Because the method proposed here is new and exploratory in nature, the committee does not recommend it as the primary approach to the MeHg risk assessment at the present time. Indeed, there are a number of questions associated with the approach that would require further exploration before it could be used as the basis of a definitive analysis. For example, one concern is the relatively small number of studies (three) available for the MeHg study. The Dominici et al (2000) analysis involved a relatively large number of studies, and therefore, does not have the same concern.

REFERENCES

Daniels, M.J., and R.E. Dass. 1998. A note on first-stage approximation in two-stage hierarchical models. Sankhya B 60(1):19-30.

Dominici, F., J.M. Samet, and S.L. Zeger. In Press. Combining evidence on air pollutoin and daily mortality from the largest 20 US cities: A hierarchical modeling strategy. Royal Statistical Society, Series A, with discussion.

Laird, N.M. 1990. Analysis of linear and non-linear growth models with random parameters. Pp. 329-343 in Advances in Statistical Methods for Genetic Improvement of Livestock, D. Gianola and K. Hammond, eds. Berlin: Springer-Verlag.

Spiegelhalter, D.J., A. Thomas, N.G. Best, and W. R. Gilks. 1996. BUGS: Bayesian Inference Using Gibbs Sampling, Version 0.5, (version ii). Online. Available: http://www.mrc-bsu.cam.ac.uk/bugs/

GLOSSARY

Absorbed dose – The amount of a substance that penetrates an exposed organism's absorption barriers (e.g., skin, lung tissue, gastrointestinal tract) through physical or biological processes. The term is synonymous with internal dose.
Administered dose – The amount of a substance given to a test subject (human or animal) in determining dose-response relationships, especially through ingestion or inhalation. In exposure assessment, since exposure to chemicals is usually inadvertent, this quantity is called potential dose.
Adverse effect – Any effect that produces functional impairment and/or a pathological lesion that may affect the performance of the whole organism, or that reduces an organisms ability to respond to an additional challenge (Stara et al. 1985)
Anthropogenic – Of human origin.
Applied dose – The amount of a substance in contact with the primary absorption boundaries of an organism (e.g, skin, lung, gastrointestinal tract) and available for absorption.
Autoimmunity – A condition resulting from the production of autoantibodies, characterized by cell-mediated or humoral immunological responses to antigens of one's own body, sometimes with damage to normal components of the body.

Benchmark dose analysis – A technique for quantitative assessment of noncancer health effects.
Benchmark dose – An exposure level that corresponds to a statistical

lower bound on a standard probability of an effect, such as 10% of people affected.

Bioaccumulation - An increase in concentration in living organisms as they take in contaminated air, water, or food because the substances are very slowly metabolized or excreted.

Bias - Any effect tending to produce results that depart systematically from the true values. Two principle forms of bias in human epidemiological studies are misclassification, when there are misassignments in exposure or adverse outcome, and selection, in which subjects selected for study differ systematically from those not selected.

Bioactivation - A metabolic process wherein an inactive chemical is converted to an active one in the body.

Bioavailability - The state of being capable of being absorbed and available to interact with the metabolic processes of an organism. Bioavailability is typically a function of chemical properties, physical state of the material to which an organism is exposed, and the ability of the individual organism to physiologically take up the chemical.

Biomarker of Effect - A measurable biochemical, physiological, or other alteration within an organism that, depending on magnitude, can be recognized as an established or potential health impairment or disease.

Biomarker of Exposure - An exogenous substance, the metabolite(s) or the product of interactions between a xenobiotic agent and some target molecule or cell that is measured in a compartment within an organism.

Biologically effective dose - The amount of the deposited or absorbed contaminant that reaches the cells or target site where an adverse effect occurs or where an interaction of that contaminant with a membrane surface occurs.

Biotransformation - A series of chemical alterations within the body whereby a foreign substance is transformed to a more or less toxic substance.

Case-control study - An epidemiological study in which persons are selected because they have a specific disease or other outcome (cases) and are compared to a control (referent comparison) group without the disease to evaluate whether there is a difference in their reported frequency of exposure to possible disease risk factors. Also termed a retrospective study or case referent study.

Chronic exposure - Multiple exposures occurring over an extended

period of time or over a significant fraction of an animal's or human's lifetime (Usually seven years to a lifetime.)
Chronic toxicity - The capacity of a substance to cause long-term poisonous health effects in humans, animals, fish, and other organisms.
Cohort study - An epidemiological study in which a defined group of persons known to be exposed to a potential disease risk factor is followed over time and compared to a group of persons who were not known to be exposed to the potential risk factor to evaluate the differences in rates of the outcome. Also termed a prospective study, follow-up study, incidence study, retrospective cohort, or historical cohort study.
Concentration (C) - The total quantity of substance present in a given unit volume (of gas or liquid). It may be expressed in any unit or mass per unit of volume such as milligrams per cubic meter (mg/m^3), or as volume per volume such as parts per million (ppm).
Confidence interval (95%) - A range of values for the effect estimate within which the true value is though to lie with a 95% level of confidence.
Confounder (confounding factor) - A factor that is associated with both the exposure and outcome of interest and can distort the apparent magnitude of the effect of the study factor.

Developmental toxicity - The occurrence of adverse effects on the developing organism that may result from exposure to a chemical prior to conception (either parent), during prenatal development, or postnatally to the time of sexual maturation. Adverse development effects may be detected at any point in the life span of the organism.
Dose - The amount of a risk agent that enters or interacts with organisms. An administered dose is the amount of substance administered to an animal or human, usually measured in milligrams per kilogram of body weight; milligrams per square meter of body surface area; or parts per million of the diet, drinking water, or ambient air. An effective dose is the amount of the substance reaching the target organ.
Dose estimation - The process by which a delivered dose is estimated from an exposure dose or from a biomarker of exposures.
Dose-response assessment - The determination of the relationship between the magnitude of administered, applied, or internal dose and specific biological response. Response can be expresses as measured or

observed incidence, percent response in groups of subjects (or populations), or the probability of occurrence of a response in a population.
Dose-response curve - A graphical representation of the quantitative relationship between administered, applied, or internal dose of a chemical or agent, and a specific biological response to that chemical or agent.
Dose-response model - A mathematical description of the relationship between exposure levels and the incidence rates of an effect.
Dose-response relationship - A relationship between the amount of an agent (either administered, absorbed, or believed to be effective) and changes in certain aspects of the biological system (usually toxic effects), apparently in response to the agent.

End points of toxicity - Adverse effects elicited as a result of exposure to a substance.
Epidemiology - The core public health science, investigating the causes and risk factors of disease and injury in populations and the potential to reduce such disease burdens.
Exposure - An event that occurs when there is contact at a boundary between a human and the environment with a contaminant of a specific concentration for an interval of time; the units of exposure are concentration multiplied by time.
Exposure assessment - The determination or estimation (qualitative or quantitative) of the magnitude, frequency, duration, and route of exposure.
Exposure dose - The level of contaminant in the air, water, or soil to which people are actually exposed.

Genotoxic - Capable of altering the structure of DNA and causing mutations.

Half-Life - The time required for the elimination of half a total dose from the body.
Human health risk assessment - A process used to estimate the likelihood of adverse health outcomes of environmental exposures to chemicals.

Immediate versus Delayed Toxicity - The immediate effects that occur or develop rapidly after a single administration or exposure of substances; delayed effects are those that occur after a lapse of some time.

These effects have also been referred to as acute and chronic, respectively.
Immunological toxicity - The occurrence of adverse effects on the immune system that may result from exposure to environmental agents such as chemicals.
Ingested dose - The amount of a substance consumed by an individual, usually expressed as amount per kilogram body weight over a given time period.
Intake - The amount of material inhaled, absorbed through skin, or ingested during a specified period of time.
Internal dose - In exposure assessment, the amount of a substance penetrating the absorption barriers (e.g., skin, lung tissue, gastrointestinal tract) of an organism through either physical or biological processes.

Latency - Time from the first exposure of a chemical until the appearance of a toxic effect.
Lifetime exposure - Total amount of exposure to a substance that a human would receive in a lifetime (usually assumed to be 75 years).
Lipid solubility - The maximum concentration of a chemical that will dissolve in fatty substances. Lipid soluble substances are insoluble in water. They will very selectively disperse through the environment via intake in living tissue.
Lowest-observed-adverse-effect level (LOAEL) - The lowest exposure level at which there are statistically or biologically significant increases in frequency or severity of adverse effects between the exposed population and its appropriate control group.

Margin of exposure - A ratio defined by EPA as a dose derived from a tumor bioassay, epidemiological study, or biological marker study, such as the dose associated with a 10% response rate divided by an actual or projected human exposure.
Mechanism of action - The way in which a substance (e.g., a chemical) exerts its toxic effect(s).
Metabolism - All the biological reactions that take in a cell or an organism.

Neurotoxicity - The occurrence of adverse effects on the nervous system following exposure to chemical.
No-observed-adverse-effect level (NOAEL) - An exposure level at

which there are no statistically or biologically significant increases in the frequency or severity of adverse effects between the exposed population and its appropriate control; some effects may be produced at this level, but they are not considered as adverse, nor precursors to adverse effects. In an experiment with several NOAELs, the regulatory focus is primarily on the highest one, leading to the common usage of the term NOAEL as the highest exposure without adverse effect.

Point of departure - An estimate or observed level of exposure or dose which is associated with an increase in adverse effect(s) in the study population. Examples of points of departure include NOAELs, LOAELs, BMDs, and BMDLs.
Population at risk - a population subgroup that is more likely to be exposed to a chemical, or is more sensitive to the chemical, than is the general population.
Power - The probability of detecting a specified difference in effect between experimental and control groups.
Probability - A numerical value between 0 and 1 that represents the likelihood of something.

Reference Dose (RfD) - an estimate (with uncertainty spanning perhaps an order of magnitude) of daily exposure to the human population (including sensitive subgroups) that is likely to be without an appreciable risk of deleterious effects during a lifetime.
Reproductive toxicity - The occurrence of adverse effects on the reproductive system that may result from exposure to a chemical. The toxicity may be directed to the reproductive organs and/or the related endocrine system. The manifestation of such toxicity may be noted as alterations in sexual behavior, fertility, pregnancy outcome, or modifications in other functions that are dependent on the integrity of this system.
Risk - A measure of the probability that damage to life, health, property, and/or the environment will occur as a result of a given hazard.
Risk assessment - An organized process used to describe and estimate the likelihood of adverse health outcomes from environmental exposures to chemicals. The four steps are hazard identification, dose-response assessment, exposure assessment, and risk characterization.
Risk characterization - The last phase process of the risk assessment process that estimates the potential for adverse health or ecological

effects to occur from exposure to a stressor and evaluates the uncertainty involved.
Risk management - The process of evaluating and selecting alterative regulatory and non-regulatory responses to risk. The selection process necessarily requires the consideration of legal, economic, and behavioral factors.

Solubility - The amount of mass of a compound that will dissolve in a unit volume of solution. Aqueous Solubility is the maximum concentration of a chemical that will dissolve in pure water at a reference temperature.
Standardized mortality ratio (SMR) - The ratio of observed deaths to expected deaths.
Statistical control - The process by which that variability of measurements or of data outputs of a system is controlled to the extent necessary to produce stable and reproducible results. To say that measurements are under statistical control means that there is statistical evidence that the critical variables in the measurement process are being controlled to such an extent that the system yields data that are reproducible within well-defined limits.
Susceptibility - The extent to which an individual is liable to infection or the effects of substances, such as toxicants, allergens, or other influences. The antithesis of resistance.

Toxicant - A harmful substance or agent that may injury an exposed organism.
Toxicity - A degree to which a substance or mixture of substances can harm humans or animals. Acute toxicity involves harmful effects in a organism through a single or short-term exposure. Chronic toxicity is the ability of a substance or mixture of substances to cause harmful effects over an extended period, usually upon repeated or continuous exposure sometimes lasting for the entire life of that exposed organism. Subchronic toxicity is the ability of the substance to cause effects for more than one year but less than the lifetime of the exposed organism.
Toxicokinetics - The processes of absorption, distribution, metabolism, and excretion that occur between the time a toxic chemical enters the body and when it leaves.

Uncertainty - An estimate of the extent to which a risk estimate reflects reality.

Uncertainty factor - One several, generally 10-fold factors, used in operationally deriving the Reference Dose (RfD) from experimental data. UF's are intended to account for (1) the variation in sensitivity among the members of the human population; (2) the uncertainty in extrapolating animal data to the case of humans; (3) the uncertainty in extrapolating from data obtained in a study that is of less-than-lifetime exposure; and (4) the uncertainty in using LOAEL date rather than NOAEL data.

Weight of the scientific evidence - Considerations involved in assessing the interpretation of published scientific information—quality of methods, ability of a study to detect adverse effects, consistency of results across studies, and biological plausibility of cause-and-effect relationships.